Macromolecular Symposia 213

Stereospecific Polymerization and Stereoregular Polymers

Milan, Italy
June 8–12, 2003

Symposium Editors:
A. Giarrusso, G. Ricci, I. Tritto, Milan, Italy

pp. 1–394 · June 2004
ISBN 3-527-31046-0

Macromolecular Symposia publishes lectures given at international symposia and is issued irregularly, with normally 14 volumes published per year. For each symposium volume, an Editor is appointed. The articles are peer-reviewed. The journal is produced by photo-offset lithography directly from the authors' typescripts.
Further information for authors can be found at http://www.ms-journal.de
Suggestions or proposals for conferences or symposia to be covered in this series should also be sent to the Editorial office (E-mail: macro-symp@wiley-vch.de).

Macromolecular Symposia:
Annual subscription rates 2004
Macromolecular Full Package: including Macromolecular Chemistry & Physics (18 issues), Macromolecular Rapid Communications (24), Macromolecular Bioscience (12), Macromolecular Theory & Simulations (9), Macromolecular Materials and Engineering (12), Macromolecular Symposia (14):

Europe	Euro	6.424 / 7.067
Switzerland	Sfr	11.534 / 12.688
All other areas	US$	7.948 / 8.743

print only **or** electronic only / print **and** electronic

Postage and handling charges included. All Wiley-VCH prices are exclusive of VAT. Prices are subject to change.

Single issues and back copies are available. Please ask for details at: service@wiley-vch.de

Orders may be placed through your bookseller or directly at the publishers: WILEY-VCH Verlag GmbH & Co. KGaA, P. O. Box 10 11 61, 69451 Weinheim, Germany, Tel. +49 (0) 62 01/6 06-400, Fax +49 (0) 62 01/60 61 84, E-mail: service@wiley-vch.de

For copyright permission please contact Claudia Rutz at:
Fax: +49 (0) 62 01/6 06-332, E-mail: crutz@wiley-vch.de

For USA and Canada: Macromolecular Symposia (ISSN 1022-1360) is published with 14 volumes per year by WILEY-VCH Verlag GmbH Co. KGaA, Boschstr. 12, 69451 Weinheim, Germany. Air freight and mailing in the USA by Publications Expediting Inc., 200 Meacham Ave., Elmont, NY 11003, USA. Application to mail at Periodicals Postage rate is pending at Jamaica, NY 11431, USA. POSTMASTER please send address changes to: Macromolecular Symposia, c/o Wiley-VCH, III River Street, Hoboken, NJ 07030, USA.

© WILEY-VCH Verlag GmbH & Co. KGaA, Weinheim, Germany, 2004
Printing: Strauss Offsetdruck, Mörlenbach. Binding: J. Schäffer, Grünstadt

Macromolecular Symposia

Articles published on the web will appear several weeks before the print edition. They are available through:

www.ms-journal.de

www.interscience.wiley.com

EUPOC 2003
Milan (Italy), 2003

Preface
A. Giarrusso, G. Ricci, I. Tritto

Author Index

Photograph of Giulio Natta.

Preface

The International Symposium on "Stereospecific Polymerization and Stereoregular Polymers" EUPOC 2003 was held on June 8–12, 2003 in Milan, Italy. The symposium was organized to celebrate the centenary of Giulio Natta's birth (26 February, 1903).

Fifty years have elapsed since the first revolutionary synthesis of isotactic polypropylene carried out at the Politecnico of Milan in March 1954, using the new organometallic catalysts discovered by Karl Ziegler. For this discovery Giulio Natta and Karl Ziegler were awarded the 1963 Chemistry Nobel Prize jointly. The work of Ziegler and Natta opened up the era of metal catalyzed stereospecific polymerization, which produced a dramatic impact on science and technology.

The symposium has provided a forum for scientists involved in the field to discuss new results and perspectives on novel catalysts and on synthesis, structure, characterization, and applications of polyolefins. The symposium covered the following topics:

– Traditional Catalysts
– Metallocene Catalysts
– More Recent Catalysts
– Molecular Modeling and Polymerization Mechanism
– Structural Aspects and Properties of Stereoregular Polymers
– New Products and Processes

More than 220 scientists and young researchers from academia and industrial research institutions of 22 countries (15 from all over Europe and 7 from overseas: Africa, North and South America, Asia) attended the meeting. 57 Oral contributions and 88 posters were presented on the topics listed above.

This volume is a collection of some of the contributions presented at the symposium and the opening lecture given by Prof. Lido Porri, one of Natta's most famous and brilliant coworkers at the Politecnico of Milan. This lecture enlightens Giulio Natta's profile as a scientist and as a man.

We would like to thank all those who, through their presence and excellent presentations, have contributed to the scientific success of the meeting and to create a friendly and stimulating atmosphere. Finally we would like to thank our sponsors (Politecnico di Milano, Università degli Studi di Milano, Università degli Studi di Milano-Bicocca, Provincia di Milano, Basell Polyolefins Italia SpA, Fondazione Cariplo, Italcementi, Sistema Ecodeco Ambiente e Innovazione, and Consorzio Interuniversitario Nazionale La Chimica per l'Ambiente) for their generosity, which has made a worthy celebration possible.

A. Giarrusso
Politecnico di Milano

G. Ricci, I. Tritto,
Istituto per lo Studio delle Macromolecole – CNR, Milano

Giulio Natta – His Life and Scientific Achievements

Lido Porri

Politecnico di Milano, Dip. Chimica, Materiali e Ingegneria Chimica - Piazza
Leonardo da Vinci 32 - 20133 Milano, Italy
E-mail: lido.porri@polimi.it

EUPOC 2003 has been organized by the European Polymer Federation to honor Giulio Natta, on the occasion of the centenary of his birth. Karl Ziegler and Giulio Natta, as is well known, shared the 1963 Nobel Prize in Chemistry for their discoveries in the field of chemistry and technology of high polymers. In this introduction I would like to briefly highlight the life and scientific achievements of Giulio Natta.

Giulio Natta was born in Imperia, a small city near the Italian-French border, on February 26, 1903. He attended the middle schools in Genoa, then he moved to Milan as student in Chemical Engineering at the Polytechnic of Milan. After graduating in Chemical Engineering in 1924, he remained at the Polytechnic as post-doc and then as assistant Professor. He was a brilliant student. His great passion for Chemistry is demonstrated by the fact that he set up a small chemical lab in his apartment, to do chemical experiments at home.

His academic career was rapid: He was professor at the University of Pavia (1933-35), at the University of Rome (1935-37) and at the Polytechnic of Turin (1937-38). In 1938 he was called at the chair of Industrial Chemistry of the Polytechnic of Milan, where he would remain for 35 years, until his retirement in 1973.

At the beginning of his career, his main scientific interest was the study of chemical structures by x-ray and electron diffraction techniques. Later he became interested in reactions and processes of industrial importance, such as the synthesis of MeOH with new catalysts, and the hydroformylation reaction.

At the end of the '40s, he established a collaboration with Montecatini, the biggest chemical company in Italy at the time. As a result of this collaboration about 25 young chemists were working in Natta's Institute at the beginning of the '50s. As we'll see later this collaboration was of great benefit to both Natta and Montecatini.

The turning point in Natta's scientific career began in 1952. In May of that year Natta attended

© 2004 WILEY-VCH Verlag GmbH & KGaA, Weinheim　　　　　　DOI: 10.1002/masy.200450901

a conference given in Frankfurt by Karl Ziegler, Director of the Max-Plank Institut für Kohlenforschung in Mülheim. Ziegler reported on the oligomerization of ethylene catalyzed by AlEt$_3$ (the so-called *Aufbaureaktion*). Natta was impressed by Ziegler's results and convinced Montecatini to purchase the rights of the *Aufbaureaktion*. Three young chemists of Natta's group went to Mühlheim to become familiar with the experimentation using aluminum alkyls. Natta began to work on the *Aufbau* reaction in Milan.

In the 2nd half of 1953, Ziegler's group discovered that the combination AlEt$_3$-TiCl$_4$ was capable of polymerizing ethylene under very mild conditions. At the end of 1953 Montecatini and Natta were informed by Ziegler of the new discovery.

Natta started to investigate the polymerization of propylene with the catalyst of Ziegler and obtained a polymer in March 1954. The polymerization product appeared heterogeneous, consisting of a mixture of rubbery material and a white product. It was fractionated by extraction with successive boiling solvents. Boiling acetone removed all the oligomers, that is oily products. Boiling diethylether dissolved all the amorphous polypropylene.

The residue was crystalline polypropylene, which in turn could be separated into fractions having different degrees of crystallinity by extraction with heptane, toluene, xylenes.

The crystalline structure of polypropylene was determined, and it was found that the crystallinity was due to the fact that in each chain all the tertiary C atoms had the same configuration, at least for long sections of the polymer chain. Natta called these polymers isotactic. A Patent was filed in June '54. Butene-1 and styrene were also polymerized to isotactic polymers a few weeks later.

In November 1954, the group in Milan made a significant change to Ziegler's catalyst. TiCl$_4$ was replaced by TiCl$_3$, which increased the stereospecificity from 40 to ca. 80%. This paved the way for the commercial production of polypropylene, which started in Ferrara in 1957.

At the end of 1954 a letter was submitted to the editor of the *Journal of American Chemical Society*, which caused some surprise in the chemical world.

Paul Flory, a member of the Editorial Board of *J. Am. Chem. Soc.*, who was to get the Nobel Prize in Chemistry in 1974, wrote a letter to Natta (in January '55), in which he said inter alia: " The results disclosed in your manuscript are of extraordinary interest; perhaps one should call them revolutionary in significance. The possibilities opened up by such asymmetric polymerization are of the utmost importance, I am sure."

Other groups succeeded in polymerizing propylene during 1954, but, in their patents, polypropylene characterization was made only by I.R. methods. No fractionation or x-ray examinations were reported, so that they did not recognize the importance of what they had obtained.

However, the synthesis of isotactic polypropylene, polybutene and polystyrene was only the beginning of a long story. Natta understood that the new catalysts were causing a revolution in polymer chemistry and therefore decided to extend the investigation to other classes of monomers: diolefins, cycloolefins, acetylenes.

At the beginning of 1955, butadiene was polymerized to a crystalline polymer having a 1,2 structure in which the chiral configuration of the tertiary C atoms along each chain were alternated. Natta coined the term syndiotactic to indicate polymers of this structure (*syndio* meaning "two by two" was taken from Greek).

A few months later, in October 1955, butadiene was polymerized to a polymer having a 1,2 isotactic structure, which showed for the first time that a monomer can be polymerized to iso- or syndiotactic polymer depending on the catalyst used. It now appears obvious that a monomer can give, at least in principle, iso- or syndiotactic polymers, but at that time, 1955, it was not.

Other two stereoregular polymers were obtained from butadiene (with a cis-1,4 and a trans-1,4 structure), so that all the four foreseeable stereoregular polybutadienes were obtained in a few months.

Stereoregular polymers were also obtained from other 1,3-dienes (such as isoprene, 1,3-pentadiene, several other substituted butadienes), small ring cycloalkenes (such as cyclobutene and norbornene), and from acetylenes.

Syndiotactic polypropylene was also isolated and characterized in 1958, and more or less at the same time crystalline alternating copolymers of ethylene/cyclopentene, ethylene/2-butene, ethylene/butadiene were obtained.

Various propylene homologs were polymerized to isotactic polymers, e.g. 3-Me-1-butene, 4-Me-1-pentene, 4-Me-1-hexene, 5-Me-1-hexene, 5-Me-1-heptene and vinylcyclohexane. Work on these monomers allowed one to formulate general rules determining the structure of linear macromolecules. In particular, the data on these isotactic polymers indicated that fourfold or higher order helices exist besides the threefold one observed in isotactic polypropylene.

Stereoregular polymers were also obtained from non-hydrocarbons monomers, such as vinylalkylethers, alkenylethers, O-methoxystyrene, N-vinylcarbazole, benzofuran, using either cationic initiators, Mg-alkyls or even AlEt₃.

I don't know exactly how many stereregular polymers Natta and his collaborators have synthesized, but I think more than 90. The crystal structures of these polymers were determined. Their chemical, physical, and, for some of them, mechanical properties were investigated. The relationship between physical properties and stereoregularity was studied. Suitable catalysts for the production were identified. The mechanism of stereoregulation was studied. And finally, the foundations of polymer stereochemistry were laid.

In 1961 the *Journal of Polymer Science* devoted an issue (vol. 51, issue 156) to Giulio Natta, "to express appreciation to the man who first established with vigorous scientific methods the existence of stereoregular polymers". Natta was called the father of stereoregular polymers. The editor wrote in the introduction to this issue.

> *"Seldom has a scientific contribution aroused such profound fundamental interest and has been followed by such a rapid technical development as the series of publications by Professor Giulio Natta and coworkers on the stereospecific polymerization of olefins, which started to appear in Italian journals several years ago and have continued ever since. Many prominent scientists in very large research laboratories have become interested in the new technique and have focused their interests and efforts on its promotion. Yet Professor Natta has succeeded in maintaining undisputed leadership and continues to surprise his colleagues by new and unexpected discoveries along the general principles of stereregulation."*

In my opinion there are two reasons for this success:

1. The intuition of Natta, who was the first to realize that Ziegler catalysts were causing a revolution in polymer science. He gave a name to the stereoregular polymers, isotactic and syndiotactic, and one gives a name when he realizes that something new and important has been created.

2. The organization that Natta gave to his Institute, where chemists, physicists, spectroscopists, crystallographers were working together on the same project, which allowed an interdisciplinary approach to the problems. Most of the people working in

Natta's Institute were Montecatini employees and without their work and the financial support of Montecatini it would have been impossible to maintain for several years the leadership in polymer science.

At the end of this introduction I wish to cite what Sir Robert Robinson said when he introduced Natta who was awarded in London in 1961 the First International Synthetic Rubber Award (Rubber and Plastics Age, 1961, p.1195):

"Natta has developed the theme of polymerization as a grandiose fugue. The successful initiation, prosecution, and completion of so much and so varied research is the result of his most unusual originality, drive and power of sustained work."

A few words about Natta as a man.

He was a very simple person, who dedicated all his life to science. He was the opposite of what we call an ambitious person. When he received the Nobel Prize, he attributed all his scientific success to luck: "We have been lucky", he said.

No doubt that luck often plays an important role in scientific discoveries. However, in science, luck is not blind. It only helps intelligent people, those who are intuitive, inquisitive, hard working, and able to quickly seize the importance of an event.

Giulio Natta was a man of this type.

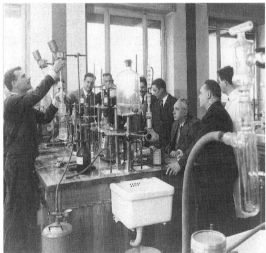

Professor Natta with some collaborators (1963).
From the left: Lido Porri, Piero Pino, Raffaele Ercoli, Enrico Mantica, Ferdinando Danusso, Giulio Natta, Gino Dall' Asta, Mario Farina.

Precise Arguments on the Distribution of Stereospecific Active Sites on MgCl$_2$-Supported Ziegler-Natta Catalysts

Boping Liu,[1] *Takashi Nitta,*[1] *Hisayuki Nakatani,*[2] *Minoru Terano**[1]

[1]School of Materials Science, Japan Advanced Institute of Science and Technology, 1-1 Asahidai, Tatsunokuchi, Ishikawa 923-1292, Japan
E-mail: terano@jaist.ac.jp
[2]Fundamental Laboratory for Engineering Education Core, Kanazawa Institute of Technology, 7-1 Ohgigaoka Nonoichi Ishikawa 921-8501, Japan

Summary: The stereospecific nature of active sites on various MgCl$_2$-supported Ziegler-Natta catalysts was investigated by stopped-flow technique combined with temperature rising elution fractionation (TREF) method. A modified three-sites model with precise description of the stereospecific nature of various types of active sites stemmed from surface titanium species, Al-alkyl compounds, Mg-compounds and electron donors has been proposed. It was demonstrated that the isospecificity of active sites strongly depends on the bulkiness of the ligands situated at the two most important ligand positions for construction of asymmetry and chirality of the active sites with steric hindrance. In general, there may exist both monometallic and bimetallic sites in heterogeneous Ziegler-Natta catalyst system. The kinds of active titanium species with different chemical structures on this catalyst system should be limited, whereas, the non-discrete distribution of isospecificity of active sites could be considered to generate from the numerous types of steric and electronic effects from the surroundings of active titanium species as well as large number of reversible and dynamic transformation reactions simultaneously occurred on the heterogeneous catalyst surface.

Keywords: MgCl$_2$-supported catalyst; polypropylene (PP); stereospecific active sites; stopped-flow method; Ziegler-Natta polymerization

Introduction

Ziegler-Natta catalyst is one of the most important discoveries in the chemistry field in the 20th century with respect to its contribution for synthesis of polyolefins at low pressure and temperature through coordination polymerization.[1] Finding the remarkable effect of MgCl$_2$ as support for Ziegler-Natta catalyst in 1960`s is the milestone for achieving super-high catalytic efficiency of this catalyst system as well as subsequent spectacular successes of large-scale commercial production of numerous polyolefin materials. In developing MgCl$_2$-supported Ziegler-Natta catalyst for propylene polymerization, electron donor compound is

 DOI: 10.1002/masy.200450902

crucial and indispensable for producing highly isotactic polypropylene (PP) with the highest commercial importance. Within the past two decades, PP product with ultra-high isotacticity has been developed based on successful innovation of electron donors in the industrial field.[2] In spite of the great industrial success and several decades of research efforts since 1953, many aspects concerning the active sites and polymerization mechanism in Ziegler-Natta catalysis still remained ambiguous and controversial. For example, the real origin of isospecificity of active sites and specific stereochemical role of electron donor are still open for discussion. The difficulties come from the complexity of this catalyst system due to the very short lifetime of growing polymer chains, low percentage of active titanium species, co-existence of multiple interactions between multi-components as well as many side reactions in the polymerization process.[3] As it has been reported in the literature, the stereospecific nature of active sites on heterogeneous Ziegler-Natta catalyst could be significantly affected not only by electron donors [4] but also by the types of titanium compounds, Al-alkyl cocatalysts and Mg-compounds,[5~10] whereas, the specific stereospecific role of each component has not been clarified yet.

The stopped-flow technique [3] has been proven to be one of the most powerful methods for studying the kinetic and stereospecific natures of active sites on heterogeneous Ziegler-Natta catalysts for propylene polymerization. By this technique, direct information corresponding to the relationship between the stereoregularity of each polymer chain and the stereospecificity of each active site can be obtained.[11] Within recent several years, we have demonstrated the combination of stopped-flow technique with TREF method can be powerful for substantial research on the variation of stereospecific nature of active sites on the MgCl$_2$-supported Ziegler-Natta catalysts.[12] From the results, modified three-sites and island model with precise description of the stereospecific nature of various types of active sites had been proposed. [12] In this paper, a brief review of this series of studies will be demonstrated in pursuing more precise understanding on the origin of isospecificity of active sites from various aspects.

Experimental

Materials and Catalysts: The specifications of various raw materials including propylene,

MgCl$_2$, Mg(OEt)$_2$, TiCl$_4$, nitrogen, triethylaluminum (TEA), heptane, ethylbenzoate (EB) and dibutylphthalate (DBP) as well as preparation procedures for three different catalysts, i.e. donor-free TiCl$_4$/MgCl$_2$ catalyst, monoester-type TiCl$_4$/EB/MgCl$_2$ catalyst and diester-type TiCl$_4$/DBP/Mg(OEt)$_2$ catalyst were previously reported.[12] Ti contents of the three catalysts were 0.50 mmol-Ti/g-cat., 0.40 mmol-Ti/g-cat and 0.54 mmol-Ti/g-cat, respectively.

Stopped-Flow Polymerization and Characterization of PP: Propylene polymerization procedures using a modified stopped-flow apparatus with each type of catalyst as well as characterization of PPs by GPC and TREF had been previously reported in detail.[12]

Results and Discussions

Donor-Free TiCl$_4$/MgCl$_2$ Catalyst

Stopped-flow polymerizations were conducted using TiCl$_4$/MgCl$_2$ catalyst pretreated by TEA cocatalyst for 0 ~ 60s. The active site concentration ([C*]), chain propagation rate constant (k_p) and weight percentage of four fractions of PPs are shown in Table 1. According to the typical distribution states of isospecificity of active sites judged from the TREF curves shown in Figure 1, each PP sample is fractionated in four different temperature ranges namely ~20°C, 20~100°C, 100~110°C and 110~140°C, which are thought to be corresponding to four kinds of active sites with different stereospecificity, here defined as aspecific sites (**AS**), poorly-isospecific sites (**IS$_1$**), the second highest isospecific sites (**IS$_2$**) and the highest

Table 1. The dependence of [C*], k_p and weight percentage of fractions of PPs on pretreatment time obtained in stopped-flow polymerization with the donor-free TiCl$_4$/MgCl$_2$ catalyst.[a]

Pretreatment time (s)	[C*] mol%	k_p L/mol·s	Fraction[b] (%)			
			F1	F2	F3	F4
0	5.2	2200	61	24	15	0
0.2	4.0	2730	48	27	23	2
2	3.0	2940	33	36	27	4
10	1.1	3340	44	19	28	9
60	0.54	2700	49	17	23	11

[a] The polymerization was carried out with TEA ([Al]=14mmol, Al/Ti=30) in heptane at 30 °C for ca. 0.15s after the pretreatment.
[b] Fractionated by TREF, weight fractions: F1 (~20°C), F2 (20~100°C), F3 (100~110°C), F4 (110~140°C).

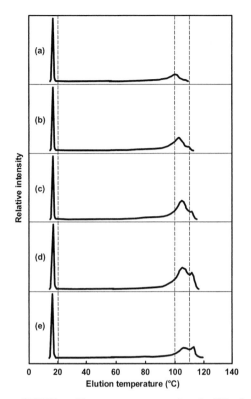

Figure 1. Dependence of TREF profiles on pretreatment time for PPs obtained in stopped-flow polymerization with TEA-pretreated TiCl$_4$/MgCl$_2$ catalysts, pretreatment time: (a) 0 s (i.e. without pretreatment), (b) 0.2 s, (c) 2 s, (d) 10 s, (e) 60 s.

isospecific sites (**IS$_3$**), respectively. There only exist three kinds of active sites namely **AS**, **IS$_1$** and **IS$_2$** on the donor-free catalyst without pretreatment (Figure 1 and Table 1). Apparently, the absence of electron donor accounts for the existence of large amount of **AS** as well as the non-existence of **IS$_3$** on the catalyst without pretreatment. The most interesting point is to find the emerging of **IS$_3$** on the donor-free catalyst after the pretreatment, which is clearly illustrated in Table 1 and Figure 1. Moreover, the amount of **IS$_3$** is obviously increasing especially up to ca. 10 s of pretreatment. These evidences clearly show **IS$_3$** can also be formed through the interaction between the catalyst and cocatalyst even in the absence of electron donor. The successive formation of **IS$_3$** during pretreatment can be ascribed to the successive transformation of some active sites with lower isospecificity (e.g. **AS**, **IS$_1$** and **IS$_2$**)

into **IS₃** by some secondary bimetallic complexing reactions between the activated titanium species and TEA (or the reaction product diethylaluminumchloride (DEAC)). The **AS** seems to be most easily deactivated up to 2s of pretreatment. The stability of the active sites increases with increasing isospecificity of the active sites probably due to the fact that the active sites with higher isospecificity are usually more sterically hindered and less acidic, thus more over-reduction-resistant.[13] The deactivation of **IS₁** and **IS₂** becomes dominant from 10s and 60s of pretreatment, respectively.

Monoester-Type TiCl₄/EB/MgCl₂ Catalyst

Stopped-flow polymerizations were conducted using TiCl₄/EB/MgCl₂ catalyst pretreated by TEA cocatalyst for 0 ~ 180s. The [C*], k_p and weight percentage of four fractions of PPs are shown in Table 2. This catalyst showed lower [C*] and higher k_p, and produced PPs with much lower amount of atactic PP and much higher amount of isotactic PP compared with the donor-free catalyst. There exist four kinds of active sites namely **AS**, **IS₁**, **IS₂** and **IS₃** on this catalyst without pretreatment. Apparently, the presence of internal donor EB accounts for the existence of much larger amount of isospecific sites (including **IS₁**, **IS₂** and **IS₃**) as well as much lower amount of **AS** sites in comparison with the donor-free catalyst. One of the most interesting and important points is the presence of **IS₃** in the EB-containing catalyst even in the absence of pre-treatment, while the results for the donor-free catalyst showed that **IS₃** was

Table 2. The dependence of [C*], k_p and weight percentage of fractions of PPs on pretreatment time obtained in stopped-flow polymerization with the TiCl₄/EB/MgCl₂ catalyst.[a]

Pretreatment time (s)	[C*] mol%	k_p L/mol·s	Fraction[b] (%)			
			F1	F2	F3	F4
0	3.5	3070	4	35	53	8
0.2	2.6	3750	4	21	51	24
2	1.9	4110	4	19	50	27
10	1.1	4580	4	18	47	31
60	0.52	3660	7	22	47	24
180	0.46	3430	-	-	-	-

[a] The polymerization was carried out with TEA ([Al]=14mmol, Al/Ti=30) in heptane at 30 °C for ca. 0.15s after the pretreatment.
[b] Fractionated by TREF, weight fractions: F1 (~20°C), F2 (20~100°C), F3 (100~110°C), F4 (110~140°C).

12

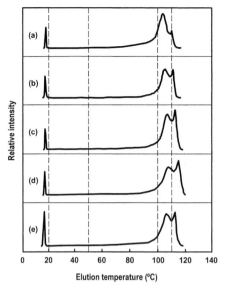

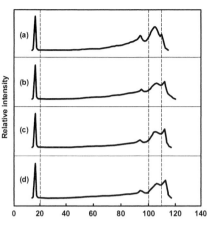

Figure 2.Dependence of TREF profiles on pretreatment time for PPs obtained in stopped-flow polymerization with TEA-pretreated TiCl₄/EB/MgCl₂ catalysts, pretreatment time: (a) 0 s, (b) 0.2 s, (c) 2s, (d) 10 s, (e) 60 s.

Figure 3. Dependence of TREF profiles on pretreatment time for PPs obtained in stopped-flow polymerization with TEA-pretreated TiCl₄/DBP/Mg(OEt)₂ catalysts, pretreatment time: (a) 10 s, (b) 60 s, (c) 180 s, (d) 600 s.

formed only after pre-treatment with the cocatalyst. This is clearly illustrated in Table 2 and also in curve (a) of Figure 2. So far, the main effects of the addition of internal donor EB to the TiCl₄/MgCl₂ catalyst can be summarized as follows: (1). to significantly restrain the formation of **AS** sites; (2). to drastically promote the formation of **IS₂** sites; (3). to construct some new **IS₃** sites. The **AS** seems to be not so easily deactivated during pretreatment compared with the **IS₁** and **IS₂** on this catalyst, as well as the **AS** on the donor-free catalyst. It means that the **AS** becomes much more stable after addition of donor EB. As for the isospecific active sites, the stability is still increasing with increase of isospecificity. Another most interesting and important point here is the successive increasing of **IS₃** sites due to transformation of active sites with lower isospecificity (e.g. **AS**, **IS₁** and **IS₂**) into **IS₃** sites up to 10 s of pretreatment derived from bimetallic complexing between the catalyst and cocatalyst. This active site transformation is especially dominant within 0.2 s of pretreatment (Table 2 and Figure 2). The extraction of EB by TEA from the catalyst surface is thought to initiate from 10s of pretreatment judging from Table 2 and Figure 2 corresponding to the dynamic process of interaction between cocatalyst and internal donor approaching to an equilibrium state as suggested by Sacchi et al.[14]

Diester-Type TiCl$_4$/DBP/Mg(OEt)$_2$ Catalyst

Stopped-flow polymerizations were conducted using TiCl$_4$/DBP/Mg(OEt)$_2$ catalyst pretreated by TEA cocatalyst for 0 ~ 600s. The [C*], k_p and weight percentage of four fractions of PPs are shown in Table 3. This catalyst shows an induction period up to 0.2s. The unique feature in terms of slow active sites formation, and slow deactivation may be mainly ascribed to the much stronger electron donating effect from DBP compared with EB. The effect from residual amount of –OC$_2$H$_5$ ligand should be negligible due to the wash by toluene followed a second TiCl$_4$ treatment during catalyst preparation. There exist four kinds of active sites namely **AS, IS$_1$, IS$_2$** and **IS$_3$** on the catalyst with 2s of TEA pretreatment. The successive formation of **IS$_3$** sites from 10s to 60s of pretreatment can be observed from Table 3 and Figure 3, which can be ascribed to similar active sites conversion through bimetallic complexing. The stability of active sites increases with the increasing isospecificity of the active sites up to 60s of pretreatment. Thereafter, all types of active sites on the catalyst become relatively stable.

Table 3. The dependence of [C*], k_p and weight percentage of fractions of PPs on pretreatment time obtained in stopped-flow polymerization with the TiCl$_4$/DBP/Mg(OEt)$_2$ catalyst.[a]

Pretreatment time (s)	[C*] mol%	k_p L/mol·s	Fraction[b] (%) F1	F2	F3	F4
0	-	-	-	-	-	-
0.2	-	-	-	-	-	-
2	0.38	1500	-	-	-	-
10	0.62	2810	24	40	29	7
60	0.41	3730	21	41	23	15
180	0.38	3840	30	32	24	14
600	0.38	3750	34	32	20	14

[a] The polymerization was carried out with TEA ([Al]=14mmol, Al/Ti=30) in heptane at 30 °C for ca. 0.15s after the pretreatment.
[b] Fractionated by TREF, weight fractions: F1 (~20°C), F2 (20~100°C), F3 (100~110°C), F4 (110~140°C).

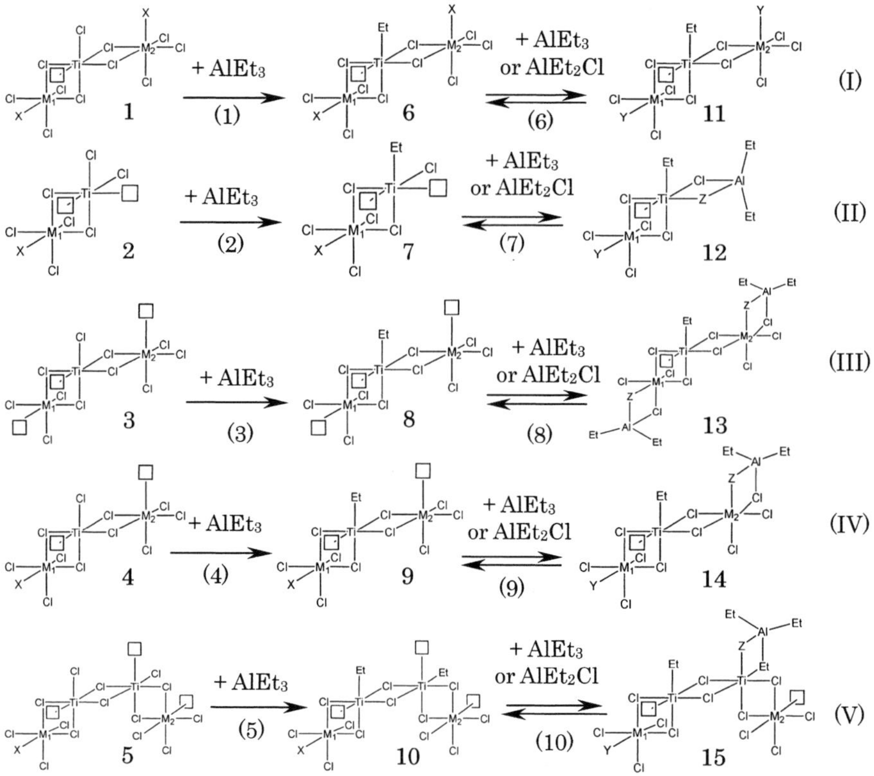

Scheme 1. Modified three-sites model in terms of formation and transformation of stereospecific active sites on heterogeneous Ziegler-Natta catalyst, M_1 and M_2: Ti or Mg, and M_1 and M_2 are bound to the catalyst substrate through chlorine bridges; X: Cl, or ED; Y: Cl, Et, or ED; Z: Cl or Et; ꞁ: coordination vacancy, for donor-free $TiCl_4/MgCl_2$ catalyst: X=Cl, Y=Cl or Et; for $TiCl_4/EB/MgCl_2$ and $TiCl_4/DBP/Mg(OEt)_2$ catalyst: X=Cl or ED, Y=Cl or ED or Et.

The extraction of DBP by TEA is thought to occur from 60s of pretreatment judging from Table 3 and Figure 3 and was observed to be much more difficult, slower and in much lower extent than the case with EB. This is consistent with some previous reports that DBP can coordinate much more strongly with catalyst surface compared with EB.[6, 13]

Modified Three-Sites Model

Busico et al. proposed a three-sites model of stereospecific active sites, which provides a reasonable explanation for the formation of stereoblock characteristics of PPs synthesized with Ziegler-Natta catalysts.[15] Whereas, the substantial difference regarding stereospecific roles between catalytic titanium species, alkyl-Al cocatalyst, $MgCl_2$ support and electron donor has not been specified in this model. According to our new understandings, a modified three-sites model of stereospecific active sites (as shown in Scheme 1) for $MgCl_2$-supported Ziegler-Natta catalysts was depicted as follows. The stereospecificity of active sites on the catalyst without TEA pretreatment was thought to originate mainly from the catalyst substrate namely neighboring $MgCl_2$, titanium chloride species and electron donor (ED). The most typical active sites formation reactions are summarized as reactions (1)~(5) in Scheme 1. Before contact with TEA cocatalyst, there exist different kinds of of Ti- precursors (**1~5** in Scheme 1) with different local steric environments. These Ti- precursors (**1~5**) were activated to form active sites (**6~10** in Scheme 1) on the catalyst after the first contact with TEA.[16] The isospecificity of these active sites is thought to be mainly determined by their local steric environments in terms of the number of coordination vacancy, pendant chlorine and ED.[17, 18] Interconversions between these active sites might be induced by ligand migration on the surface of catalyst substrate.[6] **6** with the highest steric hindrance among **6~10** is isospecific active site. For the donor-free catalyst, **6** is mostly composed of multinuclear titanium species (with X=Cl), which was previously speculated to be isospecific active sites by Soga et al.[5, 6] and Busico et al.[7]. The island model of monolayer multinuclear titanium species was further established to describe one typical existing state of titanium species in either donor-free or donor-contained Ziegler-Natta catalysts.[11d] One of the most important points is that **6** with X=Cl is only **IS₂**, which can not produce PP with the highest isotacticity. This means that the bulkiness of the chlorine atoms on the X positions in **6** is still not enough to construct **IS₃**. Further contact with alkyl-Al compounds during pretreatment (reaction (6)) gets the isospecific site **11**, which can be a **IS₃** when ethyl ligands are introduced into the X positions through ligand exchange between the catalyst and cocatalyst during pretreatment. For donor-contained catalysts, both **6** with X=ED and **11** with Y=ED are **IS₃** sites. **7** with the lowest steric hindrance around is **AS** and can not act as isospecific site even when X=ED due to the existence of two coordination vacancies. Further contact with alkyl-Al compounds during pretreatment (reaction (7)) can transfer **7** (**AS**) into12 (**IS₃**) through Al-Ti bimetallic complexing reaction. **8** is a syndiospecific site governed by chain-end control mechanism.[15] According to the reports by Doi,[9] Xu et al.[10] and Busico et al.[15], the syndiotactic-sequence-rich stereoblock PP mainly existed in the atactic PP fraction most probably due to the poor stability of **8**. Further contact with alkyl-Al compounds during pretreatment (reaction (8)) can transfer **8** into **13** (**IS₃**) through Al-Ti bimetallic complexing reaction. **9** and **10** (both **IS₁**), which can only produce poorly isotactic PP,[15] can also be converted into **14** (**IS₃**) and **15** (**IS₃**), respectively, by bimetallic complexing reactions (reactions (9) and (10)) during pretreatment. It is worth to mention that **10** is actually a twin-site involving two neighboring **IS₁** centers. The bimetallic complexing during pretreatment might deactivate one center [8, 19] and consequently the other is transferred to **15** (**IS₃**) (reaction (10)). The essential of these reversible bimetallic complexing reactions is to increase local steric hindrance with asymmetry and chirality through bridging with coordination vacancy, pendant chlorine or alkyl ligand around the activated Ti-species. The presence of much bulkier group on the two most important ligand positions on these bimetallic active sites (**12~15**) seems to be crucial for their higher isospecificity compared with those monometallic active sites (**6~10**) derived solely from the catalyst substrate. Extraction of ED by cocatalyst from those ED-contained sites (**6**, **11**, **12**, **13**, **14**, **15** with X=ED or Y=ED) will transfer these isospecific sites into active sites with lower isospecificity.

Conclusion

In this work, we have shown the recent new and precise understanding regarding the stereospecific nature of active sites obtained through the combination of TREF method with stopped-flow technique using heterogeneous Ziegler-Natta catalysts. A modified three-sites model with precise description of the stereospecific nature of various types of active sites stemmed from surface titanium species, Al-alkyl compounds, Mg-compounds and electron donors was proposed. It has been shown that isospecificity of active sites strongly depends on the bulkiness of the ligands situated at the two most important ligand positions for construction of the asymmetry and chirality of the active sites with steric hindrance. The bulkiness of Cl atoms at the positions is not enough for the construction of the highest isospecific active sites, which is necessary for the introduction of much bulkier ligands e.g. ethyl group or ED through either ligand exchanging or bimetallic complexing reactions. The existence of aspecific sites with two coordination vacancies should be also taken into consideration. These aspecific sites can be possibly transferred into the highest isospecific sites through Al-Ti bimetallic complexing reactions. In general, it can be concluded that there might exist both monometallic active sites and bimetallic active sites in the conventional heterogeneous Ziegler-Natta catalyst systems. The kinds of active titanium species with different chemical structures on the heterogeneous catalysts should be limited, whereas, the non-discrete distribution of isospecificity of active sites could be considered to generate from the numerous types of steric and electronic effects from the surroundings of the active titanium species as well as large number of reversible and dynamic transformation reactions simultaneously occurred on the heterogeneous catalyst surface with multi-components. The modified three-sites model combined with the island model gave rise to much more precise understanding of the real origin of stereospecificity of active sites on $MgCl_2$-supported Ziegler-Natta catalysts.

Acknowledgments

The authors thank to Mitsubishi Chemical Co., Mitsui Chemical Co., Toho Titanium Co., Ltd., Asahi Denka Co., Ltd., Chisso Corp., and Tosoh Akzo Corp., for their support and donation to our laboratory.

[1] [1a] K. Ziegler, *Angew. Chem.* **1955**, *67*, 426; [1b] G. Natta, P. Pino, P. Corradini, F. Danusso, E. Mantica, G. Mazzanti, G. Moraglio, *J. Am. Chem. Soc.* **1955**, *77*, 1708.
[2] E. P. Moore, Jr., "*The Rebirth of Polypropylene: Supported Catalysts*", Hanser Publishers, Munich 1998.
[3] B. Liu, H. Matsuoka, M. Terano, *Macromol. Rapid Commun.* **2001**, *22*, 1.
[4] M. C. Sacchi, F. Forlini, I. Tritto, P. Locatelli, *Macromol. Chem. Phys.* **1995**, *196*, 2881.
[5] K. Soga, J. R. Park, H. Uchino, T. Uozumi, T. Shiono, *Macromolecules* **1989**, *22*, 3824.
[6] K. Soga, T. Shiono, Y. Doi, *Makromol. Chem.* **1988**, *189*, 1531.
[7] V. Busico, P. Corradini , L. D. Martino, A. Proto, *Makromol. Chem.* **1986**, *187*, 1115.
[8] L. A. M. Rodriguez, H. M. Van Looy, *J. Polym. Sci.: Part A-1* **1966**, *4*, 1971.
[9] Y. Doi, *Makromol. Chem., Rapid Commun.* **1982**, *3*, 635.
[10] J. Xu, L. Feng, S. Yang, *Macromolecules* **1997**, *30*, 2539.
[11] [11a] H. Matsuoka, B. Liu, H. Nakatani, M. Terano, *Macromol. Rapid Commun.* **2001**, *22*, 326; [11b] B. Liu, H. Matsuoka, M. Terano, *Macromol. Symp.* **2001**, *165*, 3; [11c] H. Matsuoka,   B.   Liu,   H. Nakatani, I. Nishiyama, M. Terano, *Polym. Int.* **2002**, *51*, 781; [11d] I. Nishiyama, B. Liu, H. Matsuoka, H. Nakatani, M. Terano, *Macromol. Symp.* **2003**, *193*, 71.

18

[12] [12a] T. Nitta, B. Liu, H. Nakatani, M. Terano, *J. Mol. Catal. A: Chem.* **2002**, *180*, 25; [12b] B. Liu, T. Nitta, H. Nakatani, M. Terano, *Macromol. Chem. Phys.* **2002**, *203*, 2412; [12c] B. Liu, T. Nitta, H. Nakatani, M. Terano, *Macromol. Chem. Phys.*, **2003**, *204*, 395.

[13] V. Busico, P. Corradini, L. D. Martino, A. Proto, V. Savino, *Makromol. Chem.* **1985**, *186*, 1279.

[14] M. C. Sacchi, L. Tritto, P. Locatelli, *Prog. Polym. Sci.* **1991**, *16*, 331.

[15] V. Busico, R. Cipullo, G. Monaco, G. Talarico, and M. Vacatello, J.C. Chadwick, A.L. Segre and O. Sudmeijer, *Macromolecules* **1999**, *32*, 4173.

[16] P. Pino, G. Fochi, O. Piccolo, U. Giannini, *J. Am. Chem. Soc.* **1982**, *104*, 7381.

[17] E. J. Arlman, *J. Catal.* **1966**, *5*, 178.

[18] M. Kakugo, T. Miyatake, Y. Naito, and K. Mizunuma, *Macromolecules* **1988**, *21*, 314.

[19] V. A. Zakharov, G. D. Bukatov, Y. I. Yermakov, *Makromol. Chem.* **1975**, *176*, 1959.

Macromol. Symp. **2004**, *213*, 19-28

Recent Data on the Number of Active Centers and Propagation Rate Constants in Olefin Polymerization with Supported ZN Catalysts

V.A. Zakharov,* G.D. Bukatov, A.A. Barabanov

Boreskov Institute of Catalysis, Novosibirsk, 630090, Russia
E-mail: v.a.zakharov@catalysis.nsk.su

Summary: Data on the number of active centers (Cp) and propagation rate constants (Kp) have been obtained by means of polymerization quenching with ^{14}CO of propylene and ethylene polymerization with supported titanium-magnesium catalysts (TMC) with different composition. In the case of propylene polymerization the Cp and Kp values have been measured separately for isospecific, aspecific and low stereospecific centers. Effects of $MgCl_2$ support, internal and external donors are discussed on the basis of data obtained.
Data on the strong effect of diffusion limitation at ethylene polymerization with number of TMC have been obtained and a set of methods have been used to exclude this effect. Data on Cp and Kp values at ethylene polymerization with low stereospecific and highly stereospecific catalysts are presented.

Keywords: active centers; polyethylene; polypropylene; stereospecificity; Ziegler-Natta polymerization

Introduction

In 1959 Prof. G. Natta published a brilliant survey on the kinetics of stereospecific polymerization of olefines[1]. The author presented systematic data on the kinetics of propylene polymerization with heterogeneous catalyst [$TiCl_3$ + $AlEt_3$] and analyzed the complex process of catalytic polymerization including the reactions of initiation, propagation and chain transfer with active centers which he defined as metalloorganic complexes on the surface of $TiCl_3$. The main parameters determining the catalyst activity are the number of active centers and propagation rate constant. Approaches and concepts formulated in this paper formed the basis for further kinetic studies of catalytic olefin polymerization.

In catalytic olefin polymerization the number of active centers usually accounts for a small part of the content of transition metal in the catalyst and depends on the catalyst preparation procedure, composition and polymerization conditions. Low concentration of active centers,

DOI: 10.1002/masy.200450903

their high reactivity and lability and involvement in side reactions hamper the task of evaluation of the active centers number. The method of polymerization inhibition by radioactive CO (^{14}CO)[2,3] and stopped flow method[4-6] are considered as the most efficient for direct determination of the number of active centers (C_p) and propagation rate constant (K_p).

The present work reports the data on Cp and Kp, obtained by the method of polymerization inhibition by ^{14}CO, for the ethylene and propylene polymerization with highly active supported ZN catalysts. The method differs by some peculiarities related to the conditions of polymerization, inhibition by ^{14}CO and subsequent removal of labeled by-products.[7, 8] The presence of hydrogen is a necessary condition for correct calculation of Kp in propylene polymerization, since without hydrogen a part of active centers fall into a dormant state.[8, 9] These details were taken into account in the present work.

Experimental

Microspherical TiCl$_3$ catalyst (26 wt.% of Ti, average particle size of 15μm) is a commercial sample. TiCl$_4$/MgCl$_2$ catalysts with different Ti content (0.1 and 2.1 wt.% of Ti) have been prepared by adsorbtion of TiCl$_4$ on highly dispersed MgCl$_2$ synthesized according to.[10] TiCl$_4$/MgCl$_2$·nD catalysts (2.5 wt.% of Ti) have been prepared according to Zakharov et al.[11] [D: dibutylphthalate (DBPh) or 2,2-diisobutyl-1,3-dimethoxypropane (DBDMP)].

The procedures of polymerization, Cp and Kp determination with ^{14}CO have been described by Bukatov et al.[8, 9]

Results and Discussion

1. Propylene polymerization

We have studied the effect of catalyst composition (MgCl$_2$ support, internal and external donors) on Cp and Kp values for propylene polymerization with MgCl$_2$-supported catalysts (titanium-magnesium catalysts - TMC). There are data on the different reactivity of aspecific and isospecific centers at propylene polymerization with TMC.[6,8] We believe it is important to obtain in more detail data on the effect of catalyst composition on the number of AC with different stereospecificity and Kp values for AC with different stereospecificity.

We have performed experiments with quenching of propylene polymerization by [14]CO using catalysts with different composition. PP obtained in these experiments have been fractionated to give three fractions:

- fraction soluble in boiling pentane (PP5 fraction – atactic PP)
- fraction soluble in boiling heptane (PP7 fraction – low stereoregular PP)
- fraction insoluble in boiling heptane (IPP fraction – isotactic PP).

Then we have measured Cp and Kp values for these separate fractions obtained from propylene polymerization over all catalysts studied. Data obtained on Cp and Kp values for polymerization in the presence of hydrogen are presented in Table 1. These data allow evaluation of the combined effect of Cp and Kp values on the relative yield of different PP fractions (Y_f) with different stereoregularities. The results obtained for different catalysts are presented in Figure 1 as areas calculated according to simple equation $Y_f = C^f_p \times K^f_p$ for corresponding fractions (f).

We have done additional polymerizations without hydrogen and we used Kp values, calculated in experiments with hydrogen (Kp^{+H}) and without hydrogen (Kp^{-H}) for estimation of the portion of dormant centers (P_d) formed at polymerization without hydrogen:

$$P_d (\%) = [1 - Kp^{-H} / Kp^{+H}] \times 100 \qquad (1)$$

These data are presented in Table 1.

1.1. Effect of MgCl$_2$ support on Cp and Kp values

Data on Cp and Kp values at propylene polymerization with TiCl$_3$ and supported TiCl$_4$/MgCl$_2$ catalyst (TMC-1) are presented in Table 1. The supported catalyst has the higher activity and lower stereospecificity. Data on Cp and Kp values allow the following conclusions to be made:

(i) the higher activity of TiCl$_4$/MgCl$_2$ catalyst in comparison with TiCl$_3$ is determined mainly by a higher total Cp value;

(ii) Kp values for isospecific centers (Kp for IPP fractions) are close for TiCl$_3$ and TiCl$_4$/MgCl$_2$ catalyst, indicating similar structures of isospecific active centers for both catalysts;

(iii) Kp(PP7) and Kp(PP5) values (low stereospecific and aspecific centers) for supported TiCl$_4$/MgCl$_2$ catalyst are higher than corresponding constants for TiCl$_3$ catalyst.

Table 1. Data on Cp and Kp values at propylene polymerization over catalysts with different composition (70°C, AlEt$_3$ as cocatalyst, H$_2$/C$_3$H$_6$=0.15)

		TiCl$_3$	TiCl$_4$/MgCl$_2$ (TMC-1)	TiCl$_4$/MgCl$_2$·nDBPh (TMC-2)	TiCl$_4$/MgCl$_2$·nDBDMP (TMC-3)	TiCl$_4$/MgCl$_2$·nDBPh + DCPDMS[3] (TMC-4)
Rp[1], kg/g Ti·h·atm		1.6	16.8	26.0	25.7	37.2
Content of PP fractions, wt.%	PP5	17.1	43	10.6	3.1	1.9
	PP7	10.5	27	14.7	6.9	2.7
	IPP	72.4	30	74.7	90	95.4
Total Cp, mmol/mol Ti		2.34	11.5	23.3	18.4	21.6
Portion of Cp (%) in fractions:	PP5	45	50	41	25	14
	PP7	25	34	25	32	23
	IPP	30	16	34	43	63
Kp (L/mol·s) for fractions:	PP5	250	1230	280	170	220
	PP7	270	1120	660	300	190
	IPP	1580	2590	2480	2820	2560
Portion of dormant sites[2] (P$_d$,%) for fractions:	PP5	ca.0	53	43	71	85
	PP7	52	63	58	81	68
	IPP	60	47	73	84	72

[1]Polymerization rate at the moment of ^{14}CO addition. [2] At polymerization without hydrogen.
[3]DCPDMS-dicyclopentyldimethoxysilane as an external donor.

It means, probably, that the structures of weakly stereospecific and aspecific centers for TiCl$_4$/MgCl$_2$ catalyst differ from corresponding centers for TiCl$_3$ catalyst.

So the MgCl$_2$ support increases sharply the total number of AC and affects the structures and reactivity of weakly stereospecific and aspecific centers. The MgCl$_2$ support affects the yield of PP fractions with different stereoregularities via the combined effect of Cp and Kp values (Figure 1): the lower content of IPP fraction for the TiCl$_4$/MgCl$_2$ catalyst in comparison with TiCl$_3$ is determined by a lower portion of isospecific AC and higher Kp values for PP7 and PP5 fractions.

1.2. Effect of internal donors on Cp and Kp values at polymerization with TMC

Data on effect of internal donors (DBPh and DBDMP in composition of TMC) on Cp and Kp values are presented in Table 1 (catalysts TMC-1, TMC-2 and TMC-3).

Catalysts TMC-2 and TMC-3 with internal donors (ID) have the higher activity, higher total Cp values and higher content of IPP fraction in comparison with TMC-1 catalyst. But the procedures for preparation of TMC-2 and TMC-3 catalysts differ greatly from the procedure for the TMC-1 catalyst. We suppose the total Cp value increases for TMC-2 and TMC-3 catalysts by way of affect on the morphology of catalysts (higher surface area) on the stage of catalyst preparation. The direct and most important effect of ID is the increase of the portion of isospecific centers [Cp(IPP)] from 16% for TMC-1 catalyst up to 34 and 43 % for TMC-2 and TMC-3 catalysts respectively.

Internal donors don't affect Kp values for isospecific centers. There is a rather unexpected effect of ID on Kp values for aspecific and low stereospecific AC: addition of ID into TMC gives a decrease of Kp values for PP7 and PP5 fractions.

Finally we can conclude on the combined effect of Cp and Kp values on the stereospecificity of TMC with internal donors: the yield of IPP fraction increases mainly by way of increase of the portion of isospecific centers and a decrease of Kp values for PP7 and PP5 fractions (Figure 1).

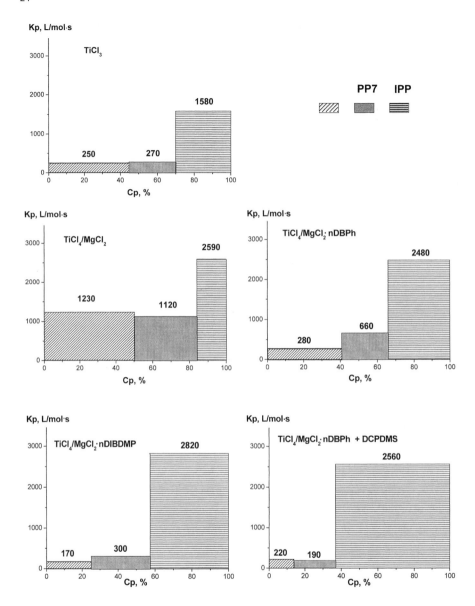

Figure 1. Relative yield of PP fractions with different stereoregularity as combined effect of C_P and K_P values for catalysts with various compositions

1.3. Effect of external donors on Cp and Kp values at polymerization with TMC

Data on the effect of external donor (DCPDMS), added into polymerization with catalyst $TiCl_4/MgCl_2 \cdot nDBPh$ (TMC-2), on Cp and Kp values are presented in Table 1. Addition of ED leads to the increase of activity and stereospecificity. External donor doesn't affect the total Cp value but increases the portion of isospecific centers from 34% up to 63% and decreases the portion of aspecific AC from 41% to 14%. We believe these results evidence the transformation of aspecific centers into isospecific AC by adsorbtion of ED on the catalyst surface.

External donor doesn't affect Kp values for isospecific centers (IPP fraction) but decreases the Kp value for PP7 fraction.

We can conclude on the combined effect of Cp and Kp values on the activity and stereospecificity of TMC with external donor:

 (i) activity increases because of the increase of the portion of Cp(IPP) with higher Kp value;

 (ii) stereospecificity increases because of the increase of the portion of Cp(IPP) and the decrease of the Kp(PP7) value (Figure 1).

The possible structures of isospecific, low stereospecific and aspecific centers are discussed in literature.[5,6,12]

2. Ethylene polymerization

As we have shown above the isospecific and aspecific centers have different reactivity (Kp values) in propylene polymerization. It is interesting to obtain data on Cp and Kp values at ethylene polymerization using weakly and highly stereospecific catalysts. We have selected for this study the following catalysts:

(1) $TiCl_4/MgCl_2$ catalyst with low content of titanium (0.1wt.% of Ti). This catalyst produces PP with a very low content of IPP fraction (10 wt.%) and contains ca. 95% of weakly strereospecific and aspecific AC.

(2) Catalyst $TiCl_4/MgCl_2 \cdot nDBPh$ + ED (ED: propyltrimethoxysilane). This catalyst produces mainly isotactic PP (96% of IPP fraction) and contains ca. 63% of isospecific AC.

But we have found that TMC with composition $TiCl_4/MgCl_2 \cdot nDBPh$ prepared for stereospecific propylene polymerization according to Ref. [11] have a very low activity in ethylene polymerization. Later we have found that activity of these catalysts at ethylene polymerization depends on porosity of catalysts. As an example data on activity at ethylene and propylene polymerization are shown in Table 2 for two catalyst samples with different porosity. These catalyst samples have a similar high activity in propylene polymerization but their activity differs greatly in ethylene polymerization. Catalyst (II) with higher porosity has much higher activity (Table 2). We propose that diffusion limitation is the possible reason of low activity in ethylene polymerization for catalysts with dense particles (with low porosity).

Table 2. Activity of catalysts $TiCl_4/MgCl_2 \cdot nDBPh$ with different porosity at ethylene and propylene polymerization (70°C, $AlEt_3$ as cocatalyst, PTMS[1])

Catalyst	Porosity	Rp kg PE/g cat·h·atm	Rp kg PP/g cat·h·atm
I	Low	0.13	0.95
II	High	0.73	0.70

[1] PTMS: propyltrimethoxysilane as an ED

The effect of diffusion limitations on the catalyst activity could be evaluated by the analysis of polymerization kinetics with catalysts differing by particle size. It is well known that as the size of catalyst particles increases, the effect of diffusion limitations on the catalyst activity becomes stronger. If polymerization proceeds within diffusion region, the propagation rate and calculated Kp values decrease with increasing catalyst particle size. For slurry polymerization, the effect of catalyst particle size on the Kp value may be estimated based on the analysis of various polymer fractions with different particle size. This approach is based on a replication phenomenon, which involves a close relationship between the catalyst particle size and the final polymer particle size. Table 3 presents the results of two experiments with the catalysts of different compositions. After polymerization inhibition by ^{14}CO, the PE powders obtained in these experiments were sieved into separate fractions and the Cp and Kp values were calculated for each fraction. It is seen that the Kp value for catalyst I depends on particle size and dramatically increases with decreasing particle size. Obviously, the low average Kp value for catalyst I ($2.3 \cdot 10^3$ L/mol·s) is dictated by diffusion restrictions in ethylene polymerization with this catalyst. The calculated Kp value for catalyst II does not depend on particle size, that is, ethylene polymerization on catalyst II proceeds in kinetic region.

An efficient method to diminish or avoid diffusion restrictions consists of a low-temperature prepolymerization at low monomer pressure. We found that the activity of catalyst $TiCl_4/MgCl_2 \cdot nDBPh$ in ethylene polymerization increased sharply after prepolymerization with propylene at 40°C. Table 4 illustrates the effect of prepolymerization on the catalyst activity and calculated Cp and Kp values. It is seen that prepolymerization causes sharp increases in both the activity and the Kp value. It seems reasonable to assume that in this case the calculated Kp ($1.2 \cdot 10^4$ L/mol·s) is close to the real Kp value.

Table 3. Ethylene polymerization over TMC with different composition: effect of particle size on Kp values (70°C, AlEt$_3$ as cocatalyst)

	(I) $TiCl_4/MgCl_2 \cdot nDBPh$ (2.3% Ti)	(II) $TiCl_4/MgCl_2$ (0.1%)
Activity, kg/g cat·h·atm	0.47	0.25
kg/g Ti·h·atm	20	250
Total Cp, mol/mol Ti	0.07	0.18
$Kp \cdot 10^{-3}$, L/mol·s:		
- for total PE	2.4	11.5
- for fraction of 1-2mm*)	1.8	12.0
- for fraction of 0.09-0.25mm*)	8.4	11.3

*) PE powders have been separated by means of sieves on fractions with different particle size

Table 4. Ethylene polymerization over TMC with composition $TiCl_4/MgCl_2 \cdot nDBPh$: effect of prepolymerization on the activity, Cp and Kp values (70°C, AlEt$_3$, DCPDMS as ED)

Exp. No.	Prepolymerization with C_3H_6 at 40°C	Rp kg/g cat·h·atm	Cp mol/mol Ti	Kp L/mol·s
1	-	0.11	0.043	900
2	+	2.3	0.063	12100

Finally we can compare data on Kp values at ethylene polymerization for weakly and highly stereospecific catalysts (Table 5). We can conclude there are no noticeable differences for Kp values in ethylene polymerization on stereospecific and nonstereospecific active centers.

Table 5. Data on Cp and Kp values at ethylene polymerization over TMC with different stereospecificity (70°C, AlEt$_3$)

Catalyst	Content of IPP fraction[1] wt.%	Portion of Cp(IPP)[1] %	Rp kg/g Ti·h·atm	Cp mol/mol Ti	Kp L/mol·s
TiCl$_4$/MgCl$_2$ (0.1% Ti)	10	5	250	0.18	11500
TiCl$_4$/MgCl$_2$·nDBPh + DCPDMS	96	63	100[2]	0.063	12100

[1] At propylene polymerization
[2] Prepolymerization with propylene at 40°C

Conclusion

In the absence of donors, a TiCl$_4$/MgCl$_2$ catalyst has a much higher active center content but a lower proportion of isospecific centers than TiCl$_3$ catalyst. Internal and external donors increase the portion of isospecific centers, and decrease Kp values for aspecific and weakly stereospecific centers. Kp values for isospecific centers are higher than for aspecific centers in propylene polymerization with all catalysts studied.

Experimental data on the strong effect of diffusion limitations in ethylene polymerization with a number of TMC have been obtained and a set of methods has been used to exclude this effect. It was found that Kp values at ethylene polymerization on isospecific and nonstereospecific active centers are close.

Acknowledgement

The authors thank *V.Goncharov, S,Sergeev and T.Mikenas* for experimental.

[1] G.Natta, I.Pasquon, .*Advances in Catalysis*, **1959**, 11, 2.
[2] G.D.Bukatov, V.A.Zakharov, Yu.I.Ermakov, *Macromol.Chem.*, **1978**, 179, 2097.
[3] V.A.Zakharov, G.D.Bukatov, Yu.I.Ermakov, *Adv. Polym. Sci.*, **1983**, 51, 61.
[4] M.Terano, T.Kataoka, T.Keii, *J.Mol.Catal.*, **1989**, 56, 203.
[5] H.Matsuoka, B.Liu, H.Nakatani, M.Terano, *Macromol. Rapid Commun.*, **2001**, 22, 326.
[6] B.Liu, T.Nitta, H.Nakatani, M.Terano, *Macromol.Chem.Phys.*, **2002**, 203, 2412.
[7] G.D.Bukatov, V.S.Goncharov, V.A.Zakharov, *Macromol.Chem.*, **1986**, 187, 1041.
[8] G.D.Bukatov, V.A.Zakharov, *Macromol.Chem.Phys.*, **2001**, 202, 2003.
[9] G.D.Bukatov, V.S.Goncharov, V.A.Zakharov, *Macromol.Chem.*, **1995**, 196, 1751.
[10] WO 95/35163 (1995), invs.: V.A.Zakharov, S.A.Makhtarulin,, *Chem.Abstr.*, **1996**, 124, 177252.
[11] WO 96/32427 (1996), invs.: V.A.Zakharov, G.D.Bukatov, S.A.Sergeev.
[12] V.Busico, R.Cipullo, *Prog.Polym.Sci.* 2001, 26, 443.

Macromol. Symp. **2004**, *213*, 29-36

Comparison of Silyl Chromate and Chromium Oxide Based Olefin Polymerization Catalysts

Kevin Cann,* Maria Apecetche, Minghui Zhang

Univation Technologies LLC, Bound Brook Technical Center, Piscataway, NJ 08854, USA

Summary: A comparison of the ethylene polymerization performance conducted with an oxo and a triphenylsilyl chromate catalyst on silica was performed. The oxo catalyst has higher activity and better comonomer response. The silylchromate catalyst has a much longer induction time and made a much broader molecular weight distribution polymer compared to the oxo analogue. Performance similar to silylchromate on silica was observed when triphenylsilanol (TPS) was added to the oxo chromium catalyst. The oxo catalyst was converted to the silyl chromate catalyst by ligand substitution. Analysis of the catalyst components when TPS was added to the oxo chrome analogue showed that bis triphenylsilyl chromate can form and be removed from the support.

Keywords: catalysts; chromium; polyethylene; polymerization; silyl chromate

Introduction

Polymerization of ethylene with supported chromium based catalysts has been known for over forty years. Most of the chromium based catalyst work has focused on oxo chromium systems, commonly referred to as the Phillips-type catalyst (Cr/SiO_2)[1,2]. However, there has also been a significant amount of work based on the organochromium systems chromocene[3] and organosilyl chromate. Of the organochrome-based systems, silyl chromate [4,5,6] derived catalysts are used extensively for commercial PE production. Organochrome-based systems may have a different ligand environment than that found with air activated oxo catalysts. A study has been conducted to better understand the relative structure and performance of the Cr^{+6} catalysts based on chromium oxide and silyl chromate.

Results and Discussion

The oxo chromium catalyst is well known and prepared by thermal oxidation of silica impregnated with a variety of chromium compounds. Activation is done in air between 600-

      DOI: 10.1002/masy.200450904

850 °C. The silyl chromate catalyst is formed by chemisorption of bis(triphenylsilyl) chromate on dehydrated silica in a hydrocarbon slurry producing triphenyl silanol (TPS) as a byproduct after the interaction (Figure 1).

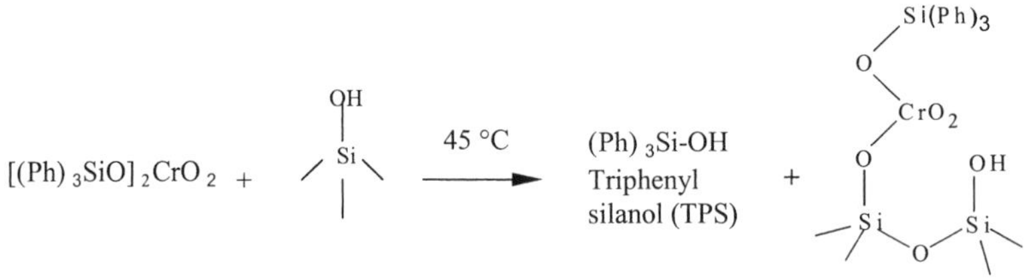

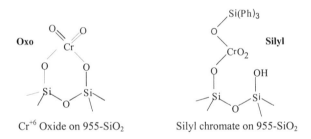

$[(Ph)_3SiO]_2CrO_2$ +   45 °C $(Ph)_3Si-OH$ +
                                      Triphenyl
                                      silanol (TPS)

Figure 1. Chemisorption of silyl chromate on silica

Using the most simplistic models for Cr^{+6} catalysts based on the oxo [7] and organosilanol [6] (triphenylsilanol) derivatives, it can be seen that the structural features of these catalyst precursors are quite similar (Figure 2).

Oxo   **Silyl**

Cr^{+6} Oxide on 955-SiO$_2$ Silyl chromate on 955-SiO$_2$

Figure 2. Possible structures of Oxo and Silyl chromate catalysts on silica

The oxo-derived system is anchored through surface inorganic siloxides. The silyl chromate is anchored by one surface siloxide with triphenyl siloxide as the other organosiloxide ligand. Although similar in structure the polymerization performances of these catalyst precursors are very different.

Ethylene polymerization comparisons (Table 1): Employing the same silica support for each system (Davison 955 silica, 1.6 cc/g, 300m^2/g) it was observed that the oxo catalyst has much higher productivity. The silyl chromate catalyst is not only less active but has a prolonged

Table 1. Ethylene Polymerization Performance of Oxo and Silylchromate Catalysts

Response Catalyst	Induction Time min	Productivity[a] gPE/g Cat	FI(I_21) gPE/10 min	Density g/cc	Bulk Density g/cc
Oxo	72	1630	13.7	0.9428	0.41
Silyl	158	302	9.8	0.9549	0.51

[a] Polymerization Conditions: 600 cc isobutane, T = 95°C, 500 cc H_2, pC_2 = 200 psi, 10 cc 1-hexene comonomer

induction period. The comonomer incorporation rate for the oxo catalyst is also much higher based on observed lower polymer density. Higher polymer bulk density and molecular weight is found with the silyl chromate catalyst compared to the oxo catalyst under these reaction conditions.

The polymer molecular weight distribution (MWD) is very different for the two systems (Figure 3). Both catalysts produce broad MWD distribution polymers, but the silyl chromate catalyst produces broader polymer distributions (broader on both the high and low molecular weight ends) with a pronounced high molecular weight tail, almost indicative of a second type of active site. It appears that one of the TPS groups remains at the site after the active site is formed.

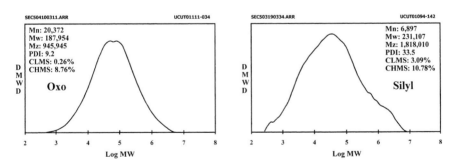

Figure 3. MWD of polymer made with oxo and silyl chromate catalysts

Conversion of Oxo to triphenylsilyl chromate catalyst: Since the catalyst structures are similar in nature it appears that it might be possible to form the silyl chromate derived structure shown in Figure 2 by ligand displacement on the oxo chromate compound. This was explored by adding triphenylsilanol (TPS) in hexane to the activated oxo chrome catalyst followed by drying under vacuum (eq. 1).

$$Cr(O)x/silica + TPS \text{ in Hexane} \xrightarrow{\text{Vacuum}} slurry \text{ TPS/Cr(O)x/silica dry catalyst} \quad (1)$$

The effect of the TPS on polymerization performance can be seen immediately (Figure 4). One equivalent of TPS results in a small decrease in activity and an increase in the initiation period. Addition of two equivalents of TPS results in a significant lowering of the catalyst activity with a large induction period. The kinetic profile produced with the oxo catalyst and two equivalents of TPS is typical of that observed with the silyl chromate catalyst. Table 2 shows the polymerization results. The catalyst productivity is lower with the TPS present due to a combination of lower inherent activity and increased induction times. The comonomer incorporation rate was also observed to decrease. This was determined by measuring the amount of 1-hexene incorporated into the polymer by ^{13}C NMR.

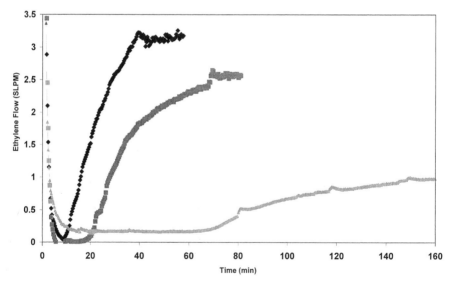

Figure 4. Effect of TPS on Oxo chromium ethylene polymerization kinetic profile; $T = 85°C$, $P = 10$ bar, 500 ml hexane slurry, 10 ml 1-hexene,
♦ = Oxo, ■ = Oxo + 1 TPS, ▲ = Oxo + 2 TPS

It is interesting that the presence of TPS inhibits the rate of active site formation (longer induction time) and lowers the rate of comonomer incorporation.

Table 2. Ethylene Polymerization Performance of Oxo Catalyst After TPS Addition

Catalyst Type	Yield g	Productivity g PE/g cat/h	FI(I$_{21}$) dg/10 min	BBF[a]	Density g/cc
Oxo	153	1,429	2.6	3.7	0.9392
Oxo + 1 eq TPS	161	607	2.1	2.9	0.9533
Oxo + 2 eq TPS	101	102	2.6	1.9	0.9517

[a] Butyl Branching frequency /1000 C as determined by ^{13}C NMR

The presence of the TPS may lower the oxidative capacity (less electro-positive) of the Cr^{+6} precursor which slows the initial reduction-oxidation reaction with monomer during site formation. The effect of TPS is still present after site formation as seen in poorer comonomer incorporation. Most likely the TPS is present as a ligand even in the reduced catalyst.

Measurement of the polymer produced in the experiments in Table 2 by size exclusion chromatography (SEC) shows that the TPS broadens the MWD of the polymer (Figure 5). One equivalent results in an increase of the dispersity index from near 10 to 13 with slight development of a high molecular shoulder.

It can be seen that the Mz component is rising faster than the Mw. The second addition of TPS results in formation of a polymer that matches very well the polymer produced by the silyl chromate catalyst.

The distribution is very broad and the high molecular shoulder is fully developed. Based on

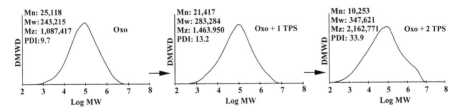

Figure 5. Effect of TPS on Oxo Chromium Polymer MWD

these results it looks like the silyl chromate catalyst can be formed through an oxo chrome intermediate.

Analysis of TPS addition to oxo chromium catalysts: Several reactions were conducted in which TPS was added to oxo chromium catalysts slurried in hexane under nitrogen. Thirty minutes after the TPS was added an aliquot of the liquid was taken. The remaining liquid was decanted away and the catalyst was washed three times more with hexane then dried under high vacuum. The reaction of TPS with oxo chromium catalyst was monitored by following the initial and final chromium loading on the support as well as the amount of chrome that dissolved in the hexane.

The first experiment shown in Table 3 was conducted on a chromium oxide catalyst activated at 825 °C with 0.4 wt % chrome present determined by ICP (Inductively-Coupled Plasma) elemental analysis. Measurement of the amount of Cr^{+6} present was determined by a standard starch iodine titration[6]. Addition of one equivalent of TPS to catalyst slurried in hexane resulted in a slight amount of chromium dissolving in the hexane.

Table 3. Effect of TPS on Oxo Chrome (Cr on Davison-955) Activated at Different Temperatures

$Cr^{a)}$ on support	Activation temperature	$Cr^{+6\,b)}$ on support	TPS/Cr	$Cr^{a)}$ on support after wash	Fraction of Cr on support after wash	Fraction of Cr lost from support	Fraction of $Cr^{b)}$ dissolved in hexane
wt %	°C	wt %		wt %	%	%	%
0.39	825	0.40	1	0.36	92	8	<4
0.39	825	0.40	2	0.28	72	28	22
0.27	600	0.27	1	0.26	>96	<4	<2
0.27	600	0.27	2	0.26	>96	<4	<3
0.48	600	0.49	1	0.45	94	6	<3
0.48	600	0.49	2	0.44	92	8	13

a) Determined by elemental analysis
b) Determined by starch iodine titration[6]

After extensive washing 92% of the chrome was found to be remaining on the support. This shows that the chromium mostly remained on the support. However, the results for the polymerization reactions with a catalyst with the same ratio of TPS/Cr show that the TPS is affecting polymerization kinetics and polymer properties. The second experiment employs the same oxo catalyst starting material but two equivalents of TPS were used. In this experiment the hexane turned yellow. Analysis of the hexane revealed that the solution contained in the form of Cr^{+6} up to 22 % of the chrome that was on the initial oxo catalyst. The dissolved chromium appears to be bis(triphenyl silyl) chromate, which is soluble in hexane. After extensive washing only 72% of the chrome was found to be present on the solid catalyst. This result shows that it is possible to make the silyl chromate catalyst by direct conversion of the chromium oxide catalyst with TPS. The reason chromium dissolved after formation of the silyl chromate derivative, even though as shown in Figure 1 silyl chromate reacts with the surface silanols, is because the surface becomes supersaturated at this chrome loading. It has been reported elsewhere[8] that 955 silica dehydrated at 825 °C becomes saturated with silyl chromate at a loading of only 0.16 wt % Cr.

The next two experiments in Table 3 were conducted with the same TPS /Cr ratios but the initial chrome loading and dehydration temperature were lower. Here, in both experiments the chromium was not observed to come off the support. The known silyl chromate saturation level on 955 silica activated at 600 °C is 0.25 wt % $Cr^{[8]}$. Therefore even if all the chromium oxide was converted to silyl chromate, the silyl chromate could completely be consumed by reaction with the surface silanols. The last two experiments in this table employ 600 °C dehydrated silica with higher chromium loading. If silyl chromate formed it would exceed the surface reaction saturation level. One equivalent of TPS results in loss of chromium, but with two equivalents the chromium can be seen to begin to dissolve (based on color and the presence of Cr^{+6} in solution) and chrome loss is observed in the final washed product.

These experiments show that it is possible to make silyl chromate catalyst directly from the oxo analogue. It also shows that it is not possible to maintain the higher chrome loadings that are achievable with the oxo catalyst when the TPS is added.

As shown in Figure 2 only one equivalent of TPS should be needed to form the chemisorbed silyl chromate analogue. However, the polymerization data showed that two equivalents were required to get complete conversion. It is possible that the TPS may not selectively react with the oxo chromate and the second equivalent of TPS is needed to complete the reaction. It is also interesting that supersaturation of the silyl chromate formed by addition of TPS to the already chemisorbed chromium oxide was observed. If the oxo chromium species were already anchored to the surface, would the bis(triphenylsilyl) chromate be able to form and migrate off the surface upon TPS addition? If the saturation level of the large bis(triphenylsilyl) chromate molecule on silica is dictated by a steric constraint, then it seems difficult to see how the second TPS molecule could approach the mono-triphenylsilylchromate anchored on the support. It should not be able to form a molecule in a pore in which the final product cannot fit.

Conclusion

Although similar in structure the oxo and triphenylsilyl chromate catalyst perform very differently in polymerization reactions. The presence of the TPS ligand is what leads to these differences. This can be seen by the clear conversion of polymerization performance observed with oxo chromium based catalysts to that observed with silyl chromate type catalysts by the addition of TPS into the former.

Acknowledgements

The authors thank Univation Technologies LLC for the permission to publish this work. Thanks are also due to T. T. Tamargo, J.W. Nicoletti, and J.H. Moorhouse.

[1] M.P. McDaniel, *Adv. Catal.* **1985**, *33*, 47.
[2] J.P. Hogan, *J. Polym. Sci.* **1970**, *8*, 2637.
[3] F. Karol, G. Karapinka, C. Wu, A. Dow, R. Johnson, W. Carrick, *J. Polym. Sci: Part A-1*, **1972**, *10*, 2621.
[4] U.S. Patent No. 3324095, June 6, **1967**.
[5] U.S. Patent No. 3324101, June 6, **1967**.
[6] W. Carrick, R. Turbett, F. Karol, G. Karapinka, A. Fox, R. Johnson, *J. Polym. Sci., Part A-1*, **1972**,*10*, 2609.
[7] B. Liu, H. Nakatani, M. Terano, *J. Mol. Catal. A: Chem.* **2002**, *184*, 387.
[8] Internal communication.

Surface Physico-Chemical State of CO-Prereduced Phillips CrO_x/SiO_2 Catalyst and Unique Polymerization Behavior in the Presence of Al-Alkyl Cocatalyst

Boping Liu,[1] Yuwei Fang,[1] Hisayuki Nakatani,[2] Minoru Terano[1]*

[1] School of Materials Science, Japan Advanced Institute of Science and Technology, 1-1 Asahidai, Tatsunokuchi, Ishikawa 923-1292, Japan
E-mail: terano@jaist.ac.jp
[2] Fundamental Laboratory for Engineering Education Core, Kanazawa Institute of Technology, 7-1 Ohgigaoka Nonoichi Ishikawa 921-8501, Japan

Summary: In this work, a CO-prereduced Phillips CrO_x/SiO_2 catalyst (PC600/CO) was characterized by X-ray photoelectron spectroscopy (XPS) and temperature programmed desorption with mass spectrometer (TPD-MS) in comparison with a calcined catalyst (PC600). It was found that surface chromate Cr(VI) species had not been completely reduced into Cr(II) species, meanwhile, CO and CO_2 still strongly adsorbed on the PC600/CO catalyst. Subsequently, the unique real-time profiles of ethylene polymerization rate using PC600/CO in the presence of TEA indicated the existence of two types of active sites: the first type derived from the desorption of CO or CO_2 from Cr(II) species by alkyl-Al and featured with fast activation, higher activity and fast decay; the second type generated from the further reduction of residual chromate Cr(VI) species by alkyl-Al and featured with slow activation, lower activity and slow decay.

Keywords: ethylene polymerization; Phillips CrO_x/SiO_2 catalyst; polyethylene; temperature programmed desorption (TPD); X-ray photoelectron spectroscopy (XPS)

Introduction

As one of the most important industrial catalysts for olefin polymerization, Phillips CrO_x/SiO_2 catalyst is still producing several million tons of high density polyethylene (HDPE) per year in the world. This catalyst system can be very easily prepared through a simple calcination process but shows quite unique polymerization behavior compared with Ziegler-Natta and metallocene catalysts. For example, it can be activated for ethylene polymerization solely by ethylene monomer or CO, as well as alkyl-Al cocatalyst, furthermore, its HDPE products usually featured with ultra-broad molecular weight distribution, unsaturated chain end and

DOI: 10.1002/masy.200450905

long chain branching.[1] All these above-mentioned properties have been proven to be very difficult to realize utilizing either Ziegler-Natta or metallocene catalyst within the past 50 years. In spite of its spectacular success in the industrial field and great research effort since early 1950s, its main mechanistic aspects concerning active sites and polymerization still remain mysterious.[1, 2]

Our recent studies on the induction period of ethylene polymerization using Phillips catalyst activated by ethylene monomer have elucidated some important mechanistic points concerning the induction period and subsequent typical accelerating-type polymerization behavior.[3, 4] The induction period is corresponding to the reduction of surface chromate Cr(VI) species into Cr(II) species. Whereas, the Cr(II) species during induction period can only act as ethylene metathesis active precursor due to the adsorption of formaldehyde, a byproduct of the redox reaction between chromate species and ethylene. The subsequent accelerating-type polymerization period is considered to be resulted from a gradual transformation of metathesis active sites into polymerization active sites due to the desorption of formaldehyde molecules.[4] It has also been shown that desorption of formaldehyde from the Cr(II) species became very difficult at even high temperature (e.g. up to 500°C) when ethylene monomer is absent.[3, 4] Therefore, for the CO-prereduced Phillips catalyst, it can be expected that CO, which present similar electron property with formaldehyde molecule, could strongly adsorbed to the surface Cr(II) species and can not be easily desorbed at even high temperature (typically 350°C) in N_2 atmosphere. This may explain why only the induction period can be eliminated, while similar accelerating-type polymerization kinetic can still be observed for the CO-prereduced Phillips catalyst,[1] which can be most probably interpreted in terms of similar active sites transformation due to the gradual desorption of CO from the Cr(II) species in the presence of ethylene monomer. To confirm these speculations, a CO-prereduced Phillips catalyst was prepared and characterized by XPS and TPD-MS in this study. It was found that the surface chromate Cr(VI) species on the calcined Phillips catalyst cannot be completely reduced to Cr(II) species even after 1.5h`s calcination in CO flow under 350°C, which are typical conditions for preparing CO-prereduced Phillips catalyst in the literature. Moreover, the existence of residual CO and $CO_2$ adsorbed on the CO-prereduced Phillips catalyst was confirmed by the TPD-MS method even after 1.5h`s further calcination in N_2 flow under the same temperature (350°C). Based on these preliminary evidences, ethylene polymerization tests using the CO-prereduced Phillips catalyst were designated to combine with the use of alkyl-Al cocatalyst, which has never been studied before. The alkyl-Al cocatalyst was expected to facilitate the formation of active sites for the CO-prereduced Phillips catalyst through both instant removing the CO or CO_2 from the Cr(II) species and further reduction of the residual chromate Cr(VI) species. Subsequently, a unique polymerization behavior was demonstrated in these ethylene polymerization tests using the CO-prereduced Phillips catalyst in the presence of alkyl-Al cocatalyst indicating the existence of two types of active sites: the first type featured with instant site formation, higher activity and fast decay, which was considered to be derived from the desorption of CO or CO_2 from Cr(II) species by alkyl-Al; the second type featured with slow site formation, lower activity and very slow decay, which was postulated to be generated from the further reduction of residual chromate Cr(VI) species by alkyl-Al.

Experimental

Raw Materials

Nitrogen of G2-grade (total impurity < 2ppm, in which O_2 < 0.3ppm, CO < 0.3ppm, CO_2 < 0.3ppm, CH_4 < 0.1ppm, NO_x < 0.1ppm, SO_2 < 0.1ppm, dew point of H_2O < −80°C), high purity helium gas (total impurity < 1ppm, in which O_2 < 0.05ppm, N_2 < 0.1ppm, CO < 0.02ppm, CO_2 < 0.02ppm, CH_4 < 0.01ppm, dew point of H_2O < −80°C), pure air of G1-grade (total impurity < 1ppm, CO < 0.1ppm, CO_2 < 0.1ppm, THC < 0.1ppm, NO_X < 0.01ppm, SO_2 < 0.01ppm, dew point of H_2O < −80°C) and high purity CO gas (CO > 99.95%, O_2 < 75ppm, N_2 < 175ppm, CO_2 < 30ppm, H_2 < 15ppm, THC < 0.5ppm, dew point of H_2O < −70°C) were purchased from Uno Sanso Co. Ethylene of research grade (C_2H_4 > 99.9%, air < 0.03%, methane < 0.01%, ethane < 0.05%, propane < 0.01%, analyzed by Gas Chromatography method) was donated by Mitsubishi Chemical Co. Molecular sieves 4A and molecular sieves 13X, which were purchased from Wako Pure Chemical Industries, Ltd., were used as moisture scavenger for gas purification. Q-5 reactant catalyst (13 wt.% of copper (II) oxide on alumina), which was purchased from Aldrich, was used as oxygen scavenger for gas purification. A catalyst precursor of Crosfield ES370X with 1.0wt% of Cr loading and surface area of 280~350 m^2/g, which was donated from Asahi Kasei Co., was used for preparation of Phillips CrO_x/SiO_2 catalysts.

 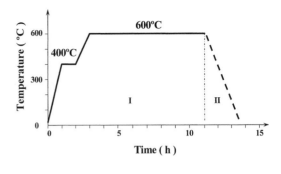

Figure 1. Schematic diagram of the calcination device of Phillips catalyst, a: spouted fluidized-bed quartz reactor; b: electronic heater with temperature-program-controller; c: catalyst powder.

Figure 2. Temperature program for the preparation of the calcined Phillips catalyst, gas flow rate: 200ml/min, stage I: in dry air; stage II: in N_2.

Catalyst Preparation

About 15g of catalyst precursor was added into a spouted fluidized-bed quartz reactor (see Figure 1). Thereafter, a calcination process was performed to obtain a 600°C-calcined Phillips catalyst (named as PC600) according to a temperature program as shown in Figure 2. During the first stage (I) of this calcination process, pure air with a flow rate of 200ml/min was used as gas media. It was switched to nitrogen (200ml/min) during the second stage (II) of cooling. Another CO-prereduced PC600 catalyst (named as PC600/CO) was prepared following a typical temperature-controlling program as shown in Figure 3. Up to the second stage (I to II), the calcination process is almost the same as the PC600 catalyst.

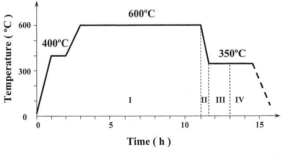

Figure 3. Temperature program for the preparation of the CO-prereduced Phillips catalyst, gas flow rate: 200ml/min, stage I: in dry air; stage II: in N_2; stage III: in CO; stage IV: in N_2.

After being cooled down to 350°C, the gas media was switched to CO (200ml/min) and kept at the same temperature for 1.5 h followed by the final stage (IV) at 350°C in N_2 for 1.5 h before cooling down to room temperature. After preparation, both catalyst samples were distributed and sealed into several large glass tubes within N_2 for storage. Finally, each catalyst sample was distributed and sealed in small glass ampoule bottles under nitrogen atmosphere. The amount of catalyst in each small glass ampoule bottle could be controlled at ca. 100mg and precisely weighed.

XPS and TPD-MS Measurements

The specific procedures and instrumental conditions of XPS and TPD-MS measurements for Phillips catalyst have been described in detail in our previous report.[3~5] Hereafter only a simplified introduction was given. XPS data were obtained on a Physical Electronics Perkin-Elmer Model Phi-5600 ESCA spectrometer with monochromated Al Kα radiation (1486.6eV) operated at 300W. A vacuum transfer vessel (Phi Model 04-110, Perkin-Elmer Co., Ltd.) was used for sample transformation from catalyst storage vessel into the XPS instrument in N_2 atmosphere. XPS high resolution scan measurements for Si 2p, Cr 2p regions of each sample was performed within 10min in order to minimize the X-ray induced reduction to the surface Chromate species during XPS measurement. All spectra were internally referenced to the Si 2p peak from silica gel at 103.3eV to correct for the charging effect during acquisition. The objective of the TPD-MS measurements of the PC600/CO catalyst is to check if there still exist residual CO and CO_2, which were considered to be very difficultly desorbed from the highly-coordinative-unsaturated Cr(II) species even at 350°C under N_2 atmosphere for 1.5 h. Firstly the PC600/CO catalyst sample (ca.100mg) was set into a quartz micro-reactor for the TPD measurement in a glove box. Then TPD-MS spectra were obtained from a Multi-task TPD instrument produced by Bel Japan, Inc. equipped with a quadrupole mass spectrometer (MS). The typical TPD-MS conditions are shown as follows: helium gas flow rate 50ml/min, mass spectrometer sampling time interval 15s, bridge current 2mA, vacuum for MS at ca. $2.5×10^{-8}$ Torr, etc. The temperature was elevated at a linear rate of 2 °C/min from room temperature (RT) to 600°C. The gaseous components of CO (m/e=28) and CO_2 (m/e=44) were continuously monitored.

Ethylene Polymerization

Schematic diagram of the experimental system for semi-batch slurry ethylene polymerization using the PC600/CO catalyst was shown in Figure 4. N_2 was further purified by passing through a Q-5 catalyst column and a 13X molecular sieve column before entering the polymerization system. Heptane was passed through a column of 13X molecular sieve followed by 24h`s bubbling by high purity N_2 before use. One ampoule bottle with ca. 100mg of PC600/CO was set in the top part of the glass polymerization reactor (volume ca.100ml). Then, the reactor system was vacuumed for 2 h before introduction of heptane solvent, TEA cocatalyst and ethylene. The polymerization was initiated after breaking of the catalyst ampoule bottle by a steel bar. Ethylene monomer was further purified by passing through a 4A molecular sieve column, a Q-5 catalyst column and a 13X molecular sieve column before entering the polymerization reactor. The polymerization conditions are shown as follows: ethylene pressure 0.13MPa, polymerization temperature: 60°C, 20 ml purified heptane, Al/Cr molar ratio at 7.5~22.5 using 1M TEA heptane solution. The real-time ethylene consumption was continuously monitored by a on-line mass flowmeter. The polymerization was stopped by adding 20ml ethanol/HCl after 1.5 h. The polymer was washed and dried in vacuum at 60°C for 6 h.

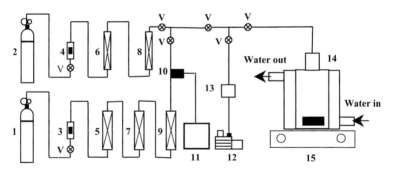

Figure 4. Schematic diagram of the experimental system for ethylene homopolymerization using the CO-prereduced Phillips CrO_x/SiO_2 catalyst, 1: ethylene gas cylinder; 2: N_2 gas cylinder; 3, 4: gas flow meter; 5: 4A molecular sieve column, 6, 7: Q-5 catalyst column; 8, 9: 13X molecular sieve column; 10: mass flow meter; 11: monitor and recorder of mass flow meter; 12: vacuum pump; 13: liquid N_2 trap; 14: polymerization reactor with water-jacket and magnetic stirrer; 15: controller of magnetic stirrer; V: valve.

Table 1. Oxidation states of surface Cr species on the calcined and CO-prereduced Phillips catalyst measured by XPS method.

Sample	BE [a] (eV)	Percentage (%)	Assignment	BE [a] (eV)	Percentage (%)	Assignment
PC600	581.0	96.9	Cr^{6+}	577.6	3.1	Cr^{3+}
PC600/CO	581.0	36.7	Cr^{6+}	576.7	63.3	Cr^{2+}

[a] BE: binding energy of Cr2p (3/2)

Results and Discussions

The oxidation states of surface Cr species on PC600 (calcined) and PC600/CO (CO-prereduced) catalysts measured by XPS method are shown in Table 1.

As it can be seen, about 97% (atomic percentage) surface Cr species is chromate species (monochromate, dichromate or polychromate species) with a binding energy (BE) of 581eV (Cr2p(3/2)) in PC600 catalyst. Only ca. 3% surface Cr species was thermally reduced into surface stabilized Cr(III) species with a BE of 577.6eV.[5] For PC600/CO catalyst, about 63% of surface Cr species was reduced into surface stabilized Cr(II) species with a BE of 576.6eV by CO.[6] It was interesting to find that there still exists ca.37% of residual chromate species, which has not been reduced by CO in PC600/CO catalyst. The TPD-MS evolution curves of CO (m/e=28) and CO_2 (m/e=44) from the PC600/CO catalyst were shown in Figure 5 and Figure 6, respectively. It was clearly confirmed that CO and CO_2 were still strongly adsorbed on the surface Cr(II) species and cannot be desorbed at even high temperature (up to 600°C)

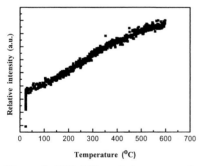

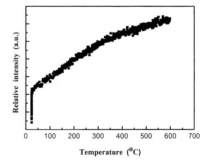

Figure 5. TPD-MS evolution curve of CO (m/e=28) from the CO-prereduced Phillips catalyst, TPD conditions: temperature elevation rate 2°C/min, RT ~ 600 °C, helium gas flow rate 50ml/min.

Figure 6. TPD-MS evolution curve of CO_2 (m/e=44) from the CO-prereduced Phillips catalyst, TPD conditions: temperature elevation rate 2°C/min, RT ~ 600 °C, helium gas flow rate 50ml/min.

in the absence of ethylene monomer. The style of the evolution curves of CO and CO_2 was similar with that of formaldehyde as reported in our previous work.[3, 4] The evolution of N_2 (also with m/e=28) from the PC600/CO was considered to be negligible due to its much weaker electron donating power and thus much weaker adsorption to Cr(II) species compared with CO.

So far, two experimental facts have been demonstrated as follows: firstly, considerable amount of surface chromate Cr(VI) species still remained on the PC600/CO catalyst. Secondly, CO and CO_2 still strongly adsorbed on the PC600/CO catalyst and can not be completely removed at even 600°C without ethylene. Based on these experimental evidences, ethylene polymerization tests using the PC600/CO catalyst in the presence of alkyl-Al cocatalyst were carried out. The alkyl-Al cocatalyst was expected to enhance the formation of active sites through fast removing CO or CO_2 from Cr(II) species and further reduction of residual chromate Cr(VI) species on the PC600/CO catalyst. Three semi-batch slurry ethylene homopolymerization experiments were performed in heptane at 60°C for 1.5 h with Al/Cr molar ratio at 7.5, 15.0 and 22.5 (using TEA as cocatalyst), respectively. The average activity of PC600/CO catalyst was found to decrease from 44.0 to 37.5 and 16.4 kg-PE/mol-Cr·h when Al/Cr molar ratio was increased from 7.5 to 15.0 and 22.5, respectively. The most interesting and important point disclosed in this study is the quite unique polymerization kinetics which has never been reported for the Phillips Cr-based catalysts up to now. As shown in Figure 7, the real-time polymerization kinetic curves seem to be a hybrid type comprised of two basic types of polymerization kinetics: one is a fast formation and fast decay type and the other is a slow formation and slow decay type. According to the surface physico-chemical state of the PC600/CO catalyst clarified by XPS and TPD-MS, it is reasonable to ascribe the origins of the two basic types of polymerization kinetics to two different types of active sites produced on PC600/CO by TEA namely Cr(II) species after desorption of CO or CO_2 by TEA (Site-A in Scheme 1) and Cr(II) species from reduction of residual chromate Cr(VI) species by TEA (Site-B in Scheme 1). Site-A is very exposed and can be easily deactivated due to over-reduction from further contact with TEA cocatalyt, while, Site-B might be protected and/or stabilized by the coordination with Al-alkoxyl groups. This is also consistent with previous polymerization kinetics reports relating to organometal-modified Phillips CrO_x/SiO_2 catalysts by McDaniel et al.[7] The Al-alkoxyl ligands coordinated on Site-B

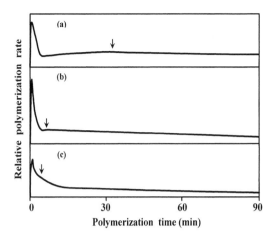

Figure 7. Polymerization kinetics profiles of ethylene homopolymerization using the CO-prereduced Phillips catalyst in the presence of TEA cocatalyst, polymerization conditions: catalyst amount 100mg, polymerization temperature 60°C, polymerization time 1.5 h, ethylene pressure 0.13 MPa, solvent heptane (20ml), cocatalyst TEA in heptane (1M), Al/Cr molar ratio (a) 7.5; (b) 15.0; (c) 22.5.

were not poisonous by-product for polymerization as indicated by McDaniel et al..[7] Figure 7 also showed that the time to reach the maximum activity for Site-B was shortened from ca. 30min to ca 5min (as shown by arrows in Figure 7) accompanied with faster decay of Site-B when Al/Cr molar ratio was increased from 7.5 to 22.5. The decrease of average activity with

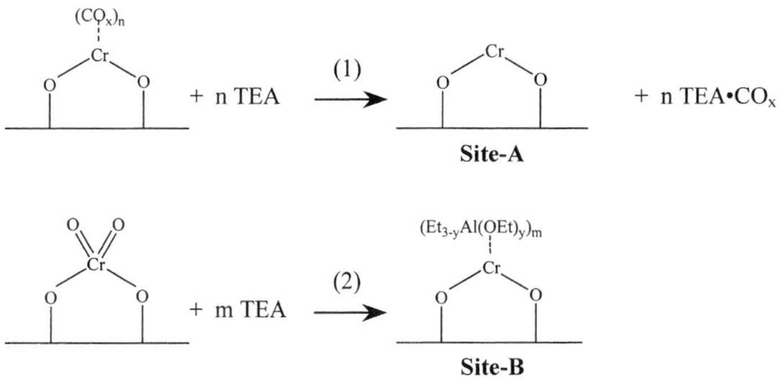

Scheme 1. Plausible mechanism of the formation of two kinds of active sites on the CO-prereduced Phillips catalyst in the presence of TEA cocatalyst for ethylene polymerization, x=1 or 2; n=1 or 2; y=1 or 2, m=1 or 2.

increasing Al/Cr molar ratio could be ascribed to enhancement of active site deactivation of both Site-A and Site-B (especially the latter).

Conclusion

It has been shown that surface chromate Cr(VI) species has not been completely reduced into Cr(II) species, meanwhile, CO and CO_2 still strongly adsorbed on the PC600/CO catalyst. Subsequent unique ethylene polymerization behavior using PC600/CO in the presence of TEA demonstrated the existence of two types of active sites: the first type derived from the desorption of CO or CO_2 from Cr(II) species by alkyl-Al and featured with fast activation, higher activity and fast decay; the second type generated from the further reduction of residual chromate Cr(VI) species by alkyl-Al and featured with slow activation, lower activity and slow decay.

Acknowledgements

The authors thank to Mitsubishi Chemical Co., Mitsui Chemical Co., Toho Titanium Co., Ltd., Asahi Denka Co., Ltd., Chisso Corp., and Tosoh Akzo Corp., Asahi Kasei Co., Japan Polyolefin Corp., for their support and donation to our laboratory.

[1] M. McDaniel, *Adv. Catal.* **1985**, *33*, 47.
[2] B. M. Weckhuysen, R. A. Schoonheydt, *Catal. Today* **1999**, *51*, 215.
[3] B. Liu, M. Terano, *J. Mol. Catal. A: Chem.* **2001**, *172*, 227.
[4] B. Liu, H. Nakatani, M. Terano, *J. Mol. Catal. A: Chem.* **2002**, *184*, 387.
[5] B. Liu, H. Nakatani, M. Terano, *J. Mol. Catal. A: Chem.* **2003**, *201*, 189.
[6] R. Merryfield, M. McDaniel, G. Parks, *J. Catal.* **1982**, *77*, 348.
[7] M. McDaniel, M. M. Johnson, J. Catal. 1986, 101, 446.

D/A-Metallocenes: The New Dimension in Catalyst Design

Aleksander Ostoja Starzewski

Bayer AG, Bayer Polymers, Innovation, Wissenschaftliches Hauptlaboratorium, 51368 Leverkusen, Germany
E-mail: Aleksander.OstojaStarzewski@Bayerpolymers.com

Summary: D/A metallocenes constitute a novel unique class of catalysts, in that they are able to express structural information within the elementary steps of the catalytic cycle out of bridged as well as out of unbridged states due to the coexistence of donor (D) and acceptor (A) atoms in complex sandwich structures. Highly polarized transannular bridging interactions as well as Lewis basic and Lewis acidic functionalities in unbridged states are the prerequisites which define a modular highly flexible catalyst system with outstanding options to tailor materials and processes.

Keywords: D/A metallocenes; long chain branching; monomodal/multimodal molecular weight distributions; polyethylene; polymerization catalysts; polypropylene; short chain branches; ultrahigh molecular weight polyolefins

Introduction

Classical metallocene catalysts are either unbridged or covalently bridged. The range of properties has been extended further by halfsandwich catalysts, in particular those with constrained geometry. Structural features have created domains of best applicability for each subclass. Intensive optimization efforts worldwide over two decades have resulted in high level catalyst profiles, but at the same time have defined system-inherent borderlines and limitations which can only be overcome by novel catalyst concepts.

The D/A Metallocene Principle

D/A metallocenes constitute a new dimension in catalyst design. [1-12] Characteristic feature is the incorporation of donor and acceptor atoms into sandwich structures, resulting in highly polarized bonding interactions, which specifically restrict the rotation of the π-ligands and widen the aperture angle. The reverse process, which eliminates the dipolar interaction, shrinks the aperture angle and favors ring rotation.(2) The energy hypersurface for the

(1)

structural changes is controlled by D and A and the nature of the donor and acceptor substituents. The actual state the system is in is temperature dependent and can be selected. It controls the architecture and the ensemble of the formed macromolecules and thereby the material properties.

(2)

The bonding D/A interaction is best described as originating from a stabilizing in-phase combination of an occupied lone pair orbital on the donor with an empty orbital of appropriate symmetry on the acceptor. The closer the two orbitals come in energy, the stronger the energy gain, which is therefore related to donor lone pair ionization energies and acceptor electron affinities. The stabilizing 2-electron/2-orbital interaction results in an electron delocalization i.e. a transfer of electron density from D to A, which in turn generates positive and negative partial charges on D and A respectively.

Thus, a dipolar interaction of variable size and orientation is introduced into the catalyst due to D/A coupling. Vice versa, the reverse process destroys the high dipole moment via a heterolytic D/A-splitting, regenerating the uncharged donor and acceptor substituents e.g. D: NR_2, PR_2, SbR_2, OR, SR, SeR, F, Cl, Br etc and A: BR_2, AlR_2, GaR_2, etc. At the same time it shrinks the aperture angle and favors ring rotation. (Breaking a covalent bridge in *ansa*-metallocenes would involve a homolytic process, which generates highly unstable radical intermediates and destroys the catalyst.)

The unbridged state is favored by weak D/A interactions, high strain energy in the bridged state (e.g. due to bulky substituents) and of course by high temperatures. Accordingly, the bridged state is favored by strong D/A interactions, low strain energy in the bridged state and by lower temperatures.

The ease, with which this bridged-to-unbridged transition takes place, can be tailored. The choice of the heteroatoms together with the nature of the donor and the acceptor substituents offer extraordinary modules for the design of novel catalysts. Macromolecular architectures obtained via D/A metallocene catalysis are controlled by the actual state(s) the chosen D/A catalyst system is in. It offers single-site, dual-site and multi-site catalyst options. The temperature dependant existence and coexistence of two stable forms (bridged/unbridged) of the same compound establishes a new dimension in catalyst design. It not only dissolves the borderline between the classical two metallocene subclasses but also, due to D/A-specific electronic consequences even the borderline between metallocenes and constrained geometry catalysts (3).

D/A-Metallocenes

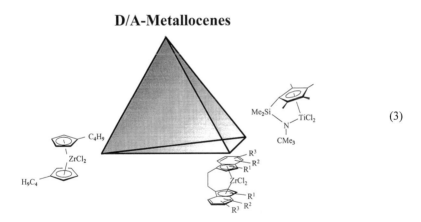

(3)

Thereby it opens the way to material properties that are at least difficult, and in some cases impossible to achieve with the traditional systems.

The novel D/A lead structures, which include reversible one-, two- and more-atom bridges, introduce in metallocene catalysis a broad range of additional and new control instruments, all of which can be used to optimize the catalytic behavior. Different donor and acceptor heteroatoms, charged and/or uncharged, together with an enormous chemical variability and selectable bulkiness of the corresponding substituents and the attached hydrocarbon or heterocyclic π-ligands provide efficient means to precisely tune stereochemical and dynamic properties and to optimize catalyst performance.

D/A Metallocene Syntheses and Structures

There are three main synthetic routes to D/A metallocenes of general applicability (1). The donor route, which is based on reacting a preformed donor halfsandwich complex with an acceptor ligand (2). The acceptor route, which is based on reacting a preformed acceptor halfsandwich complex with a donor ligand (3). The donor-acceptor route, which is based on reacting a preformed D/A ligand with a suitable metal precursor.

$$
\begin{array}{lll}
\bullet \text{ D ligand} & \xrightarrow{\text{MX}_n} \text{ D halfsandwich} & \xrightarrow{\text{A ligand}} \text{D/A metallocene} \\[2ex]
\bullet \text{ A ligand} & \xrightarrow{\text{MX}_n} \text{ A halfsandwich} & \xrightarrow{\text{D ligand}} \text{D/A metallocene} \qquad (4) \\[2ex]
\bullet \text{ D ligand +} & \longrightarrow \text{ D/A ligand} & \xrightarrow{\text{MX}_n} \text{D/A metallocene} \\
\quad \text{A ligand} & &
\end{array}
$$

Route 1 provides access to numerous D/A metallocenes even in only one set of experiments i.e. with the same central metal and the same donor substituted π-ligand. Structure/activity and structure/selectivity relationships can be elucidated in a straightforward manner, just by variing the acceptor containing moiety. Likewise route 2 provides any desired number of D/A metallocenes being derived from one and the same metal always carrying the same acceptor substituted π-ligand. The observable changes in the catalytic profile, in this case result from varying the donor containing building block. Route 3, for example, takes advantage of a preformed D/A ligand, which can be reacted with various transition metal precursors. In order to optimize the yield of a selected catalyst, one may test all three routes for the same target molecule.

A large number of D/A metallocenes has been synthesized for which crystals suitable for X-ray structure analyses have been obtained. These have provided direct proof, that the above described D/A interactions are operative. Representative D/A-bridged metallocene X-ray structures have been obtained for example for titanium and zirconium complexes with cyclopentadienyl, indenyl, fluorenyl, or phospholyl ligands, carrying dimethylamino-, dimethylphosphino-, diethylphosphino-, diisopropylphosphino-, diphenylphosphino-, dichloroboranyl-, dimethylboranyl, or bis(pentafluoropenyl)boranyl-substituents.[11,12]

For corresponding examples, see (5) and Figures 1 and 2.

In addition, multinuclear magnetic resonance methods have been used to identify the presence and the relative strength of D/A interactions of such metallocenes in solution. Especially, P-31 and B-11 NMR chemical shifts and one-bond nuclear spin-spin coupling constants are powerful means for solution studies.

Last-not-least, the polymerization results provide direct evidence if, under the activation and polymerization conditions chosen, the catalyst system under investigation is actually D/A-bridged, D/A-unbridged or both.

D/A Metallocenes, the Leading Edge in Polymerization Catalysis

The synthetic donor and acceptor „tool box" proves to be an ideal, highly flexible modular system to easily identify and develop the best catalyst for a specific application.

This way, substantial progress has been made in prominent areas of catalyst research and development: e.g. the polymerization temperature optimum of the catalyst can be adjusted to process requirements. The polymer molecular weight can be controlled at any desired level even at elevated temperatures. Molecular weight distributions can be modelled. In propene polymerization regioerrors such as 2,1- and 1,3-misinsertions can be efficiently suppressed. High molecular weight PPs are accessible with strongly differing stereochemical compositions. Sequential type homo- and copolymers with unusual microstructures can be tailored. Last-not-least, D/A-based polymers offer, when necessary, improved rheology through bimodal molecular weight distributions and/or through long chain branching.

In the following, two prominent polyolefin material areas are addressed to outline the stated merits of D/A metallocene polymerization catalysis: polyethylene and polypropylene.

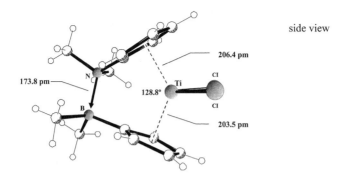

side view

Figure 1. X-ray Structure of [(cp)Me₂NBMe₂(cp)TiCl₂]

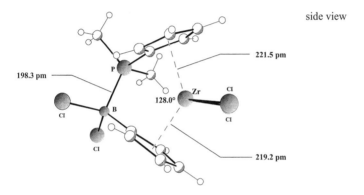

side view

Figure 2. X-ray Structure of [(cp)Me₂PBCl₂(cp)ZrCl₂] [1]

$$
\begin{array}{c}
\textbf{(cp)Me}_2\textbf{NBMe}_2\textbf{(cp)TiCl}_2], \\
\textbf{[(cp)Ph}_2\textbf{PBCl}_2\textbf{(cp)TiCl}_2], \\
\textbf{[(cp)Me}_2\textbf{PBCl}_2\textbf{(cp)ZrCl}_2], \\
\textbf{[(cp)Me}_2\textbf{PB(C}_6\textbf{F}_5)_2\textbf{(cp)ZrCl}_2], \\
\textbf{[(Me}_4\textbf{phospholyl)BMe}_2\textbf{(cp)TiCl}_2], \\
\textbf{[(ind)Et}_2\textbf{PBCl}_2\textbf{(cp)ZrCl}_2], \\
\textit{meso-}\textbf{[(ind)Ph}_2\textbf{PBCl}_2\textbf{(ind)ZrCl}_2], \\
\textit{rac-}\textbf{[(ind)Ph}_2\textbf{PBCl}_2\textbf{(ind)ZrCl}_2], \\
\textbf{[(flu)Et}_2\textbf{PB(C}_6\textbf{F}_5)_2\textbf{(cp)ZrCl}_2].
\end{array}
\qquad (5)
$$

Polyethylene

Linear high melting high density polyethylene (0.97 g/cm^3) is obtainable with the featured D/A metallocene catalysts. For example $(cp)Et_2PB(C_6F_5)_2(cp)ZrCl_2/MAO$ produces 148 tons high molecular weight HDPE per mol Zr and hour at 60 C and 10 bar ethylene with a PE melt temperature of 139 C (DSC, 2. heating) and an intrinsic viscosity of 3.56 dl/g (measured in ODCB at 140 C), which corresponds to a viscosity average molecular weight of 323 kg/mol. All PE molecular weight ranges are accessible, including ultrahigh molecular weight polyethylenes (UHMW-PE). If one uses instead of the D/A bis(cp)metallocene the corresponding D/A (flu)(cp)metallocene in an otherwise identical ethylene polymerization experiment, the intrinsic viscosity of the PE formed reaches 13.25 dl/g (!). Whereas the former PE has a polydispersity of 2.5, for the latter PD = 7.8. Monomodal but also bimodal or multimodal molecular weight distributions can be tailored. Without the need of a two-reactor/zone technology one may simply take advantage either of (1) two matched D/A metallocenes, preselected according to PE molecular weight or of (2) the coexistence of bridged and unbridged D/A states at a specific temperature, i.e. by choosing between single site, dual site, multi site options, inherent to the D/A catalyst concept.

Due to the excellent capability of D/A-bridged metallocenes to incorporate comonomers, polyethylenes with any desired amount of short chain branches along the main chain can be produced, necessary to fill the needs for any specific application. These short chain branched polyethylenes (SCB-PE) can be synthesized by copolymerizing ethylene with 1-olefins such as propene, 1-butene, 1-hexene, 4-methyl-1-pentene, or 1-octene. The different products obtained with D/A catalysts may be classified according to their density as medium density polyethylene (MDPE), linear low density polyethylene (LLDPE), very low density polyethylene (VLDPE), ultra low density polyethylene (ULDPE). With increasing amounts of incorporated 1-olefins the density drops, as does the melt temperature and the crystallinity. Relatively high amounts of uniformly incorporated 1-olefins such as 50 weight% result in high performance elastomers, characterized by high molecular weight and a low glass transition temperature e.g. –65 C for 1-hexene. Fully amorphous low-Tg-types without any melt endotherm are accessible out of bridged D/A states, whereas unbridged metallocene structures disfavor a uniform high incorporation of 1-olefins and accordingly result in semicrystalline materials due to crystallizing polyethylene sequences.

Lower T_m/moderate T_g values usually result from higher ethylene/lower 1-olefin copolymer compositions. The especially attractive high T_m/low T_g property profile of sequential thermoplastic elastomers becomes available using the D/A steering potential.

Outstanding catalytic features come from specific D/A metallocenes, which give efficient access to long chain branched polyethylene architectures (LCB-PE).

Polypropylenes

Isotactic polypropylenes (i-PP) with melt temperatures of 160 – 168 C and high to very high molecular weights can be synthesized using the D/A approach, thus taking advantage of the high regio- and stereocontrolling power of D/A systems. PP samples with 99-100 % isotactic pentads have been prepared. Conventional isospecific _ansa_-metallocenes show a pronounced correlation of the PP DSC melt temperature with the NMR _mmmm_ pentad concentration. In some D/A-catalyzed polymerizations, however, the high _mmmm_ value can be reduced considerably by raising the polymerization temperature without loosing the high melting property. This observation is consistent with the presence of a temperature window, which is defined by the coexistence of bridged and unbridged D/A states. A high melting crystalline phase formed from long isotactic sequences is produced out of the stereoselective bridged state, whereas the unbridged state generates atactic sequences forming an amorphous phase. The ratio of isospecific and aspecific sites depends on the temperature. In this way elastified isotactic polypropylenes with improved mechanical properties can be tailored. Whereas, high propylene/low ethylene compositions resulting from stereospecific catalysts show an improved low temperature behavior due to crystallizing high T_m isotactic or syndiotactic polypropylene sequences interrupted by noncrystallizing low-T_g statistical EP copolymer sequences.

Furthermore, high molecular weight essentially atactic polypropylenes with enhanced isotactic or syndiotactic contents have been accessed, which possess the attractive features of elastomeric polypropylenes (e-PP). The manifold D/A options for precisely controlling the PP microstructures allow to tailor the material properties according to the targeted applications of high or low modulus materials.

Unexpectedly, novel long chain branched polypropylene architectures (LCB-PP) highlight the D/A metallocene success story.

In summary, we have succeeded to specifically develop novel high performance D/A metallocenes polymerization catalysts (6) for a variety of target materials such as high melting crystalline thermoplastic materials with improved processability, amorphous thermoplastic materials with high Tg, as well as semicrystalline thermoplastic elastomers with tailored green strength and amorphous polyolefin elastomers with very low Tg.

D/A Metallocenes:

catalysts with added temperature stability

option for ultra high molecular weight

mono- and bimodal molecular weight distributions (6)

excellent regio-and stereocontrol

access to sequence polymers

access to long chain branching

[1] K.A. Ostoja Starzewski, W.M. Kelly, A. Stumpf, D. Freitag, *Angew..Chem.* **1999**, *111*, 2588-2592; *Angew. Chem. Int. Ed.* **1999**, *38*, 2439-2443
[2] K.A. Ostoja Starzewski, W.M. Kelly, PCT WO 98/01455 (to Bayer AG)
[3] K.A. Ostoja Starzewski, W.M. Kelly, A. Stumpf, C. Schmid, PCT WO 98/01483 (to Bayer AG)
[4] K.A. Ostoja Starzewski, W.M. Kelly, A. Stumpf, PCT WO 98/01484 (to Bayer AG)
[5] K.A. Ostoja Starzewski, W.M. Kelly, A. Stumpf, PCT WO 98/01485 (to Bayer AG)
[6] K.A. Ostoja Starzewski, W.M. Kelly, A. Stumpf, PCT WO 98/01486 (to Bayer AG)
[7] K.A. Ostoja Starzewski, W.M. Kelly, A. Stumpf, PCT WO 98/01487 (to Bayer AG)
[8] K.A. Ostoja Starzewski, W.M. Kelly, PCT WO 98/45339 (to Bayer AG)
[9] K.A. Ostoja Starzewski, M. Hoch, PCT WO 99/32532 (to Bayer AG)
[10] K.A. Ostoja Starzewski, W.M. Kelly, P. Schertl, PCT WO 99/33852 (to Bayer AG)
[11] K.A. Ostoja Starzewski, B.S. Xin, PCT WO 02/76999 (to Bayer AG)
[12] K.A. Ostoja Starzewski, B. S. Xin, to be published

Macromol. Symp. **2004**, *213*, 57-68

Microstructural Characteristics and Thermal Properties of *ansa*-Zirconocene Catalyzed Copolymers of Propene with Higher α-Olefins

Maria Carmela Sacchi,[1] *Fabrizio Fortini,*[1] *Simona Losio,*[1] *Incoronata Tritto,*[1] *Giovanna Costa,*[2] *Paola Stagnaro,*[2] *Ioan Tincul,*[3] *Udo M. Wahner*[3]

[1] Istituto per lo Studio delle Macromolecole, ISMAC - C.N.R., Via E. Bassini, 15, 20133 Milano, Italy
[2] Sezione di Genova - Via De Marini 6, 16149 Genova, Italy
[3] Sasol Technology, R&D Division, PO Box 1, Sasolburg 1947, South Africa

Summary: Copolymers of propene and several higher α-olefins, 1-butene, 1-pentene, 1-hexene, 1-heptene, 1-octene, and 4-methyl-1-pentene (4M1P), have been prepared with two different metallocene catalysts, *rac*-Et(Ind)$_2$ZrCl$_2$ (EI) and *rac*-Me$_2$Si(2-MeBenz-[e]Ind)$_2$ZrCl$_2$ (MBI), and comprehensively characterized by ^{13}C NMR spectroscopy. All copolymers produced with EI have comparable isotacticities as well as similar stereosequences. The copolymers produced with MBI significantly differ both from the reference propene homopolymer and among themselves: in fact, they are characterized by rather low isotacticities and the degree of tacticity is strongly influenced by comonomer type and concentration. A detailed ^{13}C NMR analysis revealed that these copolymers are constituted by highly isotactic and atactic sequences of various lengths. The similar thermal behavior, in terms of transition temperatures, for the homologues of the series obtained with the two different catalysts show that the capability of crystallization of these systems is not directly connected to the overall microstructural properties (e.g. the isotacticity index) but can be accounted for by the details of the microstructure.

Keywords: copolymerization; higher α-olefins; metallocene; microstructure; thermal behavior

Introduction

Recently it was revealed that a family of commercial random copolymers of propene with non conventional comonomers (e.g. odd comonomers such as 1-pentene) were characterized by an excellent combination of physical and mechanical properties compared to other commercial copolymers with commonly used comonomers, such as ethylene, 1-butene, or other even α-olefins.[1] The reason for the unique characteristics of these materials is still unknown. However these observations give clear evidence that copolymer properties depend on the

DOI: 10.1002/masy.200450907

content and distribution as well as on the type of the comonomer. Therefore we started a study aimed at investigating the effect of the kind of comonomer on microstructure and final properties of the copolymers. The research plan entailed two projects: i) study of propene/1-pentene copolymers with different comonomer content to investigate the effect of the comonomer amount; ii) study of the effect of length and branching of the comonomer through the investigation of copolymers of propene with different higher α-olefins.

The commercial copolymers are produced with traditional Ziegler-Natta catalysts. However, since multi-centered Ziegler-Natta catalysts do not allow meaningful correlations between microstructure and properties, for this investigation we used two single-site metallocene catalysts, which lead to well defined microstructures and uniform comonomer distribution:[2] the now classical and widely studied catalyst rac-Et(Ind)$_2$ZrCl$_2$, (EI), and rac-Me$_2$Si(2-MeBenz-[e]Ind)$_2$ZrCl$_2$ (MBI), which is one of the most successful isospecific catalysts. The choice was aimed at obtaining copolymers with different microstructures in order to examine the correlations between the ligand structure of the metallocene catalyst and the copolymer microstructural characteristics. A detailed ^{13}C NMR analysis of stereoregularity, and of comonomer content and distribution of all the samples was conducted. Furthermore, the investigation was addressed to the relationship between microstructural characteristics and thermal behavior of the copolymers.

Results and Discussion

Propene/1-Pentene Copolymers with Different Composition

Since 1-pentene is reported to yield superior properties to the resultant polymers, like outstanding clarity and gloss, the initial investigation focused on propene/1-pentene copolymers. Two series of propene/1-pentene copolymers with different compositions were prepared with EI and MBI. To assure that the comonomer concentration in solution remained nearly constant during the course of the reaction, the copolymerization conditions were selected to keep conversion of both comonomers lower than 5%. In Table 1 activities, feed and copolymer compositions together with some of their characteristics are reported. Moderately higher comonomer incorporation is in general observed with MBI than with EI, in line with the results from the literature.[2a] All copolymers possess narrow polydispersities, around 2. The reactivity ratio product indicates an almost random distribution of the

Table 1. Homo- and copolymerization data for different propene/1-pentene copolymers

Run	Catalyst	Feed ratio[a] C_5/C_3	1-Pe (mol-%)	Activity[b]	$Mn*10^{-3}$ (g/mol)	Mw/Mn	$r_{Pr}r_{Pe}$	[mmmm][c]
1	EI	Polypropene	-	28529	36	1.86	-	0.88
2		0.025	1.1	3881	67	1.72	n.d.	0.90
3		0.05	2.0	3639	67	1.71	n.d.	0.89
4		0.1	6.3	2706	42	3.21	1.51	0.88
5		0.2	10.6	4835	35	1.77	1.17	0.90
6		0.4	18.0	5094	34	1.71	1.16	0.91
7		1	36.4	2723	24	1.79	1.17	n.d.
8		1.5	40.7	2618	26	1.67	1.31	n.d.
9	MBI	Polypropene	-	42824	510	1.83	-	0.95
10		0.025	1.4	6734	440	1.99	n.d.	0.91
11		0.05	2.7	6926	370	1.95	n.d.	0.93
12		0.1	6.9	4223	240	2.01	1.11	0.86
13		0.2	10.3	3370	200	1.89	1.61	0.70
14		0.4	20.1	2538	170	1.76	1.12	0.53
15		1	38.6	2041	150	1.90	1.11	0.48
16		1.5	49.2	2782	140	2.32	1.07	0.37

Polymerization conditions: propene = 100 g; hydrocarbon solvent = 350 g; [Zr] = 2,5-10 · 10^{-6} mol/L; MAO/Zr = 6000 (molar ratio); T = 30°C; time = 30 min.
[a] molar ratio pentene/propene in the feed;
[b] activity expressed as: Kg polymer/mol catalyst produced in 1 hour;
[c] molar fraction of isotactic pentads determined from the methyl (G) region (runs 1-14) and from the α-methylene (D) region (runs 15-16) of the ^{13}C NMR spectra.

comonomer ($r_{Pr}r_{Pe} \approx 1$) for both catalysts, independently of the ligand structure. These properties are in accordance with the single site nature of metallocene-based systems. A surprising result concerns the influence of the ligand structure on the copolymer stereoregularity. In fact, with the moderately isospecific catalyst EI, the isotacticity value of the propene homopolymer is maintained for the entire series of copolymers, while with the highly stereospecific catalyst MBI the isotacticity of the copolymers gradually decreases with increasing comonomer content.

60

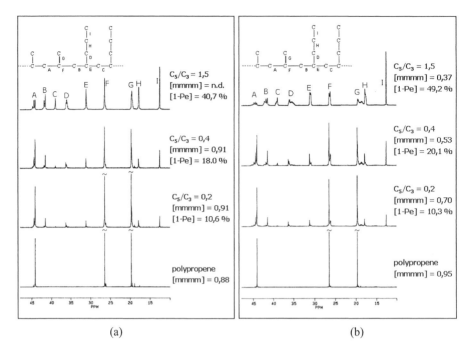

Figure 1. ^{13}C NMR spectra of polypropene and propene/1-pentene copolymer samples prepared with EI (a) and MBI (b) at different compositions

Figure 1a shows the spectra of three copolymers prepared with EI at increasing comonomer content. The spectrum of the corresponding homopolymer is reported for reference purpose. It is apparent that the three copolymers are highly isotactic. Very narrow signals are observed in all spectral regions for the three copolymer compositions. Figure 1b shows the spectra of the corresponding copolymers prepared with MBI. One can easily appreciate the progressive loss of the stereoregularity with increasing incorporation of the comonomer. A progressive broadening of the signals occurs in all the spectral regions especially in those (G and D) more sensitive to steric effects. At the high 1-pentene content of 49% MBI behaves like an almost completely aspecific catalyst.

A more detailed analysis of the tacticity was conducted on selected samples. In the series of copolymers prepared with EI only the three pentads (mmmr, mmrr and mrrm), which correspond to the isolated steric errors, are detected (Table 2). On the contrary, in copolymers

Table 2. Relative abundance of steric pentads for polypropene and some selected copolymers

Run	Cat	1-Pe (mol-%)	mmmm	mmmr	rmmr	mmrr	mrmm + rmrr	rmrm	rrrr	rrrm	mrrm
1	EI	0	0.881	0.053	-	0.045	-	-	-	-	0.021
4		6.3	0.879	0.057	-	0.037	-	-	-	-	0.027
5		10.6	0.900	0.044	-	0.036	-	-	-	-	0.020
6		18.0	0.910	0.036	-	0.036	-	-	-	-	0.018
9	MBI	0	0.950	0.023	-	0.020	-	-	-	-	0.007
12		6.9	0.865	0.031	0.010	0.022	0.023	0.013	0.008	0.016	0.012
13		10.3	0.695	0.052	0.031	0.044	0.053	0.040	0.023	0.031	0.031
14		20.1	0.528	0.040	0.052	0.082	0.105	0.081	0.044	0.033	0.035

produced with MBI (even when the comonomer content is as low as 6%) all possible stereosequences are present, including the rrrr syndiotactic pentad.

Copolymers of Propene with Higher α-Olefins

Two series of propene based copolymers with 1-butene, 1-pentene, 1-hexene, 1-heptene, 1-octene and 4-methyl-1-pentene (4M1P) were synthesized with the same catalysts EI and MBI, at a constant comonomer/propene feed ratio of 0.1 mol/mol. The selection of the comonomers allowed us to investigate the effect of gradually increasing comonomer length and of branching on the copolymer microstructure. In Table 3 activities, feed and comonomer amount together with some fundamental copolymer properties are reported. The data of the propene homopolymers are stated for reference purpose.

All copolymers feature narrow polydispersities and an almost random distribution of the comonomer (r_P r_{Cm} ≈ 1) for both catalysts. However, here we will focus mainly on the differences between the two series. A first difference is given by the comonomer content. In fact, while similar comonomer incorporation is observed in most copolymers prepared with EI (with the only exception of the copolymer with the branched 4M1P having a comonomer content of only 2.6%), with MBI the comonomer incorporation ranges from 7.5% with 1-butene to 4.9% with 1-octene for the linear 1-olefins. Again, the lowest comonomer incorporation is observed with the sterically hindered 4M1P (3%).

Table 3. Homo- and copolymerization data for propene/α-olefins copolymers

Run	Cat	Cm	Incorp. Cm (mol-%)	Activity[a]	Mn*10⁻³ (g/mol)	Mw/Mn	rₚrCm	[mmmm][b]
1	EI	Polypropene	0	28529	36	1.86	-	0.88
2		Butene	6.8	2859	38	1.79	1.23	0.86
3		Pentene	6.3	2706	42	3.21	1.52	0.88
4		Hexene	6.1	4541	37	1.79	0.85	0.84
5		Heptene	5.7	6682	37	1.74	0.98	0.85
6		Octene	6.2	4147	23	1.65	1.23	0.88
7		4M1P	2.6	41	32	1.96	n.d.	0.88
8	MBI	Polypropene	0	42824	510	1.83	-	0.95
9		Butene	7.5	4615	150	2.81	1.48	0.83
10		Pentene	6.9	4223	240	2.01	1.11	0.86
11		Hexene	5.0	2054	230	1.93	0.99	0.72
12		Heptene	6.1	3838	200	1.89	1.06	0.79
13		Octene	4.9	1870	210	1.82	1.89	0.84
14		4M1P	3.1	369	110	2.03	n.d.	0.44

Polymerization conditions: propene = 100 g; Cm/C3 = 0.1 (molar ratio); hydrocarbon solvent = 350 g; [Zr] = 2,5-4.5 · 10⁻⁶ mol/L; MAO/Zr = 6000 (molar ratio); T = 30°C; time = 30 min.
[a] activity expressed as: Kg polymer/mol catalyst produced in 1 hour;
[b] molar fraction of isotactic pentads determined from the methyl (G) region of ^{13}C NMR spectra.

However, the most notable difference between the two series of copolymers concern the stereoregularity. All copolymers produced with EI feature isotacticities comparable with those of the reference propene homopolymer. The copolymers obtained with MBI differ significantly from the reference propene homopolymer as well as among each other. Indeed, the degree of tacticity is strongly influenced by the comonomer type.

In Figure 2 the NMR spectra of propene copolymers with 1-pentene, 1-heptene and 4M1P are compared. Among these three copolymers the isotacticity decreases with the increasing bulkiness of the comonomer. In the case of the propene/4M1P copolymer, the insertion of as little as 3% of 4M1P isolated units leads to a decrease of the isotacticity from 95 (homopolymer) to 44%.

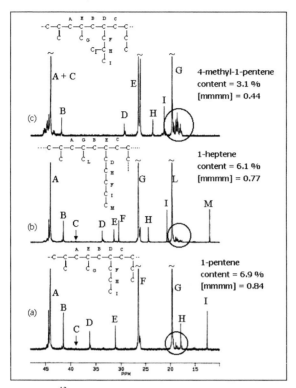

Figure 2. ¹³C NMR spectra of propene copolymers with various comonomers prepared with MBI

Therefore, when MBI is used, the isotacticity of the copolymer strongly depends not only on the comonomer content but also on its length and structure.[3] These results are rather surprising since, to our knowledge, a similar effect has never been described for copolymers of propene with higher α-olefins obtained via metallocenes.[4] The quite obvious hypothesis that MBI contained impurities of the meso form of the metallocene and consequently the material was a mixture of isotactic and atactic chains formed on the racemic and meso isomer of MBI, respectively, does not seem to be likely for a certain number of reasons. First of all reproducible results were obtained with different catalyst batches, thus making the hypothesis of the presence of impurities of the meso form in the catalyst less probable.

Moreover, if the catalyst contained a mixture of the meso and racemic isomers, in principle the ensuing copolymer should be a mixture of an isotactic and an atactic copolymer, the latter being likely characterized by higher comonomer content and lower molecular weight than the former. In fact it has been shown that, differently from propene, the higher α-olefins have remarkably greater polymerization activity with the meso form of dimethylsilyl bridged metallocenes than with the racemic one.[5] In this respect, fractionation performed with boiling n-pentane seems to indicate an uniform material although, due to the difficulty in performing an effective fractionation of this kind of copolymers, the results cannot be considered conclusive. Moreover, chain termination occurs more frequently on the meso than on the more sterically hindered racemic form.[5b] This is in contrast to the fact that all copolymers prepared with MBI, as well as those prepared with EI, are characterized by narrow molecular weight distribution and almost Bernoullian comonomer distribution, which are typical of single site catalysts.

We propose that the presence of a comonomer unit can occasionally lead to a loss of steric control of the more bulky MBI catalyst which lasts for several insertion events, thus leading to atactic blocks. At low comonomer content the atactic blocks are likely to be followed by isotactic blocks due to the recovery of the catalyst's steric control. When the comonomer incorporation reaches 30-40% there are only limited opportunities to regain the steric control prior to another comonomer insertion and the resulting copolymers are more or less atactic.

On the basis of this hypothesis catalyst EI should give a copolymer with the structure shown in Scheme 1a, i.e., a regular polypropene chain with defects due to isolated comonomeric units (for the sake of simplicity copolymers with low comonomer content are here considered) and only isolated (random) steric errors (besides other "normal" structural changes e.g. misinsertions), while with catalyst MBI (Scheme 1b) the copolymer chain would also contain more or less long atactic blocks, which might follow a comonomer unit.

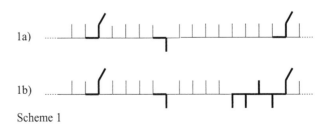

1a)

1b)

Scheme 1

Microstructure-Thermal Behavior Correlation

To ascertain the influence of the microstructure on the properties of these copolymers, an investigation on their thermal behavior was carried out by DSC analysis. After having destroyed the original crystallinity of the samples at a temperature well above the melting point of the propene homopolymers, the crystallization was performed under non-isothermal conditions (cooling rate –20°C/min) and the melting behavior was investigated during the second heating run.

Table 4 shows the melting temperatures of the two series of copolymers at increasing 1-pentene content. The thermal behavior of these copolymers is apparently contradictory since the two series prepared with the different catalysts, despite their noticeable differences in tacticity, show very similar trends of the melting temperatures. Indeed, rather similar T_m values are observed for almost all the pairs of homologues of the two series. If we compare, as a significant example, runs 6 and run 14 we notice that rather different isotacticities (0.91 and 0.53) correspond to very close melting points (60 and 63 °C).

Similar observations can be made by comparing the isotacticities and melting temperatures of

Table 4. Melting temperatures and microstructural parameters for the copolymers at increasing 1-pentene content

Run	Cat	1-Pe (mol-%)	[mmmm]	n_{iso}	T_m (°C)
1	EI	0	0.88	46.0	136
2		1.1	0.90	39.8	134
3		2.0	0.89	25.7	126
4		6.3	0.88	11.4	103
5		10.6	0.90	8.3	74
6		18.0	0.91	5.4	60
9	MBI	0	0.95	139.7	149
10		1.4	0.90	44.1	135
11		2.7	0.93	28.0	124
12		6.9	0.86	12.1	97
13		10.3	0.70	7.7	78
14		20.1	0.53	4.6	63

Table 5. Melting temperatures and microstructural parameters for the copolymers with different α-olefins

Run	Cat	Cm	Incorp. Cm (mol %)	[mmm m]	n_{iso}	T_m (°C)
1	EI	Polypropene	0	0.88	46.0	136
2		Butene	6.8	0.86	11.2	118
3		Pentene	6.3	0.88	11.4	103
4		Hexene	6.1	0.84	11.6	94
5		Heptene	5.7	0.85	12.7	92
6		Octene	6.2	0.88	12.1	88
7		4M1P	2.6	0.88	19.7	119
8	MBI	Polypropene	0	0.95	139.7	149
9		Butene	7.5	0.83	10.9	125
10		Pentene	6.9	0.86	12.1	97
11		Hexene	5.0	0.72	13.6	91
12		Heptene	6.1	0.79	12.4	86
13		Octene	4.9	0.84	16.1	98
14		4M1P	3.1	0.44	12.3	106

the two series of copolymers of propene with different 1-olefins (Table 5). By comparing the two homologues with 1-hexene, runs 4 and 11, it is evident that despite the rather different isotacticities values (0.84 and 0.72), the melting temperatures are similar (94 and 91°C).

In order to correlate the thermal behavior with the microstructural features of these copolymers, we have calculated the average lengths of the isotactic sequences, n_{iso}, according to the following equation (deduced by microstructural data in a similar way as reported in the literature[6]):

$$n_{iso} = ([mmmm] + 4 \times [mrrm]) / ([mrrm] + [c_{Cm}])$$

where c_{Cm} indicates the concentration of the constitutional errors (isolated comonomer units, comonomer diads and triads) and, for both catalysts, only the concentration of the isolated steric errors [mrrm] is considered. The calculation is in fact based on the hypothesis that all the other steric defects observed with MBI are grouped into atactic blocks eventually following a comonomer insertion (see Scheme1b).

The rather similar thermal behavior for the homologues of the series prepared with the two different catalysts can be better understood if one correlates the observed transition temperatures to the n_{iso} values rather than to the overall isotacticity: the similar n_{iso} values in runs 6 and 14, 5.4 and 4.6 respectively, (Table 4), account for similar melting temperatures. The same considerations can be made for runs 4 and 11 of Table 5. The melting temperatures *versus* the calculated n_{iso} values for the two series of copolymers with 1-pentene are given in Figure 3; the plot suggests that the thickness of crystalline lamellae is correlated with the length of isotactic sequences.

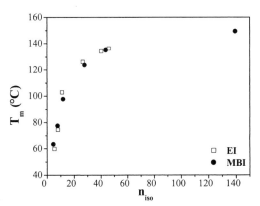

Figure 3. Melting temperatures *versus* the calculated n_{iso} values for the two series of copolymers at increasing 1-pentene content

The same trend holds for each pair of the series from the two different catalysts and suggests that the thermal behavior of the copolymers cannot be accounted for by the overall contents of isotactic pentads [mmmm], but it can be explained on the basis of the substantially equivalent average length of the isotactic sequences. At the same time the correlation observed between the thermal behavior and the n_{iso} values, calculated on the basis of copolymer structures as shown in Scheme 1, supports the reliability of the proposed structure for the copolymers from MBI as a succession of isotactic and atactic blocks, the latter likely formed by an occasional reversible loss of catalyst selectivity after a comonomer insertion.

Conclusions

In this paper several aspects relating microstructural characteristics and thermal properties of propene based copolymers to the kind and amount of the comonomer have been evidenced by introducing a series of comonomers with gradual increase of chain length and bulkiness. In

68

particular, we observed that in the presence of higher α-olefins the catalyst MBI leads to copolymers whose microstructures seem to indicate that MBI shifts reversibly from a stereoselective to a non-stereoselective state. This leads to copolymers with new and interesting microstructures extremely useful for a fundamental study of the correlations between microstructural characteristics and final properties. In fact, by properly varying the type of comonomer and its concentration in the feed, it is possible to modulate the comonomer content and distribution as well as the length and frequency of the atactic blocks. Further studies are currently in progress to ascertain the nature of the interaction between the higher 1-olefins and the catalyst MBI and to understand the reasons why, under the chosen polymerization conditions, this phenomenon occurs.

Acknowledgements

The authors wish to thank Dr. Augusto Provasoli for helpful discussions, Giulio Zannoni for performing NMR spectra and Vincenzo Trefiletti for doing DSC experiments. The helpful cooperation of Abey Shabangu for preparing the polymers and of Stefan de Goede for the determination of the molar masses is gratefully acknowledged.

[1] [1a] D. J. Joubert, I. Tincul *Propylene/1-Pentene Copolymers*, 6th SPSJ International Polymer Conference, 1997, Kusatsu, Japan; [1b] I. Tincul, D. J. Joubert, A. H. Potgieter *Impact Fracture Toughness of Propylene / 1-Pentene Random Copolymers*, PMSE Division, ACS Fall Meeting, 1998; [1c] I. Tincul, D. J. Joubert, A. H. Potgieter *1-Pentene Copolymers with Ethylene and Propylene*, Polymer '98, 1998, Brighton, UK; [1d] I. Tincul, D. J. Joubert, S. P. J. Smith, P. W. v Zyl *Propylene Copolymers with Fischer-Tropsch Olefins*, PMSE Division, Spring Meeting, 2001, San Diego, USA.
[2] see for instance: [2a] H.-H. Brintzinger, D. Fischer, R. Mülhaupt, B. Rieger, R. M. Waymouth, *Angew. Chem. Int. Ed. Engl.* 1995, *34*, 1143; [2b] S. Jüngling, R. Mülhaupt, U. Stehling, H.-H. Brintzinger, D. Fischer, F. Langhauser, *J. Polym. Sci, Part A: Polym. Chem.* 1995, *33*, 1305 and references therein.
[3] It is worthwhile to observe that the comparison of all the different comonomers shows that there is no correlation between the extent to which the tacticity is reduced and the carbon number of the comonomer. It is likely that for higher 1-olefins, effects such as the conformation of the branch, come into play.
[4] See for instance: [4a] M. J. Schneider, R. Mülhaupt, *Macromol. Chem. Phys.* 1997, *198*, 1121; [4b] M. Arnold, S. Bornemann, F. Köller, T. J. Menke, J. Kressler, *Macromol. Chem. Phys.* 1998, *199*, 2647; [4c] I. Kim, *Macromol. Chem. Commun.* 1998, *19*, 299; [4d] Z.-Q. Fan, T. Yasin, L.-X. Feng, *J. Polym. Sci. Part A: Polym Chem.* 2000, *38*, 4299, [4e] S. Hosoda, H. Hori, K. Yada, S. Nakahara, M. Tsuji *Polymer*, 2002, *43*, 7451.
[5] [5a] W. Kaminsky, A-M. Schauwienold, F. Freidanck, *J. Mol. Catal., Part A: Chem.* 1996, *112*, 37; [5b] M. Vathauer, W. Kaminsky, *Macromolecules* 2000, *33*, 1955.
[6] [6a] W. J. Gauthier, S. Collins *Macromolecules*, 1995, *28*, 3779; [6b] S. Mansel, E. Pérez, R. Benavente, J. M. Pereña, A. Bello, W. Röll, R. Kirsten, S. Beck, H.-H. Brintzinger, *Macromol. Chem. Phys.* 1999, *200*, 1292.

Relative Reactivity Enhancement of $(Me_5Cp)_2ZrCl_2$ in Mixture with $(1,2,4\text{-}Me_3Cp)_2ZrCl_2$ during Ethene /1-Hexene Copolymerization

Ingvild Bruaseth,[1] *Erling Rytter*[*2]

[1] Department of Chemical Engineering, Norwegian University of Science and Technology, N-7491 Trondheim, Norway
[2] Statoil Research Centre, N-7005 Trondheim, Norway
E-mail: err@statoil.com

Summary: Dual-site ethene/1-hexene copolymerizations with MAO-activated $(1,2,4\text{-}Me_3Cp)_2ZrCl_2$ and $(Me_5Cp)_2ZrCl_2$ catalysts were performed. Copolymers with narrow molecular weight distributions and bimodal short chain branching distributions could be produced. The combined catalyst system demonstrates a number of discrepancies from an expected average behavior of the individual sites. Dual-site $(1,2,4\text{-}Me_3Cp)_2ZrCl_2/(Me_5Cp)_2ZrCl_2$ systems produce copolymers with lower incorporation than expected. Clear evidences for relative activity enhancement of the $(Me_5Cp)_2ZrCl_2$ catalyst in the mixture were observed in melting endotherms and Crystaf profiles. Molecular weights obtained by the mixture were higher than for any of the individual catalysts. A similar effect is observed for a dual-site system of the $(1,2,4\text{-}Me_3Cp)_2ZrCl_2$ catalyst together with the $Me_4Si_2(Me_4Cp)_2ZrCl_2$ catalyst as an alternative to $(Me_5Cp)_2ZrCl_2$.

Keywords: bimodal; catalyst mixture; copolymerization; dual-site; metallocene catalysts

Introduction

Metallocenes are known as single-site catalysts. They produce polymers with narrow molecular weight distributions (MWD), and copolymers with narrow distributions of short chain branches (SCBD). Combination of two metallocenes with known polymerization behavior can be used to control polymer microstructure. Polyethene homopolymers with broader and sometimes bimodal MWD can be obtained.[1-4] Concerning copolymerization with a mixture of two metallocenes there are, to our knowledge, only a few reports in the literature.[5-7] Common to these studies was no observation of interaction between the two different sites, e.g. by re-adsorption of a terminated chain at the alternative site. Such reports have been issued, however, for polymerization of propene. Combinations of metallocenes with different stereospecifities have produced stereoblock polymers.[8-11] The stereoblock polymer is a cross product resulting from interaction of the two different sites during chain growth. Lieber and Brintzinger[11] have proposed that the transfer of a growing polymer chain occurs through an exchange of polymer chain between Zr centers and the Al center of the cocatalyst.

In the present investigation, we carried out ethene/1-hexene copolymerization catalyzed by the following methylaluminoxane (MAO)-activated zirconocene catalysts: (1,2,4-Me$_3$Cp)$_2$ZrCl$_2$ (abbreviated as 1,2,4-Me$_3$Cp), (Me$_5$Cp)$_2$ZrCl$_2$ (abbreviated as Cp*) and mixtures of these two catalysts. The influence of the dual-site catalyst on polymerization activity, incorporation of comonomer and polymer microstructure has been studied and compared with the individual catalysts at corresponding polymerization conditions.

Experimental

Chemicals. Bis(1,2,4-trimethylcyclopentadienyl)zirconium dichloride and bis(pentamethyl-cyclo-pentadienyl) zirconium dichloride purchased from Boulder Scientific Co., 10 wt% MAO in toluene purchased from Albemarle S.A., nitrogen (99.999%) purchased from Hydrogas, and ethene (polymerization grade) and 1-hexene (polymerization grade) donated from Borealis were all used as received. Toluene (p.a.) purchased from Merck was refluxed over sodium/ benzophenone and distilled under nitrogen atmosphere before use.

Polymerization. The polymerizations were performed in a steel autoclave that was repeatedly flushed and evacuated both with nitrogen and ethene. The reactor temperature was set (80 °C), and toluene (200 mL) was introduced. For copolymerizations, the desired amount of 1-hexene was added immediately after the toluene. The stirring rate was set (2000 rpm), and the reactor was pressured with ethene to the wanted total pressure (2 bar). Ethene was then equilibrated before the MAO solution was added. After 7 min the catalyst dissolved in toluene was injected. The total amount of catalyst was 0.23 µmol, and the Al/Zr ratio was 3000. The total pressure was held constant during the reaction, and the instantaneous consumption of ethene was measured. Polymerization time was 1 hour. The polymer product was poured into a mixture of methanol (300 mL) and hydrochloric acid (30 mL) and stirred overnight. After filtration the polymer was washed with methanol and dried in air.

Characterization. Differential scanning calorimetry (DSC) with a Perkin-Elmer 7 series thermal analyzer, gel permeation chromatography (GPC) with a Polymer Laboratories PL-210 GPC instrument and Fourier transform infrared spectroscopy (FTIR) with a Bruker IFS66V spectrophotometer were performed according to previously reported procedures.[12] Crystallization analysis fractionation (Crystaf) was performed with a model 200 Crystaf instrument manufactured by Polymer Char. The instrument is equipped with five separate stainless steel crystallization vessels for the simultaneous analyses of five different samples. The polymer samples (10.2 ± 0.8 mg) were dissolved in TCB (48 ml) at 160 °C for 60 min, and then the solution was equilibrated at 95 °C for 45 min. Subsequently, the solution was cooled at a rate of 0.10 °C/min to 30 °C. The solution was sampled 50 times at temperature intervals between 95 °C and 30 °C, and the change in polymer concentration in solution with temperature was monitored by a in-line infrared detector.

Results and Discussion

The Cp* catalyst has high starting copolymerization activity, but the activity decay is very fast. The average copolymerization activity of this catalyst is therefore low, and it is approximately six times lower than for the 1,2,4-Me$_3$Cp catalyst. The 1:1 mixture of these two catalysts has an activity in between the activities of the individual catalysts, as shown in Figure 1a. Because of the large difference in activities for the individual catalysts it might be expected that the mixture results in polymer properties close to the properties obtained by 1,2,4-Me$_3$Cp.

The incorporation was determined by FTIR. Figure 1b shows that 1,2,4-Me$_3$Cp incorporates significantly more 1-hexene than Cp*. The 1:1 mixture results in properties intermediate between the two individual catalysts, but the 1-hexene content certainly is lower than expected from the six-fold higher activity of the 1,2,4-Me$_3$Cp catalyst.

If the two catalysts in the dual-site system behave independently, the incorporation for the mixture can in principle be predicted from the incorporations for the individual catalysts taking into account their different activities. Other mixtures than 1:1 have been utilized in copolymerizations with similar initial 1-hexene concentration. Figure 2 shows the incorporation as a function of the mole fraction of Cp* in mixture with 1,2,4-Me$_3$Cp. By comparing the obtained and the predicted incorporations, it can be seen that each of the

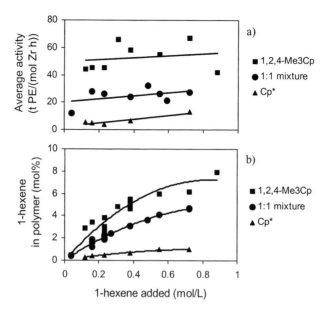

Figure 1. Average activities obtained for 1,2,4-Me₃Cp, Cp* and the 1:1 mixture of these two (a) and incorporation of 1-hexene in poly(ethene-*co*-1-hexene) (b) as a function of initial 1-hexene concentration. Cocatalyst = MAO; T = 80 °C; P_{total} = 2 bar; $n_{Zr\ tot}$ = 0.23 µmol; Al/Zr_{tot} = 3000.

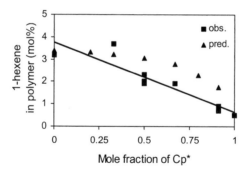

Figure 2. Observed (obs.) and predicted (pred.) incorporation of 1-hexene as a function of the mole fraction of Cp* in mixture with 1,2,4-Me₃Cp. Cocatalyst = MAO; T = 80 °C; P_{total} = 2 bar; [1-hexene]$_{initial}$ = 0.23 mol/L; $n_{Zr\ tot}$ = 0.23 µmol, Al/Zr = 3000.

mixtures (with one exception) results in poly(ethene-*co*-1-hexene) having lower comonomer content than expected. When less 1,2,4-Me₃Cp (and then more Cp*) is present in the mixture, the 1-hexene content decreases. However, the decrease is larger than expected taking into consideration the much higher activity of the 1,2,4-Me₃Cp catalyst.

The 1:1 mixture produces copolymers with two distinct melting endotherms when 0.23 – 0.72 mol/L 1-hexene was added, as illustrated in Figure 3. This is in accordance with a copolymer with two different fractions of comonomer. The high temperature peak is attributed to the Cp*-site, and the low temperature peak to the 1,2,4-Me₃Cp-site. When low amount of 1-hexene was added (0.16 mol/L), the copolymer has only one melting peak, possibly because the melting of the two fractions now occurs so close in temperature that they do not separate into two peaks. The low temperature peak moves towards lower temperature as the 1-hexene content increases, while the high temperature peak is approximately constant, or increases slightly when higher initial 1-hexene concentration was used. This may be caused by some cocrystallization of the two polymer fractions for low 1-hexene content. High comonomer concentration leads to polymer fractions with larger differences in their comonomer content, and thereby easier separation in the crystallization process.

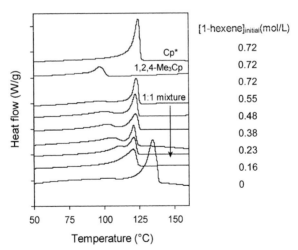

Figure 3. Melting endotherms for polyethene and poly(ethene-*co*-1-hexene) made with the 1:1 mixture and melting endotherms for poly(ethene-*co*-1-hexene) made with the individual catalysts for comparison.
Cocatalyst = MAO; $T = 80$ °C; $P_{total} = 2$ bar; $n_{Zr\,tot} = 0.23$ µmol; $Al/Zr_{tot} = 3000$.

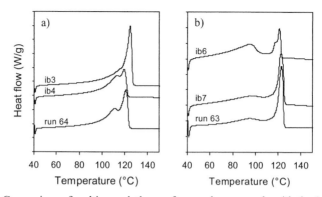

Figure 4. Comparison of melting endotherms for copolymers made with the 1:1 mixture (runs 63 and 64) and endotherms of blended copolymers made with 1,2,4-Me$_3$Cp and Cp* individually. Cocatalyst = MAO; T = 80 °C; P$_{total}$ = 2 bar; n$_{Zr\,tot}$ = 0.23 μmol; Al/Zr$_{tot}$ = 3000. (a) [1-hexene]$_{initial}$ = 0.23 mol/L; ib3 = 50/50 wt% blend; ib4 = activity ratio blend. (b) [1-hexene]$_{initial}$ = 0.72 mol/L; ib6 = activity ratio blend; ib7 = 50/50 wt% blend.

A closer look at the melting endotherms shows that the influence of the Cp*-based polymer fraction is much stronger than expected from the individual activities clearly pointing to an in-situ enhancement of the activity of the Cp*-site. To check this effect further, polymers made with the individual catalysts were blended by dissolution in hot xylene and then coprecipitated in cold methanol. The polymers were blended in accordance with the individual activity ratio and in a 50/50 wt% mixture. The melting curves of these blends are compared with the melting curves for copolymers made by the 1:1 mixture using similar 1-hexene concentrations. Figure 4 shows that the behavior of the polymer from the mixture is between what is expected from the activity ratio and a 50/50 wt% blend, and rather closer to the latter.

The weight-average molecular weights (Mw) for poly(ethene-co-1-hexene) made with 1,2,4-Me$_3$Cp, Cp* and the 1:1 mixture were determined by GPC. Figure 5 shows that Cp* and 1,2,4-Me$_3$Cp produce copolymers with quite similar Mw, but 1,2,4-Me$_3$Cp gives copolymer with rather high Mw for the high comonomer content. The 1,2,4-Me$_3$Cp catalyst has in principle many conformations due to the possibility of relative rotation of the ligands. It turns out, however, that one specific conformer is the most stable one. This structure is rather open towards coordination of the monomers, but at the same time it contains one methyl on each

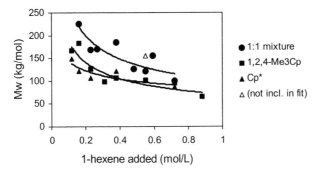

Figure 5. Mw for poly(ethene-*co*-1-hexene) as a function of initial 1-hexene concentration. Catalysts = 1,2,4-Me₃Cp, Cp*, and 1:1 mixture of these two; cocatalyst = MAO; T = 80 °C; P_{total} = 2 bar; $n_{Zr\,tot}$ = 0.23 µmol; Al/Zr$_{tot}$ = 3000.

ligand that efficiently prevents rotation between the γ-agostic and β-agostic states after 1-hexene insertion.[13] Such a rotation is a necessary step in the termination sequence. As a result, 1,2,4-Me₃Cp incorporates 1-hexene unusually easy with a simultaneous low termination frequency.

It is remarkable that Mw for copolymers made with the 1:1 mixture consistently, with only one exception, is higher than for both the individual catalysts (Figure 5). This is highly unexpected since earlier experimental evidence indicates that each metallocene in a mixture behaves as if present alone in the reactor.[1-3] This observation indicates that the two catalysts in the mixture seem to influence each other's behavior. One possible explanation for the observed high molecular weights can be deduced from previous reports stating that chain transfer to trimethylaluminum (TMA) is important for Cp*.[14-15] The chains originating at Cp* contain minimal comonomer, but terminate frequently with TMA. Thus these chains can fairly easy be incorporated into the more open 1,2,4-Me₃Cp-site and continue to polymerize. This mechanism is in analogy to the model discussed by Lieber and Brintzinger[11] for propene polymerization by mixture of two metallocenes with different stereospecifities. The Mw for polymer prepared by the 1:1 mixture followed the normal decreasing trend with higher comonomer content, while the number-average molecular weight (Mn) from GPC was more constant.[12] Consequently, the MWD became more narrow as the 1-hexene content increased. It is noteworthy that by combining two metallocenes a very narrow MWD can be observed from polymer chains that differ dramtically in their comonomer content. Chain

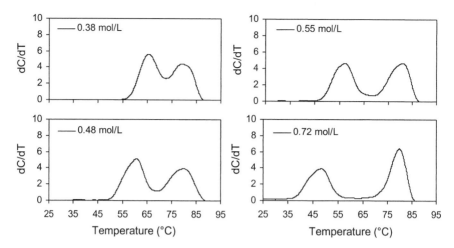

Figure 6. Crystaf profiles for poly(ethene-*co*-1-hexene) samples made with 1:1 mixture of 1,2,4-Me$_3$Cp and Cp* using different initial 1-hexene concentration. Cocatalyst = MAO; T = 80 °C; P$_{total}$ = 2 bar; n$_{Zr\,tot}$ = 0.23 µmol; Al/Zr$_{tot}$ = 3000.

transfer to TMA is an important termination mechanism for the homopolymer, but it gets less important as the 1-hexene content increases.[12] Chain transfer to TMA probably gets less important as these chains may continue to polymerize on the other site, alternatively TMA has to compete with 1-hexene for access to the active site.

In order to eliminate the possibility in the dual-site system of ligand exchange reactions, possibly resulting in a mixed ligand catalyst, we have as an alternative to Cp*, studied the Me$_4$Si$_2$(Me$_4$Cp)$_2$ZrCl$_2$ catalyst (abbreviated as aMe$_4$Cp). The geometry resembles that of Cp*, but the opening angle is 4° larger and the conformation approaches an eclipsed one.[16] The polymerization behavior is very much the same for both catalysts, although the bridged one gives increased termination probability after 1-hexene insertion, but less frequent termination after ethene insertion. The 1:1 mixture of 1,2,4-Me$_3$Cp and aMe$_4$Cp also produced poly(ethene-*co*-1-hexene) with higher Mw than for the individual catalysts at intermediate 1-hexene concentrations. In addition lower comonomer incorporation, than expected from the ca. 10:1 activity ratio between the individual catalysts, was observed. In the melting endotherms a significant overrepresentation of the high temperature melting peak, due to the

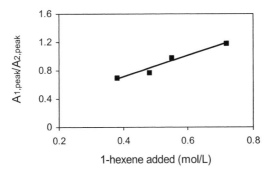

Figure 7. Ratio between weight fraction of polymer within high ($A_{1,peak}$) and low ($A_{2,peak}$) crystallization peaks for poly(ethene-*co*-1-hexene) made with the 1:1 mixture as a function of initial 1-hexene concentration.

aMe$_4$Cp catalyst, was observed.

Crystaf of poly(ethene-*co*-1-hexene) made with the 1:1 mixture of 1,2,4-Me$_3$Cp and Cp* was performed to determine the short chain branching distribution (SCBD). Examples of Crystaf profiles are given in Figure 6, and as illustrated bimodal SCBDs were obtained. The high temperature peak results from crystallization of polymer chains with few short chain branches, and these are attributed to the Cp*-site in the mixture. The low temperature peak results from crystallization of polymer chains made with the 1,2,4-Me$_3$Cp-site, and these chains contain many short chain branches. Once more, a significant contribution from the Cp*-site is observed.

The ratio between the fractions of polymer within the high and low crystallization peak temperatures ($A_{1,peak}/A_{2,peak}$) increases with increasing initial 1-hexene concentration, as illustrated in Figure 7. The relative activity of the Cp* catalyst in the mixture increases almost linearly with the 1-hexene concentration.

Conclusion

The dual-site 1,2,4-Me$_3$Cp/Cp* system demonstrates a number of discrepancies from an expected average behavior of the individual catalysts. The most striking result is Mw higher than for both the individual catalysts. Other observations were lower comonomer incorporation than expected, a much more pronounced melting/crystallization peak ascribed to Cp* than the activities predicts, and gradual elimination of chain transfer to TMA with increasing 1-hexene concentration. The data are interpreted as an enhanced activity of the Cp*-site relative to the 1,2,4-Me₃Cp-site possibly coupled with chain transfer from the Cp*-site to the 1,2,4-Me₃Cp-site. However, we do not believe that a full explanation of the unexpected relative activity enhancement of one of the sites can be given at this point. Further investigations clearly are needed.

Acknowledgement

Financial support from Borealis AS is gratefully acknowledged. We would also like to thank Professor Joao B. P. Soares at the University of Waterloo, Canada for allowing I. Bruaseth to visit his group to do Crystaf analyses.

[1]   A. Ahlers, W. Kaminsky, *Makromol. Chem., Rapid Commun.* **1988**, *9*, 457.
[2]   L. D'Agnillo, J. B. P. Soares, A. Penlidis, *J. Polym. Sci., Part A: Polym. Chem.* **1998**, *36*, 831.
[3]   T. K. Han, H. K. Choi, D. W. Jeung, Y. S. Ko, S. I. Woo, *Macromol. Chem. Phys.* **1995**, *196*, 2637.
[4]   J. Liu, E. Rytter, *Macromol. Rapid Commun.* **2001**, *22*, 952.
[5]   K. Heiland, W. Kaminsky, *Makromol. Chem.* **1992**, *193*, 601.
[6]   J. D. Kim, J. B. P. Soares, *J. Polym. Sci., Part A: Polym. Chem.* **2000**, *38*, 1427.
[7]   K.-J. Chu, J. B. P. Soares, A. Penlidis, *Macromol. Chem. Phys.* **2000**, *201*, 340.
[8]   C. Przybyla, G. Fink, *Acta Polym.* **1999**, *50*, 77.
[9]   J. C. W. Chien, Y. Iwamoto, M. D. Rausch, W. Wedler, H. H. Winter, *Macromolecules* **1997**, *30*, 3447.
[10] J. C. W. Chien, Y. Iwamoto, M. D. Rausch, *J. Polym. Sci., Part A: Polym. Chem.* **1999**, *37*, 2439.
[11] S. Lieber, H.-H. Brintzinger, *Macromolecules* **2000**, *33*, 9192.
[12] I. Bruaseth, E. Rytter, *Macromolecules* **2003**, *36*, 3026.
[13] H. Wigum, L. Tangen, J. A. Støvneng, E. Rytter, *J. Polym. Sci., Part A: Polym. Chem.* **2000**, *38*, 3161.
[14] K. Thorshaug, J. A. Støvneng, E. Rytter, M. Ystenes, *Macromolecules* **1998**, *31*, 7149.
[15] M. Ystenes, J. L. Eilertsen, J. Liu, M. Ott, E. Rytter, *J. Polym. Sci., Part A: Polym. Chem* **2000**, *38*, 3106.
[16] I. Bruaseth, J. B. P. Soares, E. Rytter, *to be published.*

Sterically Demanding Bis(2-silylindenyl)zirconium(IV) Dichlorides as Polymerisation Catalyst Precursors

Andreas Hannisdal,[1] Andreas C. Möller,*[1,3] Erling Rytter,[1,2] Richard Blom[3]

[1] Department of Chemical Engineering, Norwegian University of Science and Technology (NTNU), N-7491 Trondheim, Norway
[2] Statoil Research Centre, N-7005 Trondheim, Norway
[3] SINTEF, Department of Hydrocarbon Process Chemistry, P.O. Box 124 Blindern, N-0314 Oslo, Norway

Summary: A set of different 1- and 2-silyl-substituted zirconocene dichloride/MAO catalyst systems was investigated with respect to their performance in ethene/1-hexene copolymerisations. In-depth studies of bis(2-dimethylsilylindenyl) zirconium(IV) dichloride (**1**) revealed a multi-site behaviour, illustrating sensitivity to the reaction temperature and the comonomer mole fraction. Surprisingly, an upper limit is observed for the latter, leading to complete catalyst inhibition. Analysis of the chain termination processes implies the possibility of a predominant, although in general less favourable, β-hydride elimination route under certain polymerisation conditions.

Keywords: copolymerisation; end group analysis; ethene-*co*-1-hexene; metallocenes; multi-modal polymers

Introduction

The control of polymer microstructure by design of the catalyst ligand framework symmetry has been strongly exploited since the first discoveries in the 1980s.[1-3] Sterically rigid and bulky structures have been synthesised and successfully employed for the tailoring of tacticity and comonomer incorporation of homo- and copolymers.[4-6] When Waymouth et al.[7] introduced the concept of oscillating stereocontrol for the production of elastomeric polypropylene (ePP) as an alternative to the stereoblock ePP discovered by Kaminsky et al.,[8] conformationally flexible metallocenes attracted new attention in the community.[9] In the former system, it has been claimed that the observed elastomeric properties could be attributed to a *rac/meso* interconversion of the active site during chain growth, leading to atactic/isotactic sequences.[7] Results obtained by molecular mechanics and DFT calculations suggested that the afore mentioned *rac* and *meso* conformers represented actual energy minima for the catalyst precursors.[10,11] However, experimental evidence for the

DOI: 10.1002/masy.200450909

interconversion could not be provided, and recent studies of the polymer microstructure conducted by Busico et al. by means of advanced high-field ^{13}C NMR clearly show that atactic sequences are not contributing to the phenomenon. Isolated stereoerrors due to a *rac*/*rac** interconversion seem to be the most suitable explanation.[12]

Regardless of the validity of the suggested mechanisms of polymerisation, the idea of controlling the number and type of several possible active sites of a conformationally flexible zirconocene remains an intriguing challenge. In the present study we investigate bis-indenyl zirconocenes bearing a silyl moiety in 2-position, which can be substituted with different functionalities. The silyl group appeared to us as a natural choice for the systematic variation and study of different substitution patterns in the 2-position. To our knowledge there are only few reports on the corresponding 1-isomers and therefore these have been synthesized as well. Details on the preparation and structural as well as catalytic characterisation with respect to ethylene homopolymerisations of these zirconocenes will be reported elsewhere.[13] We herein wish to report a screening of different catalysts in the ethene/1-hexene copolymerisation and some aspects of in-depth copolymerisation studies of bis(2-dimethylsilylindene) zirconoium(IV) dichloride (**1**).

Results and Discussion

Comparative Ethene/1-Hexene Copolymerisation Studies

In order to obtain a general idea of the catalyst characteristics of the metallocenes, a comparative copolymerisation screening was run to account for activity. The bis(indenyl)zirconium(IV) dichloride/MAO catalyst was chosen as a standard, since polymerisation activities might vary due to different experimental setups and procedures. A concise overview over the catalysts employed is given in Figure 1.

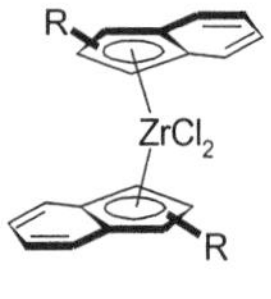

Acronym	R	2-isomer	meso 1-isomer	rac 1-isomer
DMS	Me₂HSi	1	4	-
TMS	Me₃Si	2	5	7
TBDMS	ᵗBuMe₂Si	-	-	8
DMPS	Me₂PhSi	3	6	9

Figure 1. Numbering scheme for the *rac*-like conformer of bis(1/2-silylindenyl)-zirconocene(IV) dichloride metallocenes.

The copolymerisations were run at a comonomer mole fraction of $X_H = 0.60$ in the feed at a reaction temperature of 30°C. The results presented in Table 1 indicate that for the 2-isomers, the DMS moiety (**1**) yields the highest activity. Introduction of one additional methyl group at the silyl moiety (**2**) causes a decrease of 60%, whereas total catalyst inhibition occurs for a phenyl group (**3**). As for the 1-isomers, the *rac* diastereomer of the sterically most demanding DMPS system **9** displays an activity comparable to that of **1,** the least congested 2-isomer. This is in agreement with the results obtained for the homopolymerisation.[13] A striking contrast is that **3** and **9** show complementary copolymerisation activities simply due to a change from the 2- to the 1-isomer. The unsubstituted bis(indenyl)zirconocene is slightly more active than any of the silyl substituted-indenyl zirconocenes considered here.

Table 1. Average activities with absolute standard deviations for the several zirconocene/MAO systems in ethene/1-hexene copolymerisations, using 30°C, 1.2 bar ethene, $X_H = 0.20$, $c(Zr) = 10^{-6}$ mol/L, Al/Zr = 1000, 30 min, toluene.

No.	Compound	Activity [$t/mol_{Zr} \cdot mol_{E+H} \cdot h$]
1	2-DMS	115 ± 12
2	2-TMS	47 ± 1
3	2-DMPS	0
Ref	[Ind]$_2$ZrCl$_2$	151 ± 6
4	*meso* 1-DMS	53 ± 4
5	*meso* 1-TMS	22 ± 1
6	*meso* 1-DMPS	13 ± 1
7	*rac* 1-TMS	19 ± 0
8	*rac* 1-TBDMS	62 ± 9
9	*rac* 1-DMPS	107 ± 5

Bis(2-DMS)zirconium(IV) Dichloride in Ethene/1-Hexene Copolymerisation

Activity Response on 1-Hexene Mole Fraction and Temperature

In general, zirconocene/MAO catalysts can show a polymerisation rate enhancement at certain temperatures upon the addition of comonomer to the reaction feed. In the case of **1,** the activity decreases rapidly as a function of the comonomer mole fraction in the feed at 80°C. After an initial activity drop down to 60% at $X_H = 0.20$, it stabilises between 3-5%

relative to the homopolymerisation for $0.80 \leq X_H \leq 0.90$. Complete catalyst inhibition is achieved at higher values for X_H. For the same set of experiments at a reaction temperature of 30°C, the slope of inhibition is steeper, although it is qualitatively of the same form, and complete catalyst inhibition is not achieved until X_H exceeds 0.95.

The effect of a variation in temperature is less pronounced and more complicated. In the temperature series all parameters except for the temperature were constant, but the activity ranged from 148 to 245 t/mol$_{Zr}$·mol$_{E+H}$·h without any apparent correlation. Conditions used: $c(E) = 0.14$ mol/L, $c(Zr) = 5 \cdot 10^{-7}$ mol/L, Al/Zr = 8700, 30 min polymerisation time.

Polymer Microstructure and Molecular Weights

The polymer microstructure of the LLDPE samples has been analysed by means of FT-IR and ^{13}C NMR, and the homogeneity of the products was investigated by DSC and high-temperature GPC. The polymers obtained under the various experimental conditions are strongly affected by the polymerisation temperature.

The comonomer enchainment was between 1.0 and 0.3 mol% at constant concentration in the reaction feed, $X_H = 0.20$ and a temperature range from 10 to 80°C. Based on the results from the end group analysis (Figure 2) and molecular weights (Table 2), the reason for these values becomes apparent. The ratio of vinyl to *trans*-vinylene termination is larger than unity at all temperatures. While the concentration of vinyl terminations is higher than that of vinylidene up to 30°C, a reversal of the situation is observed from 40°C upwards. Vinylidene bonds originate by hydrogen transfer to monomer or β-hydride elimination after 1-hexene enchainment. Here, the latter process is considered to be dominating for two reasons. Firstly, the activation entropy for a bimolecular hydrogen transfer to monomer is negative and this process becomes less favourable with increasing temperature. Secondly, the activation enthalpy, estimated to be on the order of 155 kJ/mol ($\pm$17 kJ/mol),[14] is in reasonable agreement with the calculated reaction energy of 113 kJ/mol, as predicted by Støvneng and Rytter.[15]

Interestingly, the concentration of vinylidene double bonds does stabilise around 0.36/1000C at 70 and 80°C (Figure 2). At 80°C the average number of comonomer units per polymer chain is about 2, based on a molecular weight of 17 kg/mol, *i.e.* $P_n \cong 600$. This implies that 32% of all chains will be terminated after comonomer enchainment or, in other words, enchainment of 1-hexene will lead to chain growth termination one out of five times. Hence,

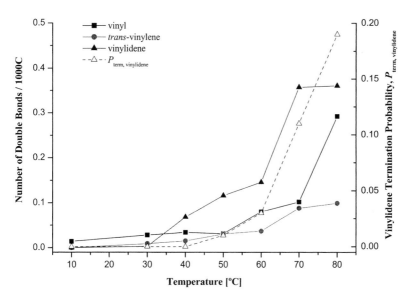

Figure 2. End group analysis as a function of the polymerisation temperature, catalysed with **1**/MAO, X_H = 0.20, 30 min polymerisation time, Al/Zr = 8700. Left ordinate shows number of double bonds / 1000C, right ordinate the calculated probabilities for chain termination after 1-hexene insertion.

the catalytic process is approaching dimensions of quantitative conversion relative to the usually rather unlikely vinyl termination process.

The increase in the termination rate to vinyl and *trans*-vinylene end groups at elevated temperatures indicates a superposition of mechanisms, *i.e.* β-hydride elimination occurs in addition to chain-transfer to monomer.[17] That is in agreement with thermodynamics which call for faster and more frequent termination, while comonomer insertion should slow down. Further, chain termination by transmetalation to aluminium loses significance with increasing temperature, since the contribution of the negative activation entropy depresses the Gibbs' free energy and makes this process unfavourable. A comparison of the molecular weights determined by means of FT-IR and GPC reveals that there is virtually no termination through transmetalation to be considered at 70 and 80°C (Table 2).

Similar end group effects have been observed for the variation of the 1-hexene mole fraction in the feed at elevated temperatures, but the inhomogeneity of the products obtained made an interpretation more ambiguous.

Table 2. Variation of temperature at constant 1-hexene concentration. The 1-hexene incorporations x_H reported here approach the detection limit for ^{13}C NMR and IR analysis. The uncertainty in these figures is difficult to estimate. We assume ±0.003 for x_H and ±10% for the M_n. PDI = polydispersity index.

T [°C]	$x_{H, NMR}$	M_n [kg/mol]$_{IR}$	M_n [kg/mol]$_{GPC}$	PDI
80	0.003	21	17	15
70	0.005	33	16	25
60	0.006	58	17	37
50	0.008	79	35	21
40	0.009	133	47	18
30	0.010	370	58	13
10	0.010	974	208	3.4

Homogeneity of Products and Dual-Site Hypothesis

Since rotationally flexible zirconocenes offer at least in theory the possibility of conformer interconversion during the polymerisation, the question of the accessibility and conversion rate of these conformers arises. Taking into account the polydispersity index (PDI) for the previously discussed copolymerisation, we observe indications for non-ideal single-site catalysis at 10°C. Quijada and Dupont reported a PDI of 3.5 for a copolymerisation with [Ind]$_2$ZrCl$_2$/MAO at 60°C,[17] while 1/MAO yields a ten-fold PDI at three times the M_n under similar reaction conditions. The response of the PDI as a function of the temperature, shown in Table 2, is rather systematic and passes a maximum around 60°C.

Complications arise when the effect of the comonomer concentration is studied at constant temperatures (Table 3). For the interval $0.00 \leq X_H \leq 0.90$, the PDI ranges between 16 and 1.9 at 80°C. After an initial increase at low concentrations of 1-hexene, the PDI undulates between 8 and 5 and finally displays single-site character at the highest comonomer concentration. In contrast to this, the PDI lies one order of magnitude higher when X_H exceeds 0.60 for the same set of experiments at 30°C and does not exhibit the initial increase at low comonomer concentrations. It has to be emphasised that these inhomogeneous products complicated sample preparation for FT-IR, DSC and GPC analysis. However, comonomer incorporations determined by means of ^{13}C NMR and FT-IR are in good agreement. A comprehensive summary of the polydispersities under the several reaction conditions is presented in Table 3.

Table 3. Polydispersities of LLDPE samples as derived from GPC. Catalyst system **1**/MAO.

	Polydispersity Index	
X_H	**30°C**	**80°C**
0.00	3.6	5.9
0.20	5.7	16.0
0.40	8.5	12.0
0.60	7.0	8.1
0.70	43	4.5
0.80	38	5.0
0.83	29	6.6
0.87	58	5.3
0.90	34	1.9
0.95	15	-

Figure 3 illustrates the superimposition of the GPC analyses for the comonomer variation at 80°C. The product distribution is clearly a function of the comonomer and multimodal. Surprisingly, a smooth transition from the rather monomodal product to at least bimodal products occurs between $0.60 \leq X_H \leq 0.90$ and implies that the different sites are either

a) discriminated with respect to population as a function of the comonomer mole fraction rather than the temperature, or

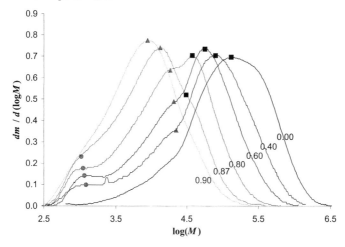

Figure 3. Superimposition of GPC elution curves for different 1-hexene mole fractions in the feed. Main fractions are marked with square, triangles and circles. T = 80°C.

86

b) respond to the change in comonomer mole fraction by a change in the propagation rate.

A less ambiguous case is presented in Figure 4, which resembles a bimodal product distribution at $X_H = 0.80$ and 30°C. The composition consists of an oligomeric fraction around $\log(M) = 3.4$ and a polymeric fraction at $\log(M) = 5.8$.

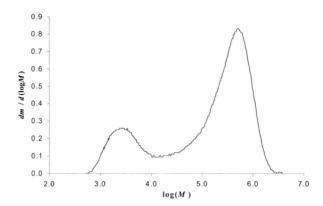

Figure 4. GPC trace of a bimodal product composition obtained with 1/MAO at $X_H = 0.80$ and 30°C.

The observed product compositions suggest more than one active site during the polymerisation process. Inspired by the works of Brintzinger et al.,[11] we applied molecular mechanics calculations to analyse the energy profile of the rotamer interconversion of the catalyst precursor. Preliminary results unambiguously show that the dichlorides have different accessible conformers, but these calculations have to be extended to active sites with attached polymer chain and associated olefins for a thorough picture to be drawn. Nevertheless, it can be pointed out for the case of compounds 1 and 2 that no stabilisation of the *meso*-like conformer is expected, accounting for the fact that our computations predict conformational transition states for the sterically less congested catalyst precursors.

Conclusions

The results presented here suggest that the enchainment of 1-hexene in the copolymerisation with ethene as catalysed by 1/MAO is hindered, although our incorporation rates are somewhat higher than those reported by Quijada for the bis-indenylzirconium(IV)-

dichloride/MAO catalyst.[18] Waymouth made similar observations for his catalyst systems in ethen-1-hexene copolymerisation.[19] While sterically demanding silyl moieties in the 2-position of an indene ring can, in the case of the DMPS group, lead to complete catalytic inhibition in the presence of 1-hexene, inhibition can be achieved at high comonomer mole fractions for even the smaller silyl moieties. Molecular weights increase by the introduction of a functional group in the 2-position. Despite the fact that the system investigated in detail here shows a temperature dependence on the active site population, there is no indication to believe that these sites interconvert rapidly enough to resemble a rotationally dynamic single-site catalyst. Quantum chemical calculations, which we are currently carrying out, suggest different, non-stabilised active sites, which matches the high polydispersity indices found by means of GPC. Still, the analysis of the polymer products remains difficult due to their inhomogeneity and display the non-SSC character of these zirconocenes.

Acknowledgement

We thank Jens Kjær Jørgensen for high-temperature GPC measurements and Aud M. Bouzga for NMR instrument support. A.C.M. gratefully acknowledges Dr.-Ing. scholarship grant no. 145544/431 by NFR (The Research Council of Norway).

[1] F.R.W.P. Wild, L. Zsolnai, G. Huttner, H.H. Brintzinger *J. Organomet. Chem.* **1982**, *232*, 233-247.
[2] J.A. Ewen *J. Am. Chem. Soc.* **1984**, *106*, 6355-6364.
[3] W. Kaminsky, K. Külper, H.H. Brintzinger, F.R.W.P. Wild, *Angew. Chem. Int. Ed. Engl.* **1985**, *24*, 507-508.
[4] W. Spaleck, M. Antberg, J. Rohrmann, A. Winter, B. Bachmann, P. Kiprof, J. Behm, W. A. Herrmann *Angew. Chem.* **1992**, *104*, 1373-1376.
[5] J.A. Ewen, J.M. Elder, R.L. Jones, L. Haspeslagh, J.L. Atwood, S.G. Bott, K. Robinson *Makromol. Chem., Macromol. Symp.* **1991**, *48-9*, 253-295.
[6] T. Uozumi, K. Miyazawa, T. Sano, K. Soga *Macromol. Rapid Commun.* **1991**, *18*, 883-889.
[7 G.W. Coates, R.M. Waymouth *Science* **1995**, *267*, 217-219.
[8] W. Kaminsky, M. Buschermöhle, in: *"Recent Advances in Mechanistic and Synthetic Aspects of Macromolecules"*, NATO ASI Series C, M. Fontanille, A. Guyot, Eds., D. Reidel Publishing Company, Dordrecht 1987, Vol. 215, 503-513.
[9] *ansa*-Metallocenes are also known to produce ePP at high activity and low deactivation rates. See U. Dietrich, M. Hackmann, B. Rieger, M. Klinga, M. Leskelä *J. Am. Chem. Soc.* **1999**, *121*, 4348-4355.
[10] M.A. Pietsch, A.K. Rappé *J. Am. Chem. Soc.* **1996**, *118*, 10908-10909.
[11] N. Schneider, F. Schaper, K. Schmidt, R. Kirsten, A. Geyer, H.H. Brintzinger *Organometallics* **2000**, *19*, 3597-3604.
[12] a) V. Busico, R. Cipullo, A.L. Segre, G. Talarico, M. Vacatello, V. Van Axel Castelli *Macromolecules* **2001**, *34*, 8412-8415. b) V. Busico, R. Cipullo, W. Kretschmer, G. Talarico, M. Vacatello, V. Van Axel Castelli *Angew. Chem. Int. Ed.* **2002**, *41*, 505-508.

[13] A.C. Möller, R. Blom, R.H. Heyn *manuscript in preparation*.

[14] This estimate assumes that the activity is constant over the whole period of polymerisation, which is not the case. The average termination rate after 1-hexene enchainment plotted versus $1/T$ yields $R = 0.97999$ and passes ANOVA at $p=0.05$ and pairwise t-testing. The pre-exponential factor A was calculated to $6 \cdot 10^{12}$ mol/s, its error of the pre-exponential calculates to zero, but the polydispersity index indicates an inhomogeneous product.

[15] J.A. Støvneng, E. Rytter *J. Organomet. Chem.* **1996**, *519*, 277-280.

[16] a) T.K. Woo, L. Fan, T. Ziegler *Organometallics* **1994**, *13*, 2252-2261. b) J.C.W. Lohrenz, T.K. Woo, L. Fan, T. Ziegler *J. Organomet. Chem.* **1995**, *497*, 91-104.

[17] The activation energy was estimated to be 92 kJ/mol, which is right in between the expected 42-58 kJ/mol for chain transfer to monomer and the reaction energy of 113 kJ/mol for β-hydride elimination.

[18] R. Quijada, J. Dupont, M.S.L. Miranda, R.B. Scipioni, G.B. Galland *Macromol. Chem. Phys.* **1995**, *196*, 3991-4000.

[19] Reybuck, S.E.; Meyer, A.; Waymouth, R.M. *Macromolecules* **2002**, *35*, 637-643.

Macromol. Symp. **2004**, *213*, 89-99

From Zirconium to Titanium: The Effect of the Metal in Propylene Polymerisation Using *Fluxional* Unbridged Bicyclic Catalysts

Eleonora Polo,[1] *Simona Losio,*[2] *Fabrizio Fortini,*[2] *Paolo Locatelli,*[2] *Maria Carmela Sacchi*[2]

[1] C.N.R., I.S.O.F. sez. Ferrara, c/o Dip. di Chimica dell'Università di Ferrara, via L. Borsari 46, 44100 Ferrara, Italy
E-mail: tr3@unife.it; Fax: +39 0532240709
[2] C.N.R., ISMAC, via E. Bassini 15, 20133 Milano, Italy
E-mail: m-c.sacchi@ismac.cnr.it; Fax: +39 022362946

Summary: Catalytic systems based on unbridged substituted indenyl systems are becoming of interest in the production of elastomeric polypropylene. A full understanding of the structural features necessary to control this kind of behaviour has not yet been achieved, since relatively slight changes in the molecular architecture can lead to polymers with remarkably different properties.
We report here our recent findings regarding the study of bicyclic zirconium and titanium complexes as fluxional catalysts in propylene polymerisation. Most of them have been synthesised according to a synthetic procedure that allowed us to prepare a series of complexes in which the ring fused to the cyclopentadienyl moiety is saturated and of different sizes, thus introducing a flexibility parameter that can be finely tuned.
The results obtained show that the stereoselectivity induced by this class of catalysts strongly depends both on the structure of the ligand and on the nature of metal atom (Zr *vs.* Ti). The titanium-based catalysts yield polypropylenes with new and interesting microstructures, in particular when an higher stability is achieved through a careful choice of the substitution pattern of the ligands.

Keywords: DSC; fluxional catalysts; metallocene catalysts; NMR; polypropylene

Introduction

Catalytic systems based on unbridged substituted indenyl systems are becoming more and more of interest in the production of elastomeric polypropylene. The possibility of preparing this kind of polymer by means of a single catalyst was introduced by the work of Waymouth and Coates in 1995.[1] They used for the first time the concept of "fluxional" catalysts to describe unbridged metallocenes which are able to interconvert between chiral and achiral rotational isomers on the time scale of the polymerisation reactions. This mechanism accounts

 DOI: 10.1002/masy.200450910

for the isotactic/atactic stereoblock structure obtained, which imparts interesting elastomeric properties to the propylene homopolymer. By a careful choice of the substitution pattern and/or the reaction conditions (temperature, pressure, solvent) it is possible to modulate the isomers interconversion speed in order to achieve the desired stereochemical control. Despite the number of papers appeared in the last years,[2] a clear structure/performance correlation is still lacking and there are several open questions.[3] This is due to the fact that even relatively slight changes in the catalyst molecular architecture are enough to disrupt the delicate balance of the rotational isomers ratio and to produce polymers with remarkably different properties.

In particular, the activity of bicyclic metallocenes, where the ring fused to the cyclopentadienyl unit is saturated, has not yet fully explored[2d] in the field of fluxional catalysts due to the lack of a general synthetic pathway. The study of these metallocenes can be very interesting, since, besides their higher stability and resistance to hydrolysis, they introduce a conformational flexibility parameter that can be finely tuned. Several attempts to synthesise this type of ligands have been reported.[4] However the overall yields were often low or the method used was of limited applicability. Recently, we have reported a new synthetic pathway[5] that has proved successful for a versatile and "tuned" synthesis of tetrahydroindenyl systems and of related ligands containing seven- and eight-membered saturated rings condensed to the cyclopentadienyl moiety. These ligands and their corresponding organometallic compounds have been synthesised in good yields, with only minor purification steps, and the entire procedure is reproducible in large-scale quantities.

We have applied this synthetic method to prepare some series of zirconium complexes[6] and the titanium analogues of the most promising zirconium catalysts (Scheme 1).

Although several zirconium and hafnium complexes have been already studied as fluxional catalysts, very little is known about the behaviour of the corresponding titanium derivatives. It is known that the substitution of zirconium for hafnium yields complexes that are isostructural with the original zirconium catalysts, but which produce polypropylenes of lower isotacticity.[2f] Since the titanium centre allows a closer interaction between the two bent cyclopentadienyl moieties, the effect of the substituents should be enhanced and could affect the interconversion speed between aspecific and isospecific isomers, thus favouring the formation of longer isotactic blocks.

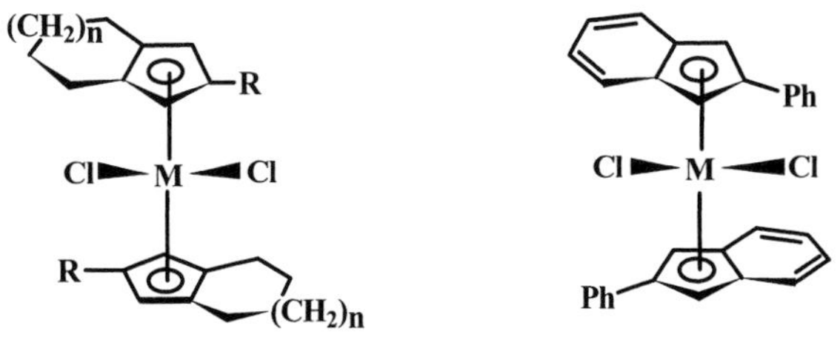

M= Ti, R= -Ph; n = 2 **(2c Ti)**

M= Zr R= -H;   n= 1-3 **(1a-c)**
    R= -Me; n= 1-3 **(2a-c)**
    R= -Ph; n= 1-3 **(3a-c)**

M= Ti **(w- 4 Ti)**
M= Zr **(w-4)**

Scheme 1

Synthesis of the Complexes

Three series of unbridged bicyclic complexes, containing a six, seven-, or eight-membered saturated ring fused to the cyclopentadienyl unit, and bearing different substituents (-H, -CH$_3$, -Ph) in position 2, were synthesised according to a new synthetic pathway (Scheme 2).

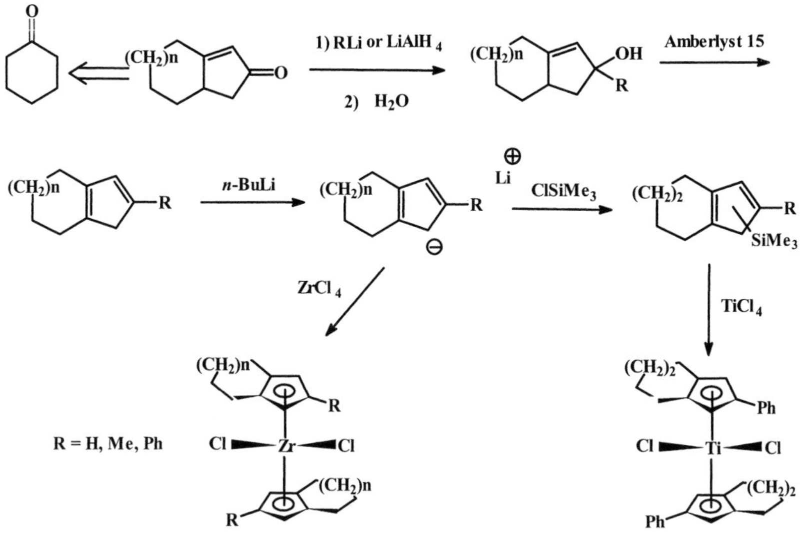

Scheme 2

We have also prepared the catalyst reported by Waymouth, (**w-4**),[1] and its corresponding titanium, (**w-4 Ti**), derivative (Scheme 3). These two catalysts have been used as a reference point under the same polymerisation conditions.

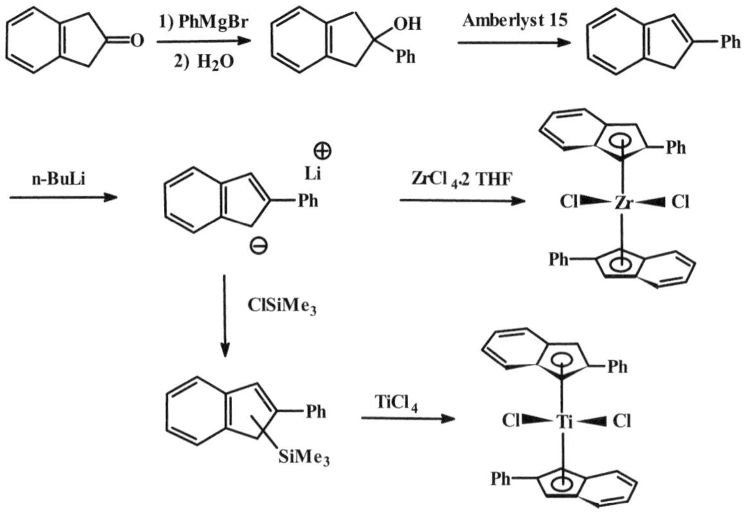

Scheme 3

Polymerisation Studies

Zirconium-Based Catalysts

First of all we have studied the polymerisation of propylene at several temperatures (Table 1) using catalysts belonging to three complete series of bicyclic cyclopentadienyl zirconium complexes bearing different substituents (-H, -CH₃, -Ph) in position 2; methylaluminoxane (MAO) was used as co-catalyst. All the new catalysts proved to be active and most of the polymers obtained were fully atactic when the polymerisations were performed at 30 °C, while they became partially isotactic at lower temperatures, with an isotactic pentad content [*mmmm*] ranging from 11% to 21%.

The conformation of the saturated ring is likely to play an important role in determining the stereochemical control and/or the rate of interconversion between isospecific and aspecific states. In fact the ^{13}C NMR analysis shows that only when a seven-membered ring is associated with a phenyl substituent in position 2 (**2-c**) the pattern of the spectrum (Figure 1L b) is similar to that of the sample produced with Waymouth's catalyst (Figure 1L a); in this

Table 1. Results of propylene polymerisation with different bicyclic zirconium catalysts[a]

Catalyst		T (°C)	Activity[b]	[mmmm][c] (%)	Mw (*10^-3)	Mw/Mn
	R= H **1-a**	30	1172	4.8	27.1	3.1
		0	1000	11.0	25.6	3.0
		-20	1200	11.4	70.2	6.4
	R = CH$_3$ **1-b**	30	558	2.4	8.6	2.8
		0	600	9.8	49.4	2.9
		-20	582	14.6	365.7	3.4
		-30	54	14.1	n.d.[d]	n.d.
	R = C$_6$H$_5$ **1-c**	30	280	2.9	17.1	3.2
		0	1010	6.4	55.5	3.4
		-20	641	8.2	230.8	2.9
		-30	76	7.9	n.d.	n.d.
	R= H **2-a**	30	1604	6.2	13.7	3.3
		0	1200	8.1	92.3	2.5
		-20	1170	8.9	43.0	3.9
	R = CH$_3$ **2-b**	30	239	6.5	30.5	2.8
		0	214	10.7	357.5	6.3
		-20	275	12.6	400.0	13.3
		-30	traces	n.d.	n.d	n.d.
	R = C$_6$H$_5$ **2-c**	30	215	6.1	26.1	2.9
		0	370	13.3	197.1	2.9
		-20	176	16.6	406.8	4.2
		-30	63	21.1	184.8	3.5
	R= H **3-a**	30	430	6.6	6.8	2.7
		0	800	11.6	114.9	2.3
		-20	567	11.7	143.8	5.9
	R = CH$_3$ **3-b**	30	1550	7.4	16.5	2.7
		0	590	11.3	461.7	2.7
		-20	288	11.6	710.9	5.2
		-30	traces	n.d.	n.d.	n.d.
	R = C$_6$H$_5$ **3-c**	30	770	7.6	16.3	2.5
		0	140	14.4	110.3	3.3
		-20	60	15.0	n.d.	n.d.
		-30	traces	n.d.	n.d.	n.d.
	w-4	30	61	8.1	n.d.	n.d.
		0	340	13.9	476.4	4.8
		-20	370	15.1	778.4	5.0
		-30	213	14.1	413.2	1.9

[a]Polymerisation conditions: toluene: 100 mL, [Zr] = 20x10^{-6} mol/L, [Al]/[Zr] = 3000, t = 30 min
[b](kg PP)•(mol catalyst)$^{-1}$•h^{-1}
[c]Isotactic pentad contents by ^{13}C NMR; n.d. = not determined

case we can observe a predominance of the isotactic heptad (indicated by an arrow in Figure 1) in the methyl region of ^{13}C NMR spectrum at 19.67 ppm from hexamethyldisiloxane (HMDS). The spectrum of the sample obtained with **3-c** (Figure 1L c) shows instead the pattern of an atactoid-like polypropylene, even if its atactic pentad content is practically the same as in the reference sample.

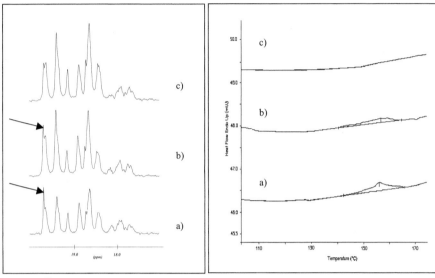

Figure 1.^{13}C NMR spectra (Left) and the corresponding scanning calorimetry heating curves (Right) of polypropylene samples prepared with **w-4**, a), **2-c**, b) and **3-c**, c) at -20°C

This result was confirmed by the thermal analysis of the samples (Figure 1R) carried out by differential scanning calorimetry (DSC). In fact, the reference sample prepared with **w-4** (a) shows a detectable melting peak with a maximum at 156°C. The polypropylene prepared with **2-c** (b) also shows a slight melting peak, similar to that observed in the sample prepared with **w-4**, which allows us to hypothesise the formation of stereoblocks. As expected, the DSC curve of the sample prepared with **3-c** (c) is typical of a completely amorphous polypropylene, i.e. no melting peak was detected.

Titanium-Based Catalysts

Titanium-based complexes have been also investigated. For this study we made a selection among the zirconium-based catalysts, choosing the complex that gave the best results, i.e. the complex bearing the seven-membered ring associated with the phenyl substituent (**2-c Ti**). The titanium derivative (**w-4 Ti**) of the reference Waymouth catalyst was also synthesised and used under the same conditions.

In Table 2 the data obtained with the new titanium catalysts (**2-c Ti** and **w-4 Ti**) are reported and compared with those obtained with the corresponding zirconium complexes (**2-c** and **w-4**). Polymerisations have been carried out at the constant pressure of 2.4 atm at any given temperature (0 and -20 °C). Indeed, the simultaneous decrease of the temperature and increase of the pressure both raises the monomer concentration in solution (with the consequent enhancement of the propagation rate) and slows down the rate of interconversion of isospecific and aspecific states according to the proposed mechanistic model.[7] Both effects should be beneficial in increasing the length of the isotactic stereoblocks.

Table 2. Results of propylene polymerisation with Ti and Zr catalysts at 2.4 atm[a]

Catalyst	T (° C)	Activity[b]	$[mmmm]$[c] %	N_{iso}[d]	T_m (°C)
2-c	0	60	14.0	5.6	134
2-c	-20	47	17.4	5.9	131
2-cTi	0	141	5.5	5.6	142
2-cTi	-20	145	7.8	6.0	142
w-4	0	497	15.9	6.2	142
w-4	-20	149	17.6	6.3	145
w-4Ti	0	44	16.4	6.1	144

[a]Polymerisation conditions: 100 mL toluene, P = 2.4 atm, [Mt] = 10x10^{-5} mol/L, [Al]/[Mt] = 1000, t = 60 min;
[b](kg PP)·(mol catalyst)$^{-1}$·h^{-1};
[c]Isotactic pentad contents by ^{13}C NMR;
[d]Average length of isotactic blocks estimated by ^{13}C NMR

The most interesting result is the high difference in stability between indenyl and tetrahydroindenyl titanium complexes compared with the corresponding zirconium analogues. In contrast to **w-4 Ti**, which is less stable with respect to **w-4**, and is consequently a less effective catalyst, **2-c Ti**, likely due to the saturated ring, is quite stable and more active with respect to the titanium reference catalyst **w-4 Ti**.

Other interesting results come from the study of the polymer microstructure by ^{13}C NMR (Figure 2L) and from the thermal behaviour by DSC (Figure 2R). Among the microstructural data the most striking result is that, although the isotactic pentad content is significantly lower with the titanium-based catalysts, the average length of the isotactic blocks (N_{iso}), estimated by ^{13}C NMR spectroscopy according to the formula $N_{iso} = 4 + 2$ ([*mmmm*]/[*mmmr*])[8], is similar to that obtained with the zirconium based homologues. This result is confirmed by the analysis of the methyl region of the spectra of the samples prepared with **2-c Ti** (Figure 2L a) and **w-4 Ti**, where it is possible to appreciate the presence of a relatively intense signal of the [*mmmmmm*] heptad.

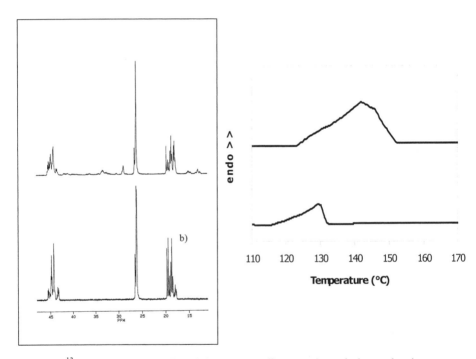

Figure 2. ^{13}C NMR spectra (Left) and the corresponding scanning calorimetry heating curves (Right) of polypropylene samples prepared with **2-c Ti**, a) and **2-c**, b) at -20 °C

Another significant difference observed in the spectrum of the titanium samples is the noticeable increase of the *rr* centred pentads, and the presence of a considerable amount of regioirregularities; these characteristics suggest a different polymerisation mechanism. Surprisingly, the DSC analysis shows a sensibly more evident melting peak in the sample from **2-c Ti**. The fractionation of this sample with boiling diethyl ether allowed the separation of two fractions of comparable amount. From the NMR spectra (Figure 3L) it is possible to conclude that the fractionation has successfully separated a fully atactic fraction (b) from a fraction (a) containing the 18% of isotactic pentad, in which the calculated average length of isotactic sequences reaches the value of 8.1.

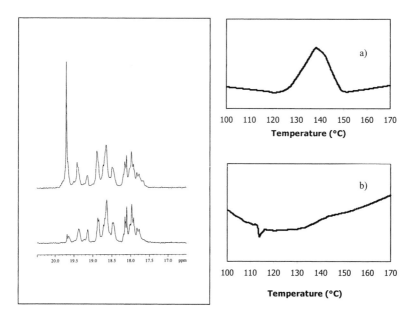

Figure 3. ^{13}C NMR spectra (Left) and DSC heating curves (Right) of ether insoluble fraction, a) and ether soluble fraction, b) of the polypropylene sample prepared with **2-c Ti** at -20 °C

Conclusions

Unbridged bicyclic metallocenes, where the ring fused to the cyclopentadienyl unit is saturated, constitute a very promising family of "fluxional" catalysts. The nature of the ligand has shown to have a marked effect on the stereoselectivity of these catalysts.[6] The present results show that the stereoselectivity is also quite sensitive to the nature of the metal atom (Ti vs. Zr). The titanium-based catalysts yield polypropylenes with new and interesting microstructures. In fact the decrease of the overall isotacticity is accompanied by a noticeable increase of the *rr* centred pentads. However, it is surprising that, despite the significantly lower isotactic pentad content [*mmmm*], the average lengths of the isotactic blocks (N_{iso}), calculated on the basis of the ^{13}C NMR analysis, are similar or even higher than those obtained with the zirconium based homologues. An ethyl ether insoluble fraction with a [*mmmm*] content of the 18% and a N_{iso} value of 8.1 can be separated, thereby demonstrating the stereoblock nature of the structures of these samples. Accordingly, the DSC analysis shows that the same samples are characterised by melting peaks which are equivalent to or more evident than those obtained by the zirconium homologues.

The present work also shows that titanium can also be used as a metal centre in "fluxional" catalysts if a saturated ring is fused to the cyclopentadienyl moiety, since this kind of ligand significantly increases the stability of its complexes. Up to now the easy decomposition of the indenyl titanium metallocenes had prevented their study and use in this field of catalysis.

Acknowledgement

We wish to thank warmly Mr. *Giulio Zannoni* for having performed all NMR spectra and Mr. *Maurizio Canetti*, who has studied the thermal behaviour.

[1] G. W. Coates, R. M. Waymouth, *Science* **1995**, *267*, 217.
[2] (a) E. Hauptman, R. M. Waymouth, J.W. Ziller, *J. Am. Chem. Soc.* **1995**, *117*, 11586; (b) S. Lin, E. Hauptman, T. K. Lal, R. M. Waymouth, R. W. Quan, A. B. Ernst, *J. Mol. Catal. A: Chem.* **1998**, *136*, 23; (c) R. Kravchenko, A. Masood, R. M. Waymouth, *Organometallics* **1997**, *16*, 3635; (d) L. Maciejewski Petoff, M. D. Bruce, R. M. Waymouth, A. Masood, T. K. Lal, R. W. Quan, S. J. Behrend, *Organometallics* **1997**, *16*, 5909; (e) P. Witte, T. K. Lal, R. M. Waymouth, *Organometallics* **1999**, *18*, 4147; (f) M. D. Bruce, G. W. Coates, E. Hauptman, R.M. Waymouth, J. W. Ziller, *J. Am. Chem. Soc.,***1997**, *119*, 11174; (g) R. Schmidt, M. Deppner, H. G. Alt, *J. Mol. Catal. A: Chem.* **2001**, *172*, 43; (h) N. E. Grimmer, N. J. Coville, C. B. de Koning, J. M. Smith, L. M. Cook, *J. Organomet. Chem.* **2000**, *616*, 112; (i) T. Dreier, R. Fröhlich, G. Erker, *J. Organomet. Chem.* **2001**, *621*, 197; (l) T. Dreier, K. Bergander, E. Wegelius, R. Fröhlich, G. Erker, *Organometallics* **2001**, *20*, 5067.
[3] (a) V. Busico, R. Cipullo, W. P. Kretschmer, G. Talarico, M. Vacatello, V. Van Axel Castelli, *Angew. Chem. Int. Ed. Engl.* **2002**, *41*, 505; (b) V. Busico, R. Cipullo, W. Kretschmer, G. Talarico, M. Vacatello, V. Van Axel Castelli, *Macromol. Symp.* **2002**, *189*, 127.
[4] R. L. Halterman, T. M. Ramsey, N. A. Pailes, M. A. Khan, *J. Organomet. Chem.* **1995**, *497*, 43; (b) Q. Yang, M. D. Jensen, *Synlett* **1996**, *2*, 563; (c) M. W. Kim, E. Hong, T. K. Han, S. I. Woo, Y. Do, *J. Organomet. Chem.* **1996**, *523*, 211; (d) R. N. Austin, T. J. Clark, T. E. Dickson, C. M. Killian, T. A. Nile, D. J. Schabacher,

A. T. McPhail, *J. Organomet. Chem.* **1995**, *491*, 11; (e) W. T. Scroggins, M. F. Rettig, R. M. Wing, *Inorg. Chem.* **1976**, *15*, 1381; (f) W. A. Herrmann, R. Anwander, H. Riepl, W. Scherer, C. R. Whitaker, *Organometallics* **1993**, *12*, 4342; (g) R. L. Halterman, T. M. Ramsey, *Organometallics* **1993**, *12*, 2879; (h) P. L. Pauson, *Tetrahedron* **1985**, *41*, 5855; (i) G. Erker, C. Psiorz, R. Z. Frölich, *Naturforsch* **1995**, *50b*, 469.

[5]   E. Polo, R. M. Bellabarba, G. Prini, O. Traverso, M. L. H. Green, *J. Organomet. Chem.* **1999**, *577*, 211.

[6]   E. Polo, S. Losio, F. Forlini, P. Locatelli, A. Provasoli, M. C. Sacchi, *Macromol. Chem. Phys.* **2002**, *203*, 1859.

[7]   (a) M. Bruce, R. M. Waymouth, *Macromolecules* **1998**, *31*, 2707; (b) S. Lin, C. D. Tagge, R. M. Waymouth, M. Nele, S. Collins, J. C. Pinto, *J. Am. Chem. Soc.* **2000**, *122*, 11275; (c) M. Nele, S. Collins, M. L. Dias, J. C. Pinto, S. Lin, R. M. Waymouth, *Macromolecules* **2000**, *33*, 7249.

[8]   S. Mansel, E. Pérez, R. Benavente, J. M. Pereña, A. Bello, W. Roll, R. Kirsten, S. Beck, H.-H. Brintzinger, *Macromol. Chem. Phys.* **1999**, *200*, 1292.

New Polymers by Copolymerization of Ethylene and Norbornene with Metallocene Catalysts

Walter Kaminsky,* Phu-Dennis Tran, Ralf Werner

Institute for Technical and Macromolecular Chemistry, University of Hamburg, Bundesstr. 45, D-20146, Germany
E-mail: kaminsky@chemie.uni-hamburg.de

Summary: To increase the thermal stability of metallocene-methylaluminoxane catalysts. pentalene bridged zirconocenes and a constrained geometry catalyst were used for the copolymerisation of ethene and norbornene. The constrained geometry catalyst is stable in solution up to 90 °C. Surprisingly the molecular weight of the copolymers increases with higher incorporation rates of norbornene. Glass transition temperatures of 120 °C are reached. The microstructure of the cyclic olefin copolymers material is more alternating.

Keywords: COC; ethene-norbornene copolymers; metallocene catalysts; olefin polymerization; pentalene-zirconocene complexes

Introduction

Cyclic olefin copolymers (COC) are a new class of polyolefins which can be produced in an easy way by metallocene/methylaluminoxane (MAO) and other single-site catalysts.[1-9] These catalysts are able to copolymerize ethylene and strained cyclic olefins such as cyclopentene, norbornene, and substituted norbornenes. While polymerisation of cyclic olefins by Ziegler-Natta catalysts is accompanied by ring opening,[10] homogeneous metallocene and palladium catalysts achieve exclusive double bond opening polymerisation.[11-15] The obtained copolymers have special properties and a high potential as engineering plastics.

COCs are characterized by excellent transparency and very high long-life service temperatures. They are soluble, chemically resistant and can be melt-processed. Due to their high carbon/hydrogen ratios, these polymers feature a high refractive index, e.g. 1.53 for ethylene-norbornene copolymer at 50 mol% norbornene incorporation. Their stability against hydrolysis and chemical degradation, in combination with their stiffness, let them become desirable materials for optical applications, e.g. for compact discs, lenses, optical fibers and

© 2004 WILEY-VCH Verlag GmbH & Kgaa, Weinheim   DOI: 10.1002/masy.200450911

films.[16,17] Ticona GmbH runs a commercial COC plant with a capacity of 30,000 tons a year since 2000 in Oberhausen, Germany.

The first metallocene-based COC material was synthesized from ethylene and cyclopentene.[18] While homopolymerisation of cyclopentene results in 1,3-enchainment of the monomer units,[19] isolated cyclopentene units are incorporated into the ethylene-cyclopentene copolymer chain by 1,2-insertion. Ethylene is able to compensate the steric hindrance at the α-carbon of the growing chain after and before the insertion of cyclopentene.[20]

The microstructure of ethene norbornene polymers can be tailored in a wide range by different zirconium and titanium complexes. Chiral C_2-symmetric zirconocenes produce random copolymers while C_s-symmetric complexes produce more alternating structures (Table 1).

Table 1. Copolymerization of ethylene/norbornene by different metallocene/MAO catalysts by 30 °C. Norbornene ratio in feed:= 80 mol%, Al:Zr = 4000

Catalyst	Activity (kg/mol m·h)	r_E	$r_E \cdot r_N$	X_N (polymer) (mol%)	M_w (g/mol)
$[C_2H_4(Ind)_2]ZrCl_2$	9 080	3.6	0.9	28	220 000
$[Me_2Si(Flu)(Cp)]ZrCl_2$	25 000	4.4	0.1	32	1 010 000
$[Me_2C(Flu)(t\text{-}BuCp)]ZrCl_2$	430	6.0	0	30	80 000
$[Me_2Si(t\text{-}BuCpo)(N\text{-}t\text{-}Bu)]TiCl_2$	1 500	15.1	>0.1	19	270 000

Most active is the complex $[Me_2Si(Flu)(Cp)]ZrCl_2$ which gives also a copolymer with the highest molecular weight and a more alternating structure. Due to different incorporation values of the cyclic olefin in the copolymer, the glass transition temperature can vary over a wide range for a random copolymer independently from the used catalysts. A copolymer with 50 mol% of norbornene yields a material with a glass transition point of 145 °C. A Tg of 205 °C can be achieved at higher incorporation rates. The metallocene $[Me_2C(t\text{-}BuCp)(Flu)]ZrCl_2$ shows not only high activities for the copolymerisation of ethylene with norbornene, and gives alternating structure, too. The melting points and the glass transition temperatures of the alternating copolymer depend on the molar ratio of norbornene in the polymer, the microstructure, and the catalyst. A maximum melting point of 320 °C can be reached.

Palladium diimine complexes also show high norbornene incorporation and give more alternating microstructures.[21]

Results and Discussion

Most metallocene/MAO catalysts are instable at temperatures higher than 60 °C and the activity decreases drastically. We synthesized new zirconocenes (Figure 1) with pentalene ligands to stabilize the bridge in ansa metallocenes.

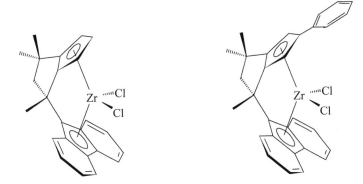

I [1-(Fluorenyl)-1,3,3-trimethyl tetrahydro-trimethyl-pentalenyl]zirconiumdichloride

II [1-(Fluorenyl)-1,3,3-5-phenyl-tetrahydro-pentalenyl]-zirconiumdichloride

Figure 1. Structures of used zirconocenes

Zirconocenes I and II were prior used to polymerise propene and tailor the microstructure.[22] Because of their open wedge, the catalysts 1 and 2 are also very active in ethylene/norbornene copolymerisation (Figure 2).

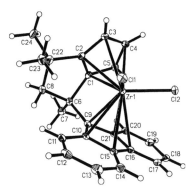

Figure 2. X-ray structure of I

The activities for the two catalysts are shown in Table 2.

Table 2. Copolymerization of ethylene and norbornene by 30 °C in 200 ml toluene. Ethylene pressure = 2 bar, zirconocene concentration : 10^{-7} - $5 \cdot 10^{-6}$ mol/L, MAO: 500 mg, reaction time 15 – 60 min, activity: kg Polymer(Ethylene)/mol Zr·h·C_E. M_v: viscosity average of molecular weight, M_w: molecular weight, measured by GPC

Zirconocene	x_N Norb. (in feed)	Activity (in copo)	X_N Norb. (kg/mol)	M_v (kg/mol)	M_w
I	0	2 700	0	490	520
I	0.19	3 700	0.07	360	400
I	0.39	3 800	0.16	230	340
I	0.59	3 500	0.23	160	330
I	0.80	3 000	0.34	190	390
I	0.90	600	0.42	120	370
II	0	3 700	0	630	660
II	0.20	5 780	0.05	660	-
II	0.40	7 600	0.09	540	670
II	0.59	6 700	0.18	470	690
II	0.80	2 500	0.29	280	830
II	0.90	100	0.36	300	830

For both catalysts we can see an increase of the activity if norbornene is present in the starting reaction mixture (feed). Up to 59 mol% of norbornene the activity is still high but decreases by higher norbornene ratios. Zirconocene II is more active for the ethylene homopoly-merisation and the ethylene/norbornene copolymerisation at low norbornene concentration but incorporates less amounts of norbornene in the copolymer. The phenyl substitution of the ligands causes steric hindrance to the cycloolefin incorporation. The calculation of the copolymerisation parameter r_1 by a first order Markov-model gives a value of:

$r_1 = 3.24$ for I/MAO

$r_1 = 5.52$ for II/MAO

$r_1 = 4.43$ for [Me$_2$C(PhCp)(Flu)]ZrCl$_2$/MAO.

For comparison also the r_1 value is given of another C_1-symmetric catalyst. The values of the copolymerisation parameters r_1 show that both catalysts can incorporate norbornene in a relatively good way. For I the norbornene incorporation is only slower by the factor 3.2 than the ethene incorporation. With both catalysts the molecular weights of the obtained copolymers are high and reach values of 800 000 (see Table 2). While in most

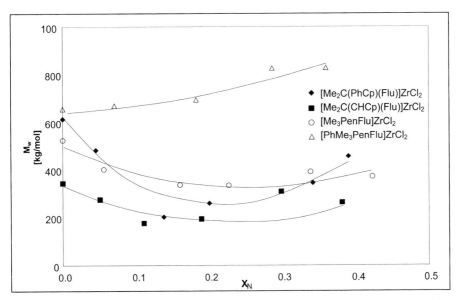

Figure 3. Molecular weights of ethylene/norbornene copolymers in dependence of the molar fraction X_N of norbornene in the polymer

copolymerisation cases the molecular weight decreases with increasing norbornene incorporation in opposite with zirconocene II/MAO the molecular weights increase (Figure 3). This is an unusual observation. The high molar masses obtained with II seem to be due to the fact that the phenyl substitution hinders β-hydrogen transfers in ethene polymerisation.

In the presence of norbornene they are further reduced. Even after an incorporation of 36 mol% of norbornene, the molecular weight of the copolymer is at the highest point of 830,000 g/mol. The microstructure is a more alternating one (Figure 4).

The calculation (lines) was made by a chain retention mechanism after each insertion step. The alternating structure is not perfect and near 70 %. Glass transition and melting temperatures are shown in Table 3.

Copolymers with norbornene parts of less than 10 % are partially crystalline and the others are amorph. The alternating structure is not high enough to form crystalline parts again. Copolymers with an incorporation of only 4 or 5 mol% of the cyclic olefin decrease the melting point to less than 100 °C. We have carried out also copolymerizations of ethylene and

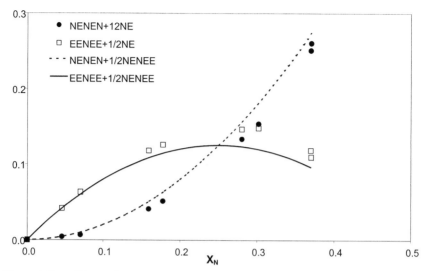

Figure 4. Part of the norbornene centralyzed pentads as a function of the incorporation rate

Table 3. Glass transition temperatures (Tg), melting points (Tm), melting entropies (ΔHm) and crystallinity (α) of ethylene/norbornene copolymers obtained by zirconocene I and II/MAO catalysts; conditions see Table 2; n.d. = not detectable

Catalyst	x_N (copo)	Tg (°C)	Tm (°C)	ΔHm	α
I	0	n.d.	134.5	152.1	54
I	0.04	n.d.	79.0	45.1	16
I	0.16	4.6	n.d.	n.d.	0
I	0.23	35.6	n.d.	n.d.	0
I	0.34	82.6	n.d.	n.d.	0
I	0.42	115.1	n.d.	n.d.	0
II	0	n.d.	135.4	149.7	53
II	0. 05	n.d.	98.0	56.3	20
II	0.09	n.d.	57.2	41.0	15
II	0.18	16.7	n.d.	n.d.	0
II	0.29	59.3	n.d.	n.d.	0
II	0.36	90.1	n.d.	n.d.	0

norbornene using the constrained geometry complex [Me$_2$Si(3-^{tert}Bu)]TiCl$_2$ and MAO as cocatalyst. This catalyst is known to be stable at higher temperatures. The activities of the copolymerisation at reaction temperatures of 30, 60, and 90 °C are shown in Table 4.

Table 4. Copolymerization of ethylene and norbornene with [Me$_2$Si(3-^{tert}BuCp)-(N^{tert}Bu)]TiCl$_2$/MAO in 200 ml toluene with polymerisation temperature; Ti-complex: $1.3 \cdot 10^{-5}$-$2.5 \cdot 10^{-7}$ mol/L; MAO: 500 mg; ethylene pressure: 2 bar, norbornene 0-1.13 mol/L

x_N [a]	Activities in kg Copolymer / mol Ti · h		
	30°C [b]	60°C [b]	90°C [b]
0	800	2 500	12 000
0.20	1 000	5 000	20 000
0.40	2 100	9 800	27 000
0.60	2 500	10 100	29 000
0.80	1 500	5 100	18 000
0.90	1 000	2 200	13 000
0.95	300	1 300	3 500

[a] Molar fraction of Norbornene in feed;

[b] Polymerization temperature

As assumed, the activities rise with increasing temperature and the differences between the activities at different temperatures decrease for $x_N \rightarrow 0$ and $x_N \rightarrow 1$. The data show that the activities at 60 °C are approximately 5 times and at 90 °C about 15 times higher than at 30 °C depending on the norbornene molar fraction in the feed x_N. The maximum of activities lies between $x_N = 0.4$ and $x_N = 0.6$. That maximum of activity is known in the literature as the positive comonomer effect and with 29 000 kg copolymer/mol Ti · h by 90 °C quite high. The molar masses of the copolymers are summarized in Table 5.

Table 5. Ethylene/norbornene copolymerizations by [Me$_2$Si(3-^{tert}BuCp)(N^{tert}Bu)]TiCl$_2$/MAO at 30, 60, and 90 °C, respectively. Molar masses determined by viscosimetry and molar fraction norbornene in the feed x_N

x_N	30 °C Mη (g/mol)	60 °C Mη (g/mol)	90 °C Mη (g/mol)
0.2	200 000	81 000	45 000
0.4	230 000	85 000	48 000
0.6	321 000	95 000	46 000
0.8	259 000	62 000	37 000
0.9	100 000	33 000	22 000
0.95	83 000	23 000	20 000
0.99	76 000	28 000	20 000

The molar masses increase with increasing norbornene fraction in the feed until $x_N = 0.6$. At higher norbornene fractions the molar masses decrease. For COCs prepared at 30 °C, the molar masses increase from 200 000 to 320 000, and then decrease to 76 000 g/mol. They are much lower than the COCs prepared by the zirconocenes I and II with pentalene ligands. The incorporation of norbornene into the polymer chain is lower for the constrained geometry catalyst compared with the zirconocenes. At 30 °C and x_N = 0.6, the molar fraction of norbornene in the copolymer is 0.09 and by x_N = 0.9 it is 0.22.

Conclusions

It could be shown that there are metallocene catalysts which can produce ethylene norbornene copolymers even at temperatures up to 90 °C. High molecular weights of the copolymers are obtained by tailored zirconocenes even at high norbornene concentrations in the feed. The norbornene incorporation, copolymerisation parameters, microstructures, and glass transition temperatures are barely influenced by the temperature. C_1-symmetric catalysts give copolymers with more alternating structures and the maximum norbornene incorporation is near 50 mol%.

[1] W. Kaminsky, *Macromol. Chem. Phys.* **1996**, 197, 3907.
[2] M. Arndt, W. Kaminsky, *Macromol. Symp.* **1995**, 95, 167.
[3] W. Kaminsky, A. Bark. R. Steiger, *J. Mol. Catal.* **1992**, 74, 109
[4] W. Kaminsky, I. Beulich, M. Arndt-Rosenau, *Macromol. Symp.* **2001**, 173, 211.
[5] W. Kaminsky, A. Bark. M. Arndt, *Makromol. Chem., Macromol. Symp.* **1991**, 47, 83.
[6] N. Herfert, P. Montag, G. Fink, *Makromol. Chem.* **2001**, 94, 3167.
[7] R.A. Wendt, G. Fink, *Macromol. Chem. Phys.* **2001**, 202, 3490.
[8] I. Tritto, L. Boggioni, M.C. Sacchi, P. Locatelli, *J. Mol. Catal. A: Chem.* **1998**, 133, 139.
[9] I. Tritto, C. Marestin, L. Boggioni, L. Zetta, A. Provasoli, D.R. Ferro, *Macromolecules* **2000**, 33, 8932.
[10] J. Boor, *Ziegler-Natta Catalysis and Polymerizations,* AP Press, New York 1979.
[11] W. Kaminsky, M. Arndt, *Adv. Polym. Sci* **1997**, 127, 143.
[12] T.J. Demming, B.M. Novak, *Macromolecules* **1993**, 26, 7089.
[13] B.L. Goodall, D.A. Barnes, G.M. Benedict, L.H. McIntosh, L.F. Rhodes, *Polym. Mater. Sci. Eng.* **1997**, 76, 56.
[14] C. Mehler, W. Risse, *Makromol. Chem. Rapid Commun.* **1991**, 12, 255.
[15] W. Heitz, T.F.A. Haselwander, *Macromol. Rapid Commun.* **1997**, 18, 689.
[16] H. Cherdron, M.-J. Brekner, F. Osan, *Angew. Makromol. Chem.* **1994**, 223, 121.
[17] W. Kaminsky, M. Arndt-Rosenau, in *Metallocene-based Polyolefins,* J. Scheirs, W. Kaminsky (eds.), Wiley Series in Polymer Science, Chichester, Vol. 2, p. 91, 2000.
[18] W. Kaminsky, R. Spiehl, *Makromol. Chem.* **1989**, 190, 515.
[19] S. Collins, W.M. Kelly, *Macromolecules* **1992**, 25, 233.
[20] W. Kaminsky, R. Engehausen, J. Kopf, *Angew. Chem.* **1995**, 107, 2469; *Angew. Chem.Int. Ed. Engl.* **1995**, 34, 2273.
[21] J. Kiesewetter, W. Kaminsky, *Chem. Eur. J.* **2003**, 9, 1750.
[22] W. Kaminsky, R. Werner, in *Metalorganic Catalysts for Synthesis and Polymerization* (W. Kaminsky, ed.), Springer Press, Berlin 1999, p. 170.

Macromol. Symp. **2004**, *213*, 109-121

On the Ethylene-Norbornene Copolymerization Mechanism

Incoronata Tritto, * *Laura Boggioni, Maria Carmela Sacchi, Paolo Locatelli, Dino R. Ferro*

Istituto per lo Studio delle Macromolecole, Consiglio Nazionale delle Ricerche
Via E. Bassini, 15, 20133 Milano, Italy
E-mail: i.tritto@ismac.cnr.it

Summary: Results of our studies on polymerization kinetics and tests of copolymerization statistical models of ethylene-norbornene (E-N) copolymers obtained on the basis of microstructures determined by ^{13}C NMR analysis are reported. Ethylene-norbornene (E-N) copolymers were synthesized by catalytic systems composed of racemic isospecific metallocenes, i-Pr[(3Pr^i-Cp)(Flu)]ZrCl$_2$ or a constrained geometry catalyst (CGC) and methylaluminoxane.
Polymerization kinetics revealed that E-N copolymerization is quasi living under standard polymerization conditions. Calculations of the number of active sites and of chain propagation and chain transfer turnover frequencies indicate that the metal is mainly in the Mt-N* state, while the Mt-E* state contributes more to transfer and propagation rates.
The first-order and the second-order Markov statistics have been tested by using the complete tetrad distribution obtained from ^{13}C NMR analysis of copolymer microstructures. The root-mean-square deviations between experimental and calculated tetrads demonstrate that penultimate (second-order Markov) effects play a decisive role in E-N copolymerizations. Results show clues for more complex effects depending on the catalyst geometry in copolymers obtained at high N/E feed ratios.
Comonomer concentration was shown to have a strong influence on copolymer microstructure and copolymer properties.
The copolymer microstructure of alternating isotactic copolymers obtained with i-Pr[(3Pr^i-Cp)(Flu)]ZrCl$_2$ have been described at pentad level. Second-order Markov statistics better describes also the microstrucure of these copolymers.

Keywords: ^{13}C NMR analysis; copolymer microstructure; ethylene-norbornene copolymers; living polymerization; metallocene catalysts; statistical models

Introduction

Interest in addition cycloolefin polymerization has increased tremendously over the last decade, due to the discovery that *ansa*-bridged metallocenes effectively copolymerize ethylene and norbornene.[1]-[8] The importance of this new class of polymers arises from their

   DOI: 10.1002/masy.200450912

special thermoplastic properties, high transparency, and high glass transition temperatures, imparted by the norbornene component. This interest led to commercialization of products such as TOPAS obtained with metallocene catalysts.

The metallocene structure plays a crucial role in the copolymerization process, the copolymer microstructure, and the copolymer properties.[9] The copolymerization mechanism consists of initiation, propagation, and termination or chain-transfer steps. It is reasonable that E-N copolymerizations begin in a manner similar to that of other α-olefin polymerizations. Detailed studies on chain termination and transfer reactions have been carried out in Prof. Brintzinger's group.[10] In order to gain deeper insights into the mechanism of these copolymerizations we have concentrated on the variables that influence the copolymer microstructure, the propagation and transfer turnover frequencies, and the presence of short and long chain branches in E-N copolymers. The present paper will review the results of our studies obtained from:

i) Polymerization kinetics

ii) Testing copolymerization statistical models from copolymer microstructures determined by ^{13}C NMR analysis.

Most recent results regarding the microstructural analysis of alternating isotactic copolymers at pental level and the influence of monomer concentration will be presented and discussed.

Results

Polymerization Kinetics

We have become interested in following the polymerization kinetics in order to understand the influence of reaction conditions and polymerization time on polymer properties. A simple sampling of the reaction mixture at different reaction times proved to be an efficient technique for studying the copolymerization kinetics, under usual conditions, that is, by using MAO and [Al]/[Zr] molar ratios as high as 2000. Both yield and molar masses grow with time. The molecular mass of E–N copolymers, obtained at low temperatures (T = 30-50°C) and high norbornene feed fractions, increases with time for up to one hour. The polydispersity can be as narrow as 1.1 at [N]/[E] feed ratios as high as 28.[11] This indicates that very little chain transfer occurs and that E-N copolymerizations are quasi-living under these conditions. Chain growth over one hour is unusual for olefin polymerization, the average lifetime of a growing ethylene and propylene polymer chain is typically less than seconds, and growth of polymer chain can only be observed with stopped-flow techniques. The type of catalyst used has a strong influence upon the quasi-living character of the reaction. The linearity of the growth of molar mass with time allows us to gain insight into the factors influencing the propagation and chain transfer reactions by using standard equations for treatment of the experimental data of the polymerization kinetics, derived from those introduced by Natta long ago. [12]

$$\frac{1}{\overline{P}_n} = \frac{M_0}{\overline{M}_n} = \frac{C^*}{R_p t} + \frac{\sum R_t^i t}{R_p t} \qquad \textbf{Eq. 1}$$

where $\overline{P}_n$ is the number average degree of polymerization, C^* is the active site concentration, R_p is the polymerization rate, and R_t^i are the rates of different chain termination reactions. For a copolymerization as a first approximation, we take the weighed average of the monomer mass by using the following relationship:

$$M_0 = x_E \cdot M_E + x_N \cdot M_N \qquad \textbf{Eq. 2}$$

where x_E and x_N are the molar fractions and M_E and M_N are the molar masses of ethylene and norbornene, respectively. The calculated number-average degree of polymerization $\overline{P}_n$ can also be expressed in terms of the average turnover frequencies of monomer insertion $<f_p>$ and chain transfer or termination $<f_t>$.[13]

$$\frac{1}{\overline{P}_n} = \frac{M_0}{\overline{M}_n} = \frac{1}{<f_p>} \cdot \frac{1}{t} + \frac{<f_t>}{<f_p>} \qquad \textbf{Eq. 3}$$

The calculated [Zr*]/[Zr] and $<f_t>/<f_p>$ values for E-N copolymerization obtained with three metallocenes are summarized in Table 1.[14]

By comparing the results obtained (Table 1) for catalyst **1** at two different N/E ratios, we observe that an increase of the N/E ratio clearly gives a decrease of the monomer insertion turnover frequency. At the same time the transfer turnover frequency decreases so that the ratio $<f_t>/<f_p>$ remains constant (Table 1). Such similarity is worth noting since **1-3** catalysts produce completely different polymers: the extremes, **1**, a norbornene-rich copolymer at high N concentration and **3**, a virtually pure polyethylene over the whole feed composition range.

Table 1. Kinetic parameters of the ethylene-norbornene copolymerization.

Catalyst		N/E ratio	feed N pol (mol-%)	$<f_p>$ (min^{-1})	$<f_t>$ (min^{-1})	$<f_t>/<f_p>$ $\cdot 10^4$	Zr*/Zr
rac-Et(Ind)$_2$ZrCl$_2$	(1)	28.4	59	346	0.062	1.79	0.68
rac-Et(Ind)$_2$ZrCl$_2$	(1)	12.5	52	634	0.109	1.72	0.66
rac-Et(4,7-Me$_2$Ind)$_2$ZrCl$_2$	(2)	12.5	36	185	0.032	1.73	0.94
90% rac /10% meso- H$_2$C(3-tert-BuInd)$_2$ZrCl$_2$	(3)	12.5	0.5	381	0.076	1.99	0.56

Polymerization conditions: solvent = toluene, MAO, [Al]/[Zr] = 2000; T = 30 °C, Pc$_2$ = 1.01 atm.

The latter along with the activity data suggests that the decrease in activity has its origin in the interference of norbornene with the reaction mechanism.

Surprisingly, the values of $<f_t>/<f_p>$ are very close to those reported by Busico et al. [13] for the stopped-flow homopolymerization of ethylene, using rac-Me$_2$Si(4-PhInd)$_2$ZrCl$_2$/MAO, in spite of the difference of several orders of magnitude in the ethylene polymerization. The fraction of active sites is higher than the value of 0.1 mol/mol(Zr) reported for ethylene polymerization with rac-Me$_2$Si(2-methyl-4-phenyl-1-indenyl)$_2$ZrCl$_2$ at very short polymerization time under stopped flow conditions. [13] Worth to note is that under the same conditions, the number of active sites found in propene polymerization is six times higher than in ethene polymerization. The values obtained for E-N copolymerization are closer to the values recently counted by Landis (0.85-0.95 mol/mol(Zr)) [15] for the polymerization of 1-hexene with dimethylmetallocene 1 by means of a new ^{2}H–tagging strategy. Each Zr* site spends a fraction of time in Zr-E* or in Zr-N* state. This depends on the norbornene content in the copolymer and on the catalyst structure.

The relatively higher concentration of Zr* active sites found in E-N copolymerization indicates that the Zr* sites that we measure are mainly in the Zr-N* state. The similarities between $<f_t>/<f_p>$ values found in E-N copolymerization and E polymerization indicate that Zr-E* sites contribute more to chain propagation and chain transfer than the Zr-N* ones. In conclusion, the presence of norbornene both in the polymer chain and in solution plays a crucial role in the livingness of E-N copolymerization due to its steric hindrance and coordinating ability. This causes a great reduction of the propagation and chain transfer rate

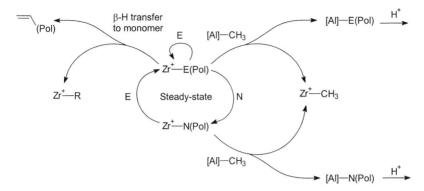

Figure 1. Reaction paths proposed to take part in the catalytic cycle during ethene-norbornene copolymerization catalyzed by metallocenes.

compared to that in the homopolymerization of α-olefins and makes the Zr* state more or less "dormant".

This agrees with the findings that the long chain branches decrease with the increasing N content in the copolymer, when using a catalyst like $Me_2Si(Me_4Cp)(N^tBu)TiCl_2$[16] which allows for incorporation of long chain branches in polyethylene: chain transfer to the monomer and the formation of the vinyl terminated polymer chain are possible only at Mt-E* sites (Figure 1).

Copolymer Microstructure and Polymerization Statistics

The scheme in Figure 1 is quite a simplified picture since penultimate inserted monomer units can be important in E-N polymerisation. When we consider second-order Markov statistics we have four copolymerization parameters linked to the probability of formation of sequences depicted in Figure 2.

A detailed knowledge of the effectiveness of catalysts in producing a given polymer structure can be achieved provided that a methodology for the detailed determination of the polymer microstructure is available. In the last few years, we have concentrated on the investigation of the microstructure of ethylene-norbornene copolymers by combining the use of NMR techniques with computational methods based on relationships between NMR spectra and conformations and an *ab initio* chemical shifts calculations.[17-20] Although some assignments

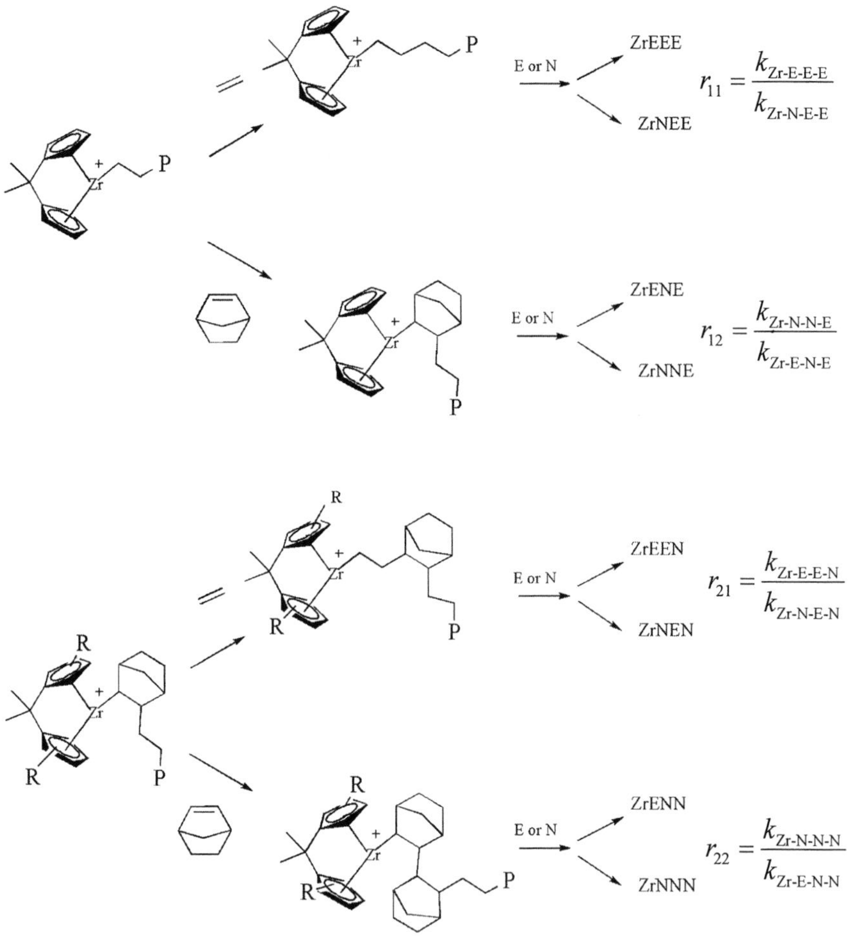

Figure 2. E-N Copolymerization reactions and reactivity ratios in the second-order Markov model.

are still controversial, there is currently a good understanding of the rather complex nature of the ^{13}C NMR spectra. This includes the assignments and quantification of isolated, alternating, and blocky norbornene sequences comprising the tacticity. Our current understanding of E-N copolymer NMR spectra yields complete tetrad sequence distributions. The analysis at tetrad level allows one to test both the first-order and the second order statistics, whilst triad distributions could allow only the fitting of the 2nd order Markov model (Figure 2). Thus, it is possible to determine the copolymerization parameters and clarify possible statistical models of copolymerization, discriminating between ultimate and penultimate effects.[22]

Reactivity Ratios and Statistical Models

The results on reactivity ratios regarding one sample per catalyst, having similar norbornene contents, are summarized in Table 2.

The C2-symmetric bridged metallocene *rac*-Me$_2$Si(2-Me-[e]-benzindenyl)$_2$ZrCl$_2$ (**5**) used in the present work is a catalyst with two homotopic sites and in principle both monomers should have the same probability of being inserted at both sides. Indeed, other C2-symmetric bridged metallocenes such as *rac*-Me$_2$Si(indenyl)$_2$ZrCl$_2$ (**4**) give rise to more random copolymers under the same conditions. However, with respect to **4**, **5** has substitutions in indenyl positions 2, 4 and 5, and the steric hindrance of the ligand structure seems to forbid two consecutive norbornene insertions at the two homotopic sites, so that a mainly alternating copolymerization occurs. However, as soon as two successive norbornene units are added some changes in the interactions occur - probably to avoid strong non-bonded interactions between the hindered growing polymer chain and the indenyl substituents - which favour the insertion of a third norbornene unit. Strong non-bonded interactions due to steric hindrance have been invoked as the cause of the formation of *rrrr* errors in isotactic polypropylene with Me$_2$C-(Flu)(3-tert-butylCp)ZrCl$_2$ based catalysts. [21]

Table 2. E-N copolymerization parameters calculated from tetrad analysis.

Catalyst		N pol (mol-%)	1st Fit	2nd Fit	r_{11}	r_{21}	r_{22}	r_{12}
rac-Et(Ind)$_2$ZrCl$_2$	(**1**)	40.16	0.010	0.004	3.041	2.191	0.018	0.030
rac-Me$_2$Si(Ind)$_2$ZrCl$_2$	(**4**)	40.73	0.017	0.001	4.414	2.339	0.026	0.066
rac-Me$_2$Si(2-Me-benzindenyl)$_2$ZrCl$_2$	(**5**)	44.1	0.034	0.027	3.373	9.931	0.039	0.005
Me$_2$Si(Me$_4$Cp)(NBu^t)TiCl$_2$	(**6**)	43.6	0.009	0.006	4.117	7.204	0.013	0.001

In general better fits are obtained with the second-order Markov statistics and this model appears to be valid. Thus, it is possible to state that the next-to-last E or N monomer unit exerts an influence on the reactivity of the propagating Mt-E* or Mt-N* species. Such an influence seems to be contingent upon the catalyst structure. The second-order Markov model must be used to describe E-N copolymerizations promoted by most of the metallocenes used.

In contrast to these results, the fits for the alternating copolymer produced from rac-Me$_2$Si(2-Me-benzindenyl)$_2$ZrCl$_2$ are not satisfactory either with first or with second order statistical models. A third-order or a more complex model may be required to fit the experimental data obtained with this catalyst, where more sterically hindered indene substitutions are dominant. This appears to confirm our hypothesis that various mechanisms are at work with this catalyst. Furthermore, the analysis of copolymers obtained at very high norbornene concentrations yields different values of copolymerization parameters. Thus, copolymerization obtained with all catalysts may require more complex models.

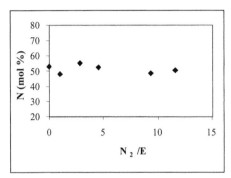 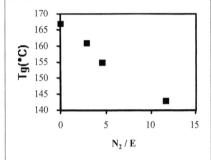

Figure 3. Plots of norbornene copolymer content of E-N copolymers obtained with 1/MAO, at constant comonomer content, as a function N$_2$/E.

Figure 4. Plots of glass transition temperature of the E–N copolymers, obtained with 1/MAO, as a function N$_2$/E.

Monomer Concentration

One important variable clearly influencing the copolymerization kinetics is the comonomer concentration, thus we tested the influence of comonomer concentration on the copolymer microstructure and the copolymer properties in copolymerizations performed with catalyst rac-Et(indenyl)$_2$ZrCl$_2$ (**1**). We varied both comonomer concentrations and kept constant the comonomer feed ratio at about 28. Ethylene pressure varied from 1,1 to 0,078 atm by adding N$_2$ to reach 1,1 atm total pressure. By decreasing both comonomer concentrations, E-N copolymer samples with very similar N content (about 50 mol%) (Figure 3), showing differences in T_g values up to 25 °C (Figure 4), are obtained.

The analysis of copolymer spectra reveals large differences in the microstructure of these copolymers as visible in Figure 5, where the experimental tetrad distribution and those calculated according to second-order Markovian models for copolymers prepared at 1,1 and 0.078 atm ethylene pressure are compared. The sample obtained at lower comonomer concentration is more alternating and shows a much lower amount of NN blocks. Both copolymers obey second order Markov statistics.

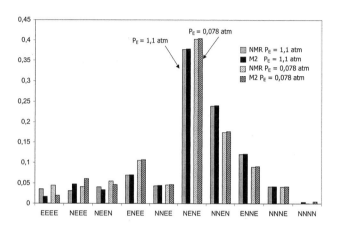

Figure 5. Tetrads distributions for E/N copolymer samples prepared with *rac*-Et(indenyl)$_2$ZrCl$_2$ (**1**), at N/E feed ratio of 28.9, ethylene pressure of 1,1 (full) and 0,078 atm (dashed), respectively; light coloured experimental data; dark coloured calculated according to second-order Markov model.

The second order Markov statistics well reproduces the major microstructural differences between the microstructure of the two samples. The lower ENNE blocks and the higher alternating NENE sequences present in the copolymer, obtained at lower E and N concentration, may explain the significantly lower T_g values observed in this sample. Indeed, the T_g differences should not arise from differences in M_n since all the copolymer samples of Figures 3 and 4 have rather high M_n values (> 70.000).

Lower PE → lower NN blocks → lower Tg values

Alternating Isotactic Copolymers

Typical catalysts which produce isotactic alternating E-N copolymers contain high steric hindered ligands such as *i*-Pr[(3-R-Cp)(Flu)]ZrCl$_2$ (R= Me, Pr^i, Bu^t).[23][24] The analysis of stereoregular alternating E-N copolymers has been used to elucidate polymerization mechanistic details such as the importance of *chain migration mechanism* vs. *chain retention mechanism*. In the case of a *chain migration mechanism* the monomer insertion occurs alternatively at one or the other coordination site. Norbornene units can be inserted only at the more open coordination site, while the small E can be inserted at both sites. This mechanism should allow only odd numbered ethylene blocks. In the case of *retention mechanism* E and N units are inserted at the same coordination site of the catalysts. This mechanism allows for the formation of odd and even numbered ethylene blocks. Arndt [23] used a statistical model which differentiates the two heterotopic sites of the catalyst *i*-Pr[(3-Bu^t-Cp)(Flu)]ZrCl$_2$: one site, always in a Zr-E* state, follows first order Markov statistics, the second can be in a Zr-N* state and follows second order Markov statistics. He concluded that E-N copolymerization follows a chain migration mechanism. On the other hand Fink *et al.* [24] have recently analyzed E-N copolymer spectra obtained with *i*-Pr[(3-Pr^i-Cp)(Flu)]ZrCl$_2$ by using their own assignments.[25] They did observe even numbered ethylene sequences and concluded that the microstructure of the copolymer is consistent with *a retention mechanism* and follows first order Markov statistics.

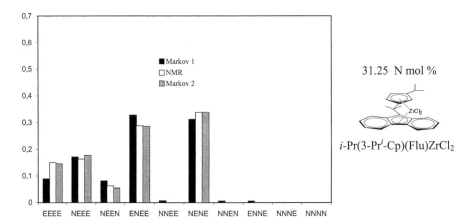

Figure 6. Experimental and calculated tetrads distributions for an E/N copolymer sample prepared with *i*-Pr[(3-Pr^i-Cp)(Flu)]ZrCl$_2$, at N/E feed ratio of 4, ethylene pressure of 1,1 atm. Black experimental data; white according to first-order Markov model; dashed according to second-order Markov model.

We have analyzed the spectra of alternating isotactic copolymers obtained with i-Pr[(3-Pr^i-Cp)(Flu)]ZrCl$_2$ at tetrad level (Figures 6 and 7). The comparison of experimental and calculated tetrads reveals that second Markov statistics reproduces the microstructure of this isotactic alternating copolymer more accurately than first order Markov statistics. Anyway, NEEN sequences are clearly visible in a copolymer sample containing 31.25 N-mol% (Figure 6).

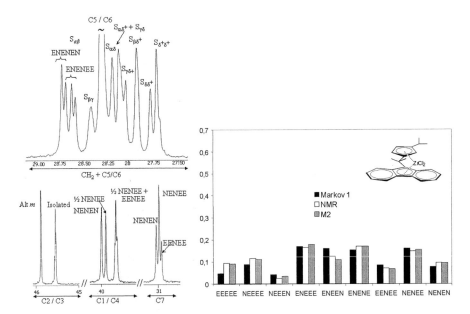

Figure 7. ^{13}C NMR spectrum performed at 100 MHz of the copolymer sample of Figure 6.

Figure 8. Experimental and calculated pentads distributions for the E/N copolymer sample of Figures 6 and 7.

Pentad Analysis

The better quality of recent NMR spectra achieved by having access to an NMR instrument at higher field (Figure 7), gives us pentad level information, which allows for even safer testing of statistical models.

This is clear in this graph where we have determined nine pentads (Figure 8), five of which are independent (only two triads and three tetrads independent). Even though there are a couple of controversies in the assignments of the CH$_2$ region, the results concerning the validity of second order statistics and the values of copolymerization parameters are practically independent from such controversies. Extension of the pentad analysis to series of E-N copolymers with a wide range of norbornene content will allow us to definitely select the best statistical model to describe E/N copolymerization with each catalyst.

Conclusions

The quasi living nature of E-N copolymerization at temperatures as low as 30 °C and high norbornene concentration in the feed allowed us to calculate the number of active sites and propagation and chain transfer turnover frequencies. We found a relatively high concentration of Zr^* active sites found in E-N copolymerization. This indicates that Zr^* that we measure are mostly in the in $Zr-N^*$ state. The $<f_t>/<f_p>$ values found are similar to those found for E polymerization. This is a sign that $Zr-E^*$ sites are those that contribute more to chain propagation and chain transfer turnover values. Moreover, the presence of norbornene both in the polymer chain and in solution causes a great reduction of the propagation and chain transfer rate compared to that in the homopolymerization of α-olefins and makes the Zr^* state more or less "dormant".

Our understanding of E-N copolymer NMR spectra and our methodology for their analysis allows us to select the best statistical model describing E/N copolymerization with a given catalyst and to study the influence of various parameters including the catalyst symmetry and ligand substitution and the monomer concentration on the polymerization mechanism.

[1]   W. Kaminsky, A. Bark. M. Arndt, *Makromol. Chem., Macromol. Symp.* **1991**, *47*, 83.
[2]   M. Arndt, W. Kaminsky, *Macromol. Symp.* **1995**, *95*, 167.
[3]   W. Kaminsky, I. Beulich, M. Arndt-Rosenau, *Macromol. Symp.* **2001**, *173*, 211.
[4]   H. Cherdron, M.J. Brekner, F. Osan, *Angew. Makromol. Chemie* **1994**, *223*, 121.
[5]   M. J. Brekner, F. Osan, J. Rohrmann, M. Antberg, (Hoechst AG) U.S. Patent 5 324 801, **1994.**
[6]   (a) Ruchatz, D., Fink, G. *Macromolecules* **1998**, 31, 4669-73; (b) Ruchatz, D.; Fink, G. ibid., 4674-80; (c) Ruchatz, D.; Fink, G. ibid., 4681-83; (d) Ruchatz, D.; Fink, G. ibid., 4684-86.
[7]   I. Tritto, L. Boggioni, M. C. Sacchi, P. Locatelli, D.R. Ferro, A. Provasoli, *Macromol. Rapid Commun.* **1999**, *20*, 279.
[8]   I. Tritto, L. Boggioni, M.C. Sacchi, P. Locatelli, D.R. Ferro, and A. Provasoli, in *Metallorganic Catalysts for Synthesis and Polymerization* W. Kaminsky Ed. Springer **1999**, 493.
[9]   J. Forsyth, J.M. Pereña, R. Benavente, E. Perez, I. Tritto, L. Boggioni, H.-H Brintzinger, *Macromol. Chem. Phys.* **2001**, *202*, 614.
[10] N. Ni Bhriain PhD thesis (**2003**)
[11] J.C Jansen, R. Mendichi, P. Locatelli, I. Tritto, *Macromol. Rapid Commun.*, **2001**, *22*, 1394.
[12] G. Natta, I. Pasquon, *Adv. Catal.* **1959**, *11*, 1-66.
[13] V. Busico, R. Cipullo, V. Esposito, *Macromol. Rapid. Commun.*, **1999**, *20*, 116

[14] J.C. Jansen, R. Mendichi, M.C. Sacchi, I. Tritto, *Macromol. Chem. Phys.* **2003**, *204*, 522.

[15] Z. Liu, E. Somsook, C. R. Landis, *J. Am. Chem. Soc.* **2001**, *123*, 2915.

[16] K. Thorshaug, R. Mendichi, I. Tritto, S. Trinkle, C. Friedrich, R. Mülhaupt, *Macromolecules* **2002**, *35*, 2903

[17] A. Provasoli, D. R. Ferro, I. Tritto, L: Boggioni, *Macromolecules* **1999**, *32*, 6697.

[18] I. Tritto, C. Marestin, L. Boggioni, L. Zetta, A. Provasoli, D.R. Ferro, *Macromolecules* **2000**, *33*, 8931

[19] I. Tritto, C. Marestin, L. Boggioni, M.C. Sacchi, H.-H. Brintzinger, D. R. Ferro, *Macromolecules* **2001**, *34*, 5770.

[20] M. Ragazzi, P. Carbone, D.R. Ferro, *Int. J. Quantum Chem.* **2002**, *88*, 663.

[21] A. Razavi, V. Bellia, Y. De Brauwer, K. Hortmann, M. Lambrecht, O. Miserque, L. Peters, S. Van Belle, in *Metallorganic Catalysts for Synthesis and Polymerization* W. Kaminsky Ed. Springer **1999**, 236.

[22] I. Tritto, L. Boggioni, J.C. Jansen, K. Thorshaug, M.C. Sacchi, D.R. Ferro, *Macromolecules*, **2002**, *35*, 616

[23] M. Arndt, I. Beulich, *Macromol. Chem. Phys.,* **1998**, *199*, 1221

[24] N. Herfert, P. Montag, G. Fink, *Makromol. Chem.,* **2001**, *94*, 3167.

[25] R.A. Wendt, R. Mynott, G. Fink, *Macromol. Chem. Phys.* **2002**, *203*, 2531.

Macromol. Symp. **2004**, *213*, 123-129

Living Random Copolymerization of Ethene and Norbornene Using *ansa*-Fluorenylamidodimethyltitanium Complex

Tariqul Hasan, Takeshi Shiono,* Tomiki Ikeda

Chemical Resources Laboratory, Tokyo Institute of Technology, Nagatsuta-cho 4259, Midori-ku, Yokohama 226-8503, Japan
E-mail: tshiono@res.titech.ac.jp

Summary: Ethene-norbornene copolymerizations were conducted by (*t*-BuNMe$_2$SiFlu)TiMe$_2$ (**1**) activated with dried methylaluminoxane (MAO) (free from Me$_3$Al) at 0 °C and 20 °C under atmospheric pressure of ethene at various norbornene concentrations. At 0 °C and high norbornene concentration, the molecular weight of the copolymer increased linearly against polymerization time while keeping narrow molecular-weight distribution. The ^{13}C NMR spectrum revealed that random copolymer with 53 mol-% of norbornene was obtained. The glass transition temperature was above 150 °C.

Keywords: amorphous; copolymerization; living polymerization; norbornene; Ziegler-Natta polymerization

Introduction

Ethene-norbornene (E-N) copolymers are high performance materials due to their high glass transition temperature (T_g) and optical transparency.[1] The recent development of single-site catalysts has brought several catalytic systems which conduct E-N copolymerization with considerable activities.[2] Most of them gave the copolymers with isolated norbornene unit or almost alternate structure. Ni-based catalysts, which are active for norbornene polymerization, also produced almost alternating E-N copolymers.[3] Constrained geometry catalysts (CGCs) are highly active for copolymerization of ethene and higher α-olefins to give random copolymers.[4] Most of the CGC derivatives, however, produce E-N copolymers with isolated or alternating norbornene sequences depending on the norbornene content.[5] A few *ansa*-zirconocenes like *rac*-Et(Ind)$_2$ZrCl$_2$, ^{i}Pr[FluCp]ZrCl$_2$, ^{i}Pr[IndCp]ZrCl$_2$, and MeCH[Cp]$_2$ZrCl$_2$ were reported to produce random copolymers containing more than 50 mol-% of norbornene.[6-8]

The living polymerization is a useful tool for the synthesis of tailor-made polymers. So far a few

          DOI: 10.1002/masy.200450913

examples of the living E-N copolymerization have been reported. Tritto *et al.* reported the quasi-living polymerization of ethene and norbornene with C_2-symmetric zirconocene catalysts[9, 10] and (t-BuNMe$_2$SiC$_5$Me$_4$)TiCl$_2$[11] using MAO as cocatalyst. Fujita *et al.* recently reported that the Ti complexes containing pyrrolide-imine ligand also produced alternating E-N copolymer in a living manner at 30 °C.[12] In this study, (t-BuNMe$_2$SiFlu)TiMe$_2$ (**1**) activated with dried methylaluminoxane (MAO) (free from Me$_3$Al), which we have previously reported as an excellent catalyst for homopolymerization of norbornene,[13] was applied for the copolymerization of ethene and norbornene, and was found to produce random copolymer in a living manner.

Experimental

Materials. The preparation and handling of the complex **1** was operated under nitrogen atmosphere with Schlenk techniques. The titanium complex **1** was synthesized according to the literature, and the references therein.[14] Dried MAO was prepared with the same procedure reported previously.[15] Norbornene (Aldrich) was purified by stirring it over calcium hydride at 60 °C for 24 h, and then distilled. The stock solution of norbornene was prepared in toluene (5.14 M). Research grade ethene (purchased from Takachiho Chemicals Co.) was purified by passing it through columns of NaOH, P$_2$O$_5$, and molecular sieves 3A. All solvents were commercially obtained and dried by standard methods.

Polymerization Procedure. Polymerizations were performed in a 100-mL glass reactor equipped with a seal septum and a magnetic stirrer. At first, the reactor was charged with 0.464 g (8.0 mmol of Al) of dried MAO and a prescribed volume of the stock solution of norbornene was added under nitrogen gas flow. The solvent toluene was added to make the total volume 30 mL, and ethene gas was charged at atmospheric pressure after the reactor had been evacuated. The reactor was kept in an ice or water bath to obtain desired temperature, and the reaction mixture was saturated with ethene. A solution of catalyst (1 mL) in toluene was added to start the polymerization. The temperature was kept constant during the polymerization. The polymerization was terminated with methanol and precipitated with acidic methanol. The polymer obtained was adequately washed with methanol, and dried under vacuum at 60 °C for 6 h.

Analytical Procedures. Molecular weight (M_n) and molecular weight distribution (M_w/M_n) of copolymer were measured by gel permeation chromatography (GPC, Waters 150C) at 140 °C using o-dichlorobenzene as solvent and calibrated by polystyrene standards. The ^{13}C NMR spectrum of copolymer was recorded at 120 °C on a JEOL GX 500 spectrometer operated at 125.65 in pulse Fourier-Transform mode in tetrachloroethane-d_2 as solvent with a 4-s delay between the pulses. Differential scanning calorimetry (DSC) was investigated under nitrogen from 20 to 240 °C with heating and cooling rate of 10 °C/min with a Seiko DSC-220 instrument. T_g values were determined from the second heating run.

Results and Discussion

E-N copolymerizations were carried out with complex **1** using dried MAO as a cocatalyst at 0 °C and 20 °C at different norbornene concentration under atmospheric pressure of ethene. The results of the polymerization are shown in Table 1. The catalyst system showed higher activity at higher temperature. The polymerization activity increased with increasing norbornene ratio in feed at 20 °C whereas the opposite trend was observed at 0 °C.

Table 1. Copolymerization of ethene and norbornene with **1** using dried MAO as cocatalyst.[a]

Temp.	N/E	Activity[b]	Nor. in copoly.	$M_n \times 10^{-4}$ [c]	M_w/M_n [c]	T_g [d]
(°C)	(mol/mol)		(mol-%)			
0	6	111	44	10.3	1.26	132
,,	9	101	49	9.8	1.21	153
,,	13	76	53	7.8	1.16	166
20	1.4	176	25	4.5	1.74	47
,,	5	480	44	7.1	1.34	126
,,	10	675	54	7.9	1.31	167

[a]Polymerization conditions: Ti = 20 μmol, Al/Ti =400, solvent = toluene, total vol. = 30 mL, ethene = 1 atm, time = 30 min at 0 °C and 2 min at 20 °C, ethene concentration 0.2M at 0 °C and 0.12 at 20 °C.
[b]Activity = kg$_{(poly)}$· mol^{-1}$_{(Ti)}$· h^{-1}.
[c]Molecular weight and molecular weight distributions were measured by GPC analysis using polystyrene standards.
[d]Glass transition temperature was measured with differential scanning calorimeter and the midpoint of the transition was used as the T_g value.

The ^{13}C NMR spectrum of an E-N copolymer produced with **1** is shown in Figure 1, which differs from that of the alternating structure, which should show only five signals[2, 6]. The signals of NN dyad and NNN triad reported by Tritto *et al.*[7] were observed as follows:

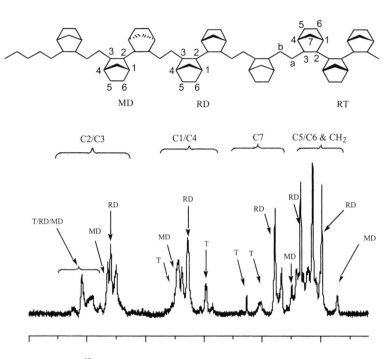

Figure 1. ^{13}C NMR spectrum of the E-N copolymer (norbornene content 53 mol-%) produced with **1** at 0°C

28.60 and 32.47 ppm, C5/C6 of *meso* ENNE; 29.91 and 31.69 ppm, racemic-ENNE; 33.8 ppm, C7 of ENNE; 35.08 and 36.32 ppm, NNN. No signal was observed above 52 ppm, where the signals of C2/C3 carbons of meso-NNN sequences should appear as reported earlier.[8] New signals appeared from 39.0 to 40.7 ppm and at 35.08 and 36.32 ppm which had never been observed in the E-N copolymers reported so far. They can be tentatively assigned to C1/C4 (39.0-40.7 ppm) and to C7 (35.08 and 36.32 ppm) carbons of racemic-NNN sequence.

The incorporation of norbornene in the copolymer was calculated by eq. 1.

$$\text{mol -\% } N = \frac{1/3(\ I_{C2/3} + I_{C1/4} + 2I_{C7})\ 100}{(\ I_{CH2}\ )} \tag{1}$$

where $I_{C2/3}$ and $I_{C1/4}$ and I_{C7} are the peak intensities of C2/C3, C1/C4 and C7 carbons in norbornene unit, and I_{CH2} is the summation of those of Ca/Cb and C5/C6 carbons as shown in Figure 1.

The norbornene content was dependent on the comonomer feed ratio and the copolymers containing ca. 53 mol-% of norbornene were obtained at the highest feed ratio employed at both temperatures. The T_g values were determined by a differential scanning calorimeter (DSC). The temperature at the midpoint of phase transition was used to determine the T_g value. The T_g values of the copolymers with 53 mol-% norbornene were around 166 ˚C (Figure 2).

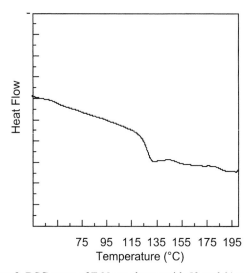

Figure 2. DSC curve of E-N copolymer with 53 mol-% norbornene

The copolymers had high M_n values and narrow M_w/M_n, which became narrower with increasing norbornene ratio in feed. Tritto *et al.* reported that high norbornene concentration and low polymerization temperature below 50 ˚C were the key factors to achieve "quasi-living" nature of E-N copolymerization with metallocene-MAO systems.[10,11] The extremely narrow polydispersity ($M_w/M_n = 1.16$) was observed at N/E = 13 feed ratio at 0 ˚C, which suggests that the living copolymerization of ethene and norbornene should proceed with these conditions. To confirm the living nature at 0 °C, the dependence of M_n on polymerization time was investigated by sampling method: 2 mL of reaction mixture was sampled with a syringe at every 10 minutes. The GPC curves of polynorbornene obtained by sampling were shifted to the higher molecular weight region with increasing polymerization time keeping narrow M_w/M_n (Figure 3). The M_n and M_w/M_n values vs time are plotted in Figure 4. The M_n values linearly increased against polymerization time. The yield of the sampled copolymer also increased linearly with polymerization time (Figure 5). These results indicate that the living E-N copolymerization proceed at 0 °C. The slightly broader M_w/M_n value at 20 °C suggests the presence of some chain transfer reaction.

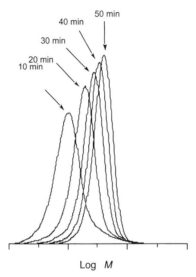

Figure 3. GPC curves of polynorbornene obtained by sampling method at 0 °C

In conclusion, complex **1** activated by dried MAO conducted living copolymerization of ethene and norbornene at 0 °C, and gave random copolymer with high T_g value.

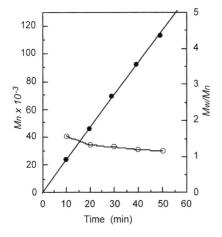

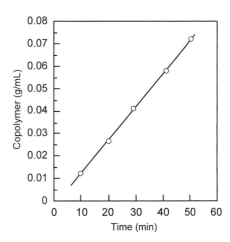

Figure 4. Plots of Mn (●) and M_w/M_n (○) vs time in E-N copolymerization with **1** at 0°C

Figure 5. Copolymer concentration in reaction mixture as a function of time

[1] (a) W. Kaminsky, A. Bark, M. Arndt, *Macromol. Chem. Macromol. Symp.* **1991**, *47*, 83. (b) H. Cherdron, M-J. Brekner, F. Osan, *Angew. Makromol. Chem.* **1994**, *223*, 121. (c) T. Rische, A. J. Waddon, L. C. Dickinson, W. J. MacKnight, *Macromolecules* **1998**, *31*, 1871 (d) W. Kaminsky, A. Noll, *Polym. Bull.* **1993**, *31*, 31.
[2] (a) D. Ruchatz, G. Fink, *Macromolecules* **1998**, *31*, 4669. (b) D. Ruchatz, G. Fink, *Macromolecules* **1998**, *31*, 4674. (c) D. Ruchatz, G. Fink, *Macromolecules* **1998**, *31*, 4681. (d) D. Ruchatz, g. Fink, *Macromoeicules* **1998**, *31*, 4684. (e) J. Forsyth, J. M. Pereña, R. Benavente, E. Perez, I. Tritto, L. Boggioni, H-H. Brintzinger, *Macromol. Chem. Phys.* **2001**, *202*, 614. (f) B. A. Harrington, D. J. Crowther, *J. Mol. Catal.* **1998**, *128*, 79. (g) P. Altamura, A. Grassi, *Macromolecules* **2001**, *34*, 9197. (h) M. Arndt, I. Beulich, *Macromol. Chem. Phys.* **1998**, *199*, 1221.
[3] G. M. Benedikt, E. Elce, B. L. Goodall, H. A. Kalamaridies, L. H. McIntosh, L. Rhodes, K. T. Selvy, *Macromolecules,* **2002**, *35*, 9878.
[4] A. L. McKnight, R. M. Waymouth, *Chem. Rev.* **1998**, *98*, 2587.
[5] A. L. McKnight, R. M. Waymouth, *Macromolecules* **1999**, *32*, 2816.
[6] I. Tritto, C. Marestin, L. Boggioni, L. Zetta, A. Provasoli, D. R Ferro, *Macromolecules* **2000**, *33*, 8931.
[7] I. Tritto, L. Boggioni, J. C. Jansen, K. Thorshaug, M. C. Sacchi, D. R. Ferro, *Macromolecules* **2002**, *35*, 616. Where the signal of carbon C5 of meso-ENNE sequence (MD) was used to adjust the chemical shifts as 28.6 ppm.
[8] (a) R. A. Wendt, G. Fink, *Macromol. Chem. Phys.* **2001**, *202*, 3490. (b) I. Beulich, M. Arndt, *Macromolecules* **1999**, *32*, 7335.
[9] J. C. Jansen, R. Mendichi, P. Locatelli, I. Tritto, *Macromol. Rapid Commun.* **2001**, *22*, 1394.
[10] (a) J. C. Jansen, R. Mendichi, M. C. Sacchi, I. Tritto, *Macromol. Rapid Commun.* **2001**, *22*, 1394. (b) J. C. Jansen, R. Mendichi, M. C. Sacchi, I. Tritto, *Macromol. Chem. Phys.* **2003**, *204*, 522.
[11] K. Thorshaug, R. Mendichi, L. Boggioni, I. Tritto, S. Trinkle, C. Friedrich, R. Mülhaupt, *Macromolecules* **2002**, *35*, 2903.
[12] Y. Yoshida, J. Saito, M. Mitani, Y. Takagi, S. Matsui, S. Ishii, T. Nakano, N. Kashiwa, T. Fujita, *Chem. Commun.* **2002**, 1298.
[13] T. Hasan, K. Nishii, T. Shiono, T. Ikeda, *Macromolecules* **2002**, *35*, 8933.
[14] H. Hagihara, T. Shiono, T. Ikeda, *Macromolecules* **1998**, *31*, 3184.
[15] T. Hasan, A. Ioku, K. Nishii, T. Shiono, T. Ikeda, *Macromolecules* **2001**, *34*, 3142.

New Organic Supports for Metallocene Catalysts Applied in Olefin Polymerizations

Markus Klapper,* Yong-Jun Jang, Kirsten Bieber, Tanja Nemnich, Nikolay Nenov, Klaus Müllen

Max-Planck-Institut für Polymerforschung, Ackermannweg 10, D-55128 Mainz, Germany
E-mail: klapper@mpip-mainz.mpg.de

Summary: Nano-sized latex particles as organic supports for metallocenes applied in olefin polymerizations are introduced. The particles are functionalized with nucleophilic surfaces such as polyethylenoxide (PEO), polypropyleneoxide (PPO) or pyridine units allowing an immobilization of the metallocene catalysts *via* a non–covalent immobilization process. The latices are obtained by emulsion or miniemulsion polymerization with styrene, divinylbenzene as the crosslinker, and either PEO or PPO functionalized styrene or 4-vinylpyridine for surface functionalization. The supported catalysts, *e.g.* [Me$_2$Si(2MeBenzInd)$_2$ZrCl$_2$/MAO] on PPO containing latices or Cp$_2$ZrMe$_2$/([Ph$_3$C][B(C$_6$F$_5$)$_4$]) on pyridine functionalized materials were tested in ethylene polymerizations. Remarkably, high activities and excellent product morphologies were obtained. The influence of the degree of surface functionalization on activity and productivity was investigated. Furthermore, the fragmentation of the catalyst was studied by electron microscopy using bismuth-labeled latex particles or by fluorescence and confocal fluorescence microscopy using dye-labeled supports.
Finally, a self-immobilizing catalyst/monomer system is presented. It is demonstrated that by using PEO-functionalized olefins, the metallocenes were immobilized on the monomers. Subjecting these mixtures to an ethylene copolymerization, again high activities and productivities as well as polyolefin beads with high bulk densities are observed, indicating that an extra supporting process for controlling the product size and shape of the polyolefins is not necessary for these monomers.

Keywords: catalyst fragmentation; ethylene polymerization; latex particles; metallocene catalyst; polymeric support; self-immobilization

Introduction

Metallocenes have proven to be attractive catalysts for olefin polymerization, as they allow a tailoring of the properties of the polyolefins, such as comonomer incorporation and stereospecificity.[1] For an industrial system, heterogenization of the metallocene catalyst is

© 2004 WILEY-VCH Verlag GmbH & KGaA, Weinheim   DOI: 10.1002/masy.200450914

necessary in order to control the morphology of products and prevent reactor fouling.[2] Inorganic supports such as silica, magnesium chloride, zeolites, clays and polymers have been used for immobilizing the catalyst in olefin polymerization. Some of these inorganic supports, however, have acidic groups on their surfaces that can cause deactivation of the catalysts.[3]-[7]

In recent years, organic supports have also been investigated as supports for metallocenes in olefin polymerization. [8], [9] Latex-particles based on polystyrene seem to be one of the most successful examples. Supports based on polystyrenes containing methoxy groups or PEO-chains, which allow an immobilization of active methylaluminoxane/metallocene complexes through non-covalent bonding with nucleophilic groups have been reported.[10], [11] These catalysts show high activities and productivities and form distinct polymer particles with a high bulk density. This concept was developed further, applying polystyrene-based nanoparticles functionalized with polyethylene oxide (PEO) chains on the surface acting as catalyst supports. In this concept, the uniform and well-defined particles (80 – 300 nm) were reversibly aggregated by the interaction of the PEO-chains with the methylaluminoxane/zirconocene clusters (Scheme 1). It was proposed that during the polymerization, the catalysts were completely and homogeneously fragmented within the final product down to the initial nanosized particles of the support, due to the formation of the polyolefins between the latex particles. Such fragmentation, proven for silica based supports, is considered essential for the control of morphology in polyolefin polymerization.[12], [13]

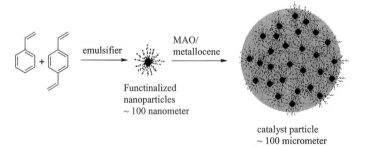

emulsifier

MAO/
metallocene

Functinalized
nanoparticles
~ 100 nanometer

catalyst particle
~ 100 micrometer

Scheme 1. PEO-functionalized nanoparticles as support for metallocenes: synthesis and supporting process

Even though it has been shown that polymeric supports are applicable supports for catalysts in olefin polymerization, there are many open questions to answer in order to fully understand the role of the supports and their effects on industrial application, such as control of the size of the supports, influence of the surface functionalization and fragmentation of the support. In this paper new approaches towards organic supports for metallocenes and their application in MAO or borate activated polymerizations are presented. The fragmentation and distribution of these supports in the polymer beads is demonstrated by several microscopic techniques using metal or dye tagged supports. Finally, a self-immobilizing process of the catalysts is discussed, which permits a morphology control in the polymerization process by using polar monomers.

Results and Discussion

New ether-functionalized supports for metallocenes

In previous work, polyether functionalized organic latices were obtained by emulsion polymerization.[18] The obtained latex particles were, however, relatively large (80-300 nm). Therefore, we applied miniemulsion polymerization for the formation of the latex particles, typically resulting in smaller and more uniform beads of 50-100 nm.[14] It has been proposed that the catalysts prepared on these nanoparticle supports can fragment to a smaller size resulting in a more uniform dispersion of the support in the polyolefin. It is expected that the PPO chains on the surface of the support coordinate to and immobilize metallocene complexes without leaching of the active metal sites during the polymerization due to their strong affinity towards aluminum compounds, e.g. MAO. To study the influence of the concentration of nucleophilic groups in the latex particles on the catalyst activities and polymer properties, the amount of block copolymer used in the miniemulsion process was varied.

PPO functionalized nanoparticles as support for Me$_2$Si(2MeBenzInd)$_2$ZrCl$_2$/(MAO activated)
Nano-sized latex particles of functionalized polypropyleneoxide (PPO) were prepared and applied in ethylene polymerization (Scheme 2). To investigate the dependence of the activity and productivity on the concentration of the PPO chains the amount of PPO on the support was increased from 0.5 to 20 mol % while keeping the PPO chain length constant (n = 15). These supports were loaded with equal amounts of Me$_2$Si(2MeBenzInd)$_2$ZrCl$_2$, activated with methylalumoxane (MAO) and applied in polymerization of ethylene.

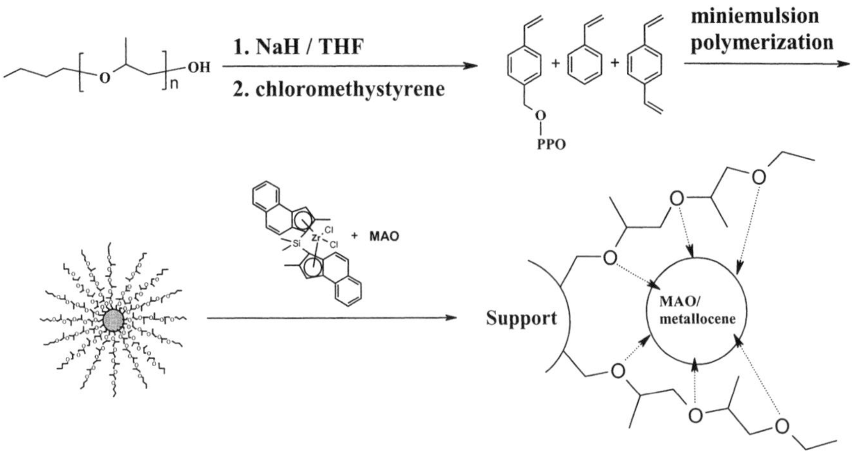

Scheme 2. Preparation of latexes and supporting process of latex and MAO/metallocene complex

In comparison to a homogeneous olefin polymerization, heterogeneous polymerization has always a higher hindrance of the diffusion of monomer gas to reach the active metal sites of the catalyst. In our case the diffusion and fragmentation should be drastically influenced by the interaction between the nucleophilic chains of the different latex particles which are reversibly crosslinked via the MAO/PPO interaction. The latter interaction between them could be strengthened drastically by increasing the amount of PPO on the particles resulting in a more stable network between the nucleophilic ether groups and active metal sites. Such a denser network could limit not only the diffusion of the monomer into the active sites of the catalyst but also the fragmentation of the catalyst.

This concept is supported by our experimental results. As the amount of PPO chains on the support increased, the activity and the productivity of the catalyst in ethylene polymerization decreased, but the bulk density increased drastically (Table 1). It is assumed that at low PPO concentration on the support, the interaction between PPO and the metallocene/MAO complex is weak and the immobilization is limited. This would explain the observed results of a more homogeneous polymerization and therefore, of higher activities but of a lower bulk density (Table 1).

Table 1. Polymerization of ethylene (catalyst: Me$_2$Si(2MeBenzInd)$_2$ZrCl$_2$, support PPO functionalized latex particle).[a]

Run	Length of PPO (units)	Amount of PPO (mol%)	Activity[b]	Productivity[c]	BD[d]
1	15	0.5	2950	4100	260
2	15	1	1800	2400	310
3	15	5	1350	2000	360
4	15	10	1250	1750	420
5	15	20	1200	1700	490

[a] Reaction conditions: 1L autoclave, isobutane 400 ml, ethylene pressure 40 bar, 70 °C, 1 hr, loading 41 μ mol/g (Zr/cat), activation: 350 MAO/Zr, catalyst 24 mg. [b] kg PE/mol Zr hr bar. [c] g PE/g cat hr. [d] BD: bulk density (g/l).

One can conclude, that there is an optimum degree of surface functionalization for obtaining good bulk density, activity and productivity. If the surface functionalization is too low, as demonstrated in previous work,[19] only fluffy materials with a low bulk density are isolated, although with high catalyst activity. On the other hand, to achieve a high bulk density essential for an industrial application, a decreased activity and productivity have to be accepted as drawbacks.

Pyridine functionalized nanoparticles as support for borate activated metallocenes

A major problem in all of the above mentioned systems is that they require high MAO-concentrations for activation in order to achieve high activities and productivities. For silica as well as for organic supports, ratios of 500 to 5000 MAO/Zr ratio are widely applied.[15]

To develop a MAO-free system which can fragment, we modified our latex particles to allow a borate activation. Bochmann *et al.* have already shown that the coordination between amines and metallocenes is strong enough to retain the cationic zirconocene without blocking the free coordinating site of the zirconocene thus not reducing the activity of the metallocene during the polymerization.[16] Fréchet *et al.* proved that amino functionalized unfragmentable polystyrene resins were able to immobilize cationic hafnocenes due to an amine hafnocene interaction and to activate the polymerisation by a borate catalyst.[17] Since pyridine moieties can easily be incorporated by radical copolymerisation and since they should be able to immobilize metallocenes we developed a synthetic approach for pyridine containing latex particles as fragmentable supports in a borate activated olefin polymerisation.

Synthesis of the supported catalysts: Pyridine (System A) or PEO/Pyridine (System B) functionalized nanoparticles / Cp_2ZrMe_2 ([Ph_3C][$B(C_6F_5)_4$])

To synthesize the pyridine containing latex-particles, an emulsion polymerization technique was applied. In this process, block copolymers and/or sodium dodecylsulfate as emulsifiers, styrene and 4-vinyl pyridine as monomers and divinylbenzene as crosslinker were chosen (Scheme 4).

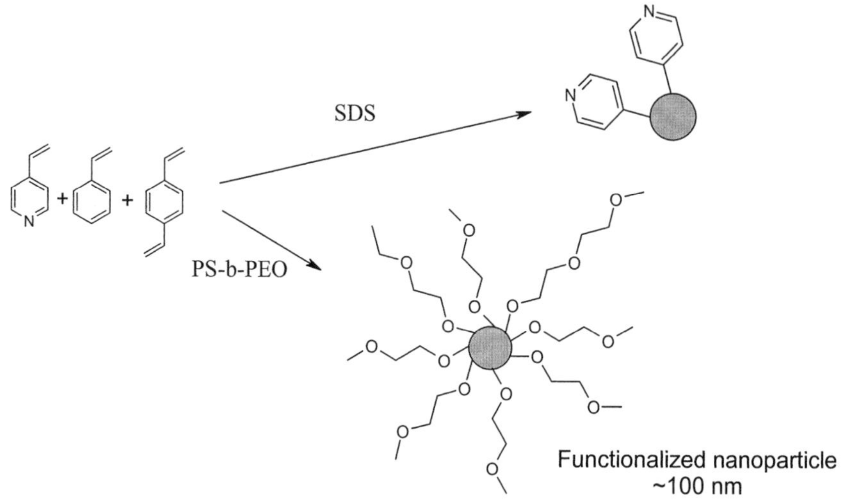

Scheme 3. Synthesis of nanoparticles *via* emulsion polymerization

For System A, latex-particles generated by emulsion polymerization of styrene, 4-vinyl pyridine as monomer and divinylbenzene as crosslinker were applied. Sodium dodecylsulfate (SDS) was added as the emulsifier. The borate activated metallocene bis(dimethyl)zirconocene (Cp_2ZrMe_2) was weakly immobilized by the pyridine ring atoms on the surface of the latex particles. The ionic surfactant was removed by dialysis.

For System B, in addition to SDS, amphiphilic block copolymers (PS-*b*-PEO) were used as emulsifiers. In this way, the surface of the latices was functionalized additionally by PEO-chains and the nanoparticles were crosslinked by the addition of trimethylaluminium (TMA) and its interactions with PEO. The particle size, after crosslinking of these so-called secondary particles was around 100 μm. The ionic surfactant was again removed by dialysis.

It is to be expected that the pyridine functions alone are enough to allow the desired reversible crosslinking of the nanoparticles. However, we could apply vinyl pyridine only in concentrations of up to 10 % in the emulsion process, otherwise no defined nanoparticles and stable dispersion are obtained. Therefore, the pyridine concentration on the surface is relatively low and the interaction between the different latex particles is too weak, which should produce a too rapid fragmentation and fluffy product beads. To enhance the interaction between the particles, we added PS-*b*-PEO for surface functionalization to the system.

Latex-particles with pyridine surfaces as support for zirconocene catalysts (System A)

The activation of Cp$_2$ZrMe$_2$ was performed in toluene by mixing the metallocene with ([Ph$_3$C][B(C$_6$F$_5$)$_4$]). After stirring at room temperature for 30 minutes, this solution was added to a suspension of latex-particles) in toluene (Scheme 4).

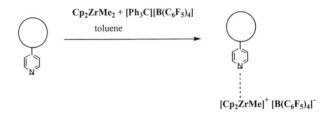

Scheme 4. Preparation of catalyst System A

After mixing this suspension for 20 minutes, the supported catalysts were dried under vacuum to achieve a free flowing powder. To obtain the catalyst particles homogeneous in size, the powder was sieved with a 50 μm and a 100 μm sieve.

Latex-particles with pyridine and PEO-functionalized surfaces as support for zirconocene catalysts (System B)

As stated above, latex-particles (primary particles) obtained by emulsion polymerization with block copolymers PS-*b*-PEO as emulsifier can form a reversible network *via* the PEO-chains on the surface. Different latex particles were generated while varying the block length and block ratio of the PS-*b*-PEO block copolymer. By adding aluminum alkyls, *e.g.* triisobutylaluminum (TIBA) or trimethylaluminium (TMA) (Al/Zr = 30), secondary particles between 50 – 100 μm were obtained in a non-covalent crosslinking process. The catalyst formation was finalized by loading the borate activated Cp_2ZrMe_2 catalyst analogous to System A (Scheme 5).

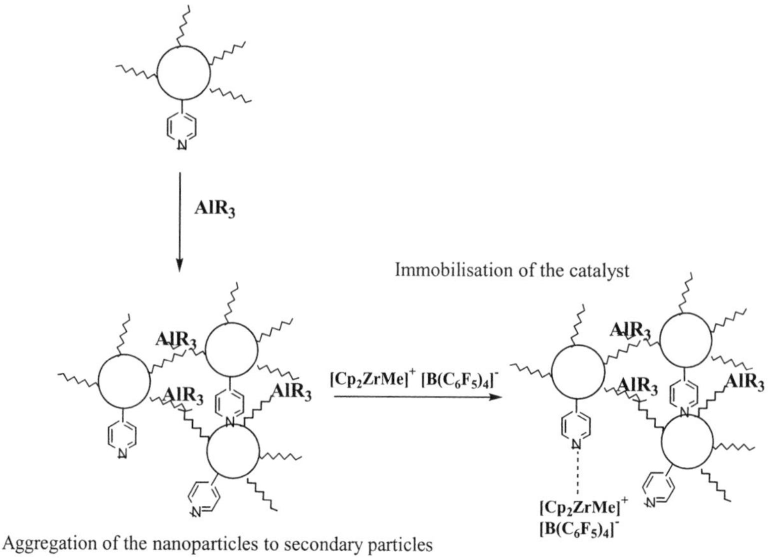

Scheme 5. Preparation of catalyst system B

After removing the solvent and sieving the particles to get a narrow size distribution of the catalyst system, the latex-particles were used in the olefin polymerization.

In Table 2 the results of some polymerization experiments using the supporting systems A and B are summarized. Even as the structure of the immobilized catalyst is not totally clear, the metallocene loaded network showed very good activities of up to 1150 kg PE/(mol Zr bar), similar to previous results for MAO-activated systems.[18] There was no adverse influence of the PEO-crosslinked system on the borate-activated metallocene. Furthermore, using trialkylaluminium compound as crosslinkers did not reduce the catalytic activity of the system.

Table 2. Results of ethylene polymerizations using latex-based supports with 4-vinyl pyridine [a)]

Run	Loading (μmol cat/ g PS)	Activity (kg PE/ mol Zr bar)	Productivity (g PE/ g cat h bar)	Bulk Density g/l
A-1	60	770	1800	140
A-2	160	860	5500	120
B-1	80	1140	3300	320
B-2	30	1150	1375	340

[a)] Reaction conditions: 1L autoclave, isobutane 400 ml, ethylene pressure 40 bar, 70 °C, 1 hr.

Remarkably, the bulk density of the polyolefin products obtained from PEO/pyridine surface functionalized particles (system B) were much higher in comparison to the pure vinylpyridine systems (system A) (Table 2). This indicates that due to the PEO chains the interaction between the latex particles is enhanced resulting in a slowing down of the fragmentation process. As demonstrated in previous publications this yields in higher bulk densities. [19]

Study of the Fragmentation Process

It has been shown for silica or $MgCl_2$ supported catalysts that fragmentation of the support is an essential prerequisite for a supported catalyst system. The first evidence that such a fragmentation also occurs for our polymeric supports is the excellent morphology control during the polymerization process. In all cases described herein, no reactor fouling was observed, but high bulk densities were achieved and millimeter sized spherical beads of polyolefin obtained. In order to prove such a fragmentation process, one has to investigate the catalyst (support) distribution within a product bead. In the case of a perfect fragmentation, a homogeneous distribution of the support in the particle is expected. However, to study this process, one has to consider that the concentration of the support in a polyolefin product bead, obtained as in our case from a very active catalyst system is extremely low. Therefore, for visualization, the supports have to be tagged either with strongly fluorescent dyes or stained with metals allowing the use of the very sensitive fluorescence, confocal fluorescence or electron microscopy techniques.

As similarly demonstrated for our PEO functionalized latex systems,[18] PPO functionalized supports catalysts were prepared, loaded with a fluorescent dye (9-styryl-N-(2,6-

diisopropylphenyl)perylene 3,4-dicarboxylimide) (Scheme 6) and applied in the polymerization of ethylene. [19]

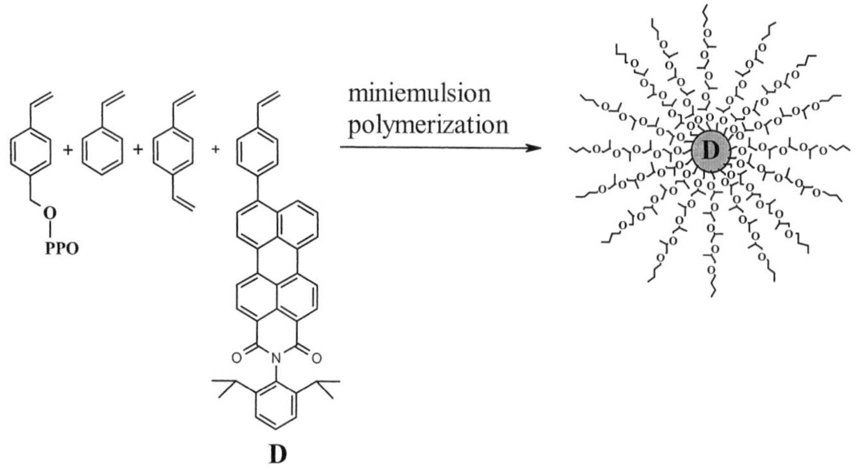

D

Scheme 6. Preparation of latex particles tagged with a fluorescent dye

To investigate the catalyst fragmentation, the polymerization was stopped at various reaction times, the polymer was isolated and the distribution of the fluorescence dye was investigated by fluorescence microscopy. As the polymerization time increased, the fluorescence distribution in the films became more and more homogenous, indicating a progressive fragmentation during the polymerization. After 60 min, the fragmentation was almost complete.[19]

These more qualitative findings can be further strengthened by studying single product beads using confocal fluorescence microscopy.[20] With this method, the fluorescence distribution can be scanned layer by layer in a polyolefin bead. The images can be combined to a three-dimensional image indicating the positions of the support by the fluorescent spots. Figure 1 shows the distribution of the visible fluorescent spots corresponding to our dye labeled latex particles, which is very homogeneous, confirming our proposal of a nearly perfect fragmentation of the catalyst.

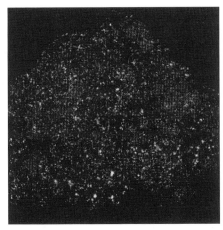

Figure 1. Fluorescence micrographs of a polyethylene particle (size 200μm) (after 60 min polymerization) obtained from a catalyst tagged with a fluorescent dye. (Laser wavelength 488 nm, emission 573 nm)

This non-invasive method does not require a slicing of the samples, therefore, the technique is very fast and easy applicable not only to organic, but also to inorganic supports such as silica. One drawback, however, is that the resolution is limited to the laser wavelength (300-600 nm). To gain a higher resolution of the structures in a polyolefin bead, electron microscopy was used. As the distinction and the selective staining of the latex particles versus polyolefin is known to be difficult, we incorporated 10 weight% of hydrophobic triphenyl bismuth into the support during the miniemulsion process[21] at the stage of the latex particle formation. Providing of the experimental procedure by K. Landfester is gratefully acknowledged.[22] This allowed us to visualize the support in the slices of the polyolefin beads in the electron microscope without any further staining.

According to Figure 2a, already after 2 minutes the bismuth containing support particles (black) were surrounded by polyethylene (white). After 7 minutes, the layer of polyolefin was relatively thick, while at longer reaction time, the support became nearly invisible due to the high dilution ,only "white images" are obtained (not shown).

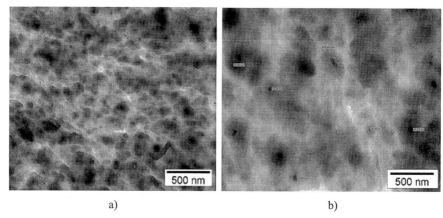

a)   b)

Figure 2. Electron microscopy: slice of a PE bead a) obtained after 2 min, b) obtained after 7 minutes (scale bar: 500nm)

Remarkably, these preliminary results suggest that the polymerization and therefore the fragmentation starts in all latex particles at the same time. This is in contrast to the model developed by Fink for silica materials describing a fragmentation of the support starting from the outer shell proceeding slowly to the inner parts.[23,24] It does correspond, however, to the mechanism proposed for Ziegler-type catalysts supported on $MgCl_2$. A more detailed study of the fragmentation process to develop a kinetic model is presently being performed in collaboration with the group of Fink.

Polyolefins by a Self-Immobilized Catalyst

In all of the previous examples, the metallocenes were supported on latex particles to control the product morphology. However, the question arises what will happen in the case of a polar monomer allowing an interaction/immobilization of the metallocene? It had already been shown by Alt et al. that by using metallocenes covalently linked to monomers, the product morphology could be controlled without supporting the catalyst.[25] The major drawback was the complicated synthesis of the polymerizable catalyst.

In our concept we applied PEO-functionalized copolymerizable monomers in the olefin polymerization. Thus is similar to the experiments using latex particles as support, but the metallocene is non-covalently bound allowing for a separate synthesis of metallocene and monomer.

PEO–functionalized norbornene and hexene derivatives were chosen as starting materials for our self-immobilizing system. Fink and coworkers have obtained copolymers with a content of comonomer of up to 23.5 mol-% were obtained using 5-norbornene-2-methanol as comonomer in an ethylene polymerization.[26] Furthermore, Marques *et al.* have reported copolymers with a 5-hexen-1-ol (TMS protected) content of up to 10 mol-%.[27]

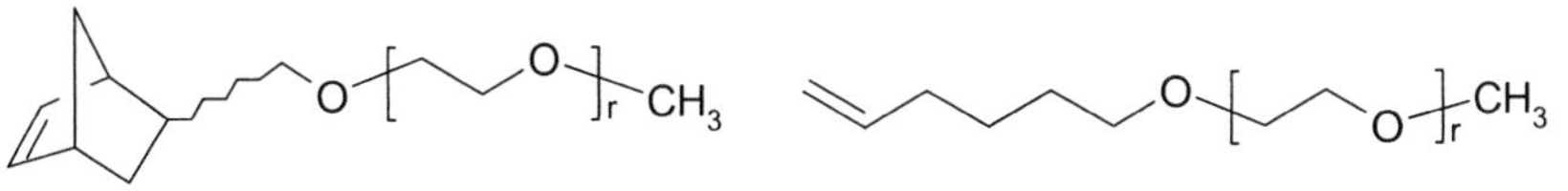

N - 1   **H - 1**

Scheme 7. Monomers used for the self-immobilizing approach: N-1: 5-norbornene-2-tetra ethylenglycol, H-1 1-hexene-6-polyethyleneglycol

In the copolymer the norbornene or hexene metallocene catalyst complex is directly attached to the polyolefin (Scheme 7). Thus, the system is self-immobilizing the catalyst during the polymerization. During the whole polymerization the MAO/metallocene complex remained attached to the PEO-chains since leaching of the catalyst was avoided as indicated by the absence of reactor fouling or dust formation (Scheme 8). Due to the interaction of the PEO chains with MAO the system started to aggregate and to form uniform particles.

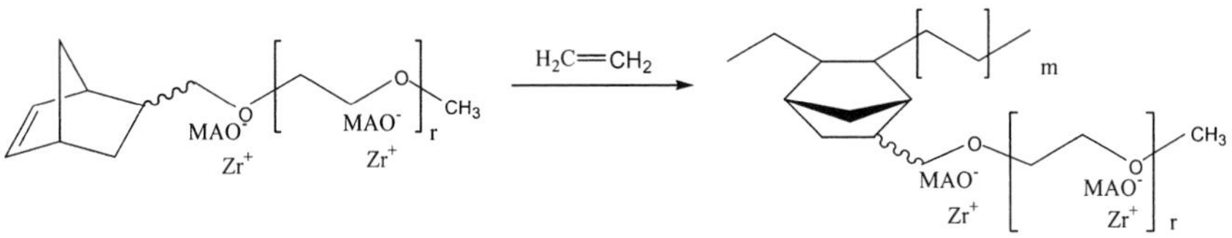

Scheme 8. Polymerization of ethylene

Table 3 shows the activities and productivities of the self-immobilizing system. The results are

Table 3. Activity and productivity of the self-immobilizing catalyst system [a]

Monomer (run)	Yield (g)	Activity (g PE/ mmol Zr·h)	Productivity (g PE/ g cat h bar) (g/g)
N-1 (1)	22	22000	39000
N-1 (2)	25	25000	44000
H-1 (3)	20	20000	35000
H-1 (4)	23	23000	40000

[a] Reaction conditions: 40 bar ethene pressure, 70 °C, 5 ml TIBA (scavenger) polymerization time: 60 min. Al/Zr: 1950, metallocene: $Me_2Si(2MeBenzInd)_2ZrCl_2$ (0.001 mmol), 0,5 mmol of N-1 or H-1 in 400 ml isobutane at 70°C.

comparable to supported systems.[8], [9] Furthermore it should be noted that the use of such polar monomers does not only allow for a control of morphology without using a support but also makes a hydrophilic polyolefin accesible shown by contact angle measurements. The contact angles films of the hydrophilic copolymers containing the PEO chains resulting from the self-immobilizing system were between 66° and 80°, for a corresponding hydrophobic polyethylene film angles of above 100° were measured .

Conclusions

It has been shown that organic particles functionalized with nucleophilic surfaces are suitable supports for metallocenes. They were successfully used in MAO and Borate activated polymerizations. Due to the variety and flexibility of the synthetic procedures for obtaining surface functionalized nanoparticles, the supports can easily be modified. Controlling the size of the particles, and of the nature and concentration of the nucleophilic groups was remarkably easy. This offers many possibilities to further optimize the catalytic system, thus improving activities and productivities. Also the fragmentation process, which directly influences product morphologies (size and shape of the product beads, bulk density), can be controlled by the concentration of nucleophilic groups on the surface of the particles. Modern fluorescence techniques such as confocal fluorescence microscopy allowed us easily to prove the fragmentation of the catalyst and will be applied in the future more extensively to elucidate the fragmentation processes of various supports as they are applicable for organic as well as for inorganic supports.

Finally, even without supports, a control of product morphologies is in principle possible as some preliminary studies demonstrated using PEO functionalized monomers. Similar to the covalently attached metallocenes to monomers, we were able to immobilize our catalyst by a non-covalent interaction to the catalyst. Therefore, the concept of self-immobilization of metallocene complexes was transferred to PEO-functionalized monomers. In all cases, products with reasonable morphologies (spherical beads, high bulk densities) were obtained and the heterogeneous manner of the polymerizations was proven as no reactor fouling was observed. One can conclude that polymeric supports are easy to synthesize and to functionalize and therefore offer many promising properties.

Acknowledgement. Financial support of the Fonds der Chemie and providing of the catalysts by Basell Polyolefins, Ludwigshafen is gratefully acknowledged.

[1] H. G. Alt, A. Köppl, *Chem. Rev.* **2000**, *100*, 1205.
[2] R. J. L. Graff, B. Kortleve, C. G. Vonk, *Polym. Lett.* **1970**, *8*, 735.
[3] S. Sensarma, S. Sivaram, *Macromol. Chem. Phys.* **1997**, *198*, 3709.
[4] M. de F. V. Maeques, C. A. Henriques, J.L.F.Monteiro, S. M. C. Menezes, F. M. B. Coutinho, *Polym. Bull.* **1997**, *39*, 567.
[5] T. Suzuki, Y. Suga, *Polym. Prepr.,* **1997**, *38*, 207.
[6] K. Czaja, L. Korach, *Polym. Bull.,* **2001**, *46*, 67.
[7] K. Czaja, L. Korach, *Polym. Bull.,* **2001**, *46*, 175.
[8] Klapper M. Koch M. Stork M. Nenov N. Mullen K., Organometallic Catalysts and Olefin Polymerisation: Catalyst for a New Millenium. **2001**, 387.
[9] Koch M. Klapper M. Mullen K. Organometallic Catalysts and Olefin Polymerisation: Catalyst for a New Millenium. **2001**, 396.
[10] M. Stork, M. Koch, M. Klapper, K. Müllen, H. Gregorius, U. Rief, *Macromol. Rapid Comm.*, **1999**, *20*, 210.
[11] M. Koch, M. Stork, M. Klapper, K. Müllen, H. Gregorius, *Macromolecules*, **2000**, *33*, 7713.
[12] G. Fink, B. Steinmetz, J. Zechlin, C. Przybyla, B. Tesche, *Chem. Rev.* **2000**, *100*, 1377.
[13] G. Fink, B. Tesche, F. Korber, S. Knoke, *Macromol. Symp.*, **2001**, *173*, 77.
[14] K. Landfester, *Macromol. Rapid. Comm.*, **2001**, *22*, 896.
[15] H. Sinn, W. Kaminsky, *Adv. Organomet. Chem.*, **1980**, *18*, 99.
[16] M. Bochmann, A.J.Jaggar, J.C.Nicholls, *Angew. Chem., Int Ed. Engl.* **1990**, *29*, 780.
[17] S. B. Roscoe, C. Gong, J. M. J. Frechet, J. F. Walzer, *J. Polym. Sci.. Part A, Chem.*, 2000, *38*, 2979-2992.
[18] M. Koch, A. Falcou, N. Nenov, M. Klapper, K. Müllen, *Macromol. Rapid. Commun.*, **2001**, *22*, 1455.
[19] Y.J. Jang , N. Nenov, M. Klapper, K. Müllen, *Polymer Bull.,* **2003**, *50*, 351.
[20] M.H. Chestnut, *Current Opinion in Colloid & Interface Science*, 1997, 2,158.
[21] F. Tiarks, K. Landfester, M. Antonietti, *Macromol. Chem. Phys.*, **2001**, *202*, 51.
[22] D. Palm, K. Landfester, publication in preparation.
[23] R. Goretzki, G. Fink, B. Tesche, B. Steinmetz, R. Rieger, W. Uzick, *J. Polym. Sci.: Part A: Polym. Chem.*, **1999**, *37*, 677.
[24] G. Fink, B. Tesche, F. Korber, S. Knoke, *Macromol. Symposia*, **2001**, *173*, 77.
[25] B. Peifer, W. Milius, H.G.Alt, *J. Organometallic Chem.* **1998**. *553*, 205.
[26] R. Goretzki, G. Fink, Macromol. *Chem. Phys.,* **1999**, 200, 881.
[27] M. M. Marques, S.G. Gorreia, J. R. Ascenso, A. F. G. Ribeiro, P. T.Gomes, A. R. Dias, P. Foster, M. D. Rausch, J., C. W. Chien, *J. Polym. Sci.: Part A*, **1999**, *37*, 2457.

Macromol. Symp. **2004**, *213*, 147-155

Heterogenization of Metalorganic Catalysts of Olefin Polymerization and Evaluation of Active Site Non-Uniformity

Lyudmila Novokshonova,[1] *Natalia Kovaleva,*[1] *Irina Meshkova,*[1]
Tatiana Ushakova,[1] *Vadim Krasheninnikov,*[1] *Tatiana Ladygina,*[1] *Ilia Leipunskii,*[2]
Alexey Zhigach,[2] *Michail Kuskov*[2]

[1] Institute of Chemical Physics, RAS, 119991 Moscow, Russia
E-mail: lnov@center.chph.ras.ru
[2] Institute of Energy Problems of Chemical Physics, RAS, 119991 Moscow, Russia

Summary: Heterogenized activators – "support-H_2O/AlR_3" (where R=Me, iBu, support=montmorillonite, zeolite), synthesized directly on the support, form with metallocenes metal alkyl complexes highly active in olefin polymerization without the use of commercial methylaluminoxane (MAO). It was shown by the method of temperature programmed desorption with the application of mass-spectrometry (TPD-MS) that the aluminumorganic compound in support-H_2O/AlR_3 is in general similar to the structure of commercial MAO. The heterogenization of Zr-cenes on support-H_2O/AlR_3 is accompanied by the appearance of the energy non-uniformity of active sites. The activation energy of thermal destruction of active Zr-C bonds in the active sites of prepared catalysts changes in the range from 25 to 32 kcal/mol.

Keywords: activator; metallocene; olefin polymerization; support

Introduction

The catalytic properties of the heterogenized metallocene catalysts of the olefin polymerization depend on many factors and first of all on the method of heterogenization and on the nature of the activator.

We synthesized directly on the surface of the support the heterogenized activator by partial hydrolysis of aluminumalkyl by water contained in a support (support-H_2O/AlR_3).[1-3] Similar way is tested now also by other investigators.[4-7] In this paper the structure of products of partial hydrolysis of aluminumalkyl by the support water was compared with the commercial MAO structure, the properties of support-$H_2O/AlR_3/Zr$-cene catalysts in the ethylene and

 DOI: 10.1002/masy.200450915

propylene polymerization were investigated, and the active site non-uniformity of supported metallocene catalysts was evaluated.

Results and Discussion

We used as supports zeolite (ZSM-5), which has very stable skeleton structure, and the layered silicate montmorillonite (MMT). The distinctive feature of these supports is a very high mobility of internal water. $AlMe_3$ and $AliBu_3$ were used for the treatment of supports. The products of partial hydrolysis of aluminumalkyl, heterogenized on the surface, activate different metallocenes in the ethylene and propylene polymerization without the use of commercial MAO:

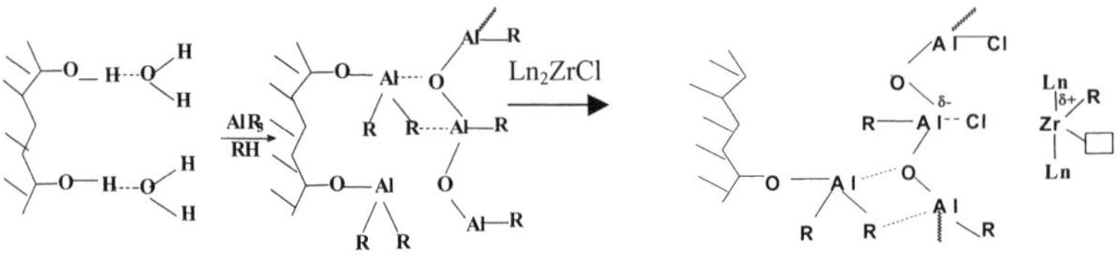

The activities of these catalysts are close to the activities of corresponding homogeneous systems with MAO. [3,8-10] The data, presented in Table 1, illustrate this for the propylene polymerization with the catalyst based on zeolite as a support and the data of Table 2 – for

Table 1. Propylene polymerization with ZSM-5(H$_2$O)/AlMe$_3$/Zr-cene and homogeneous Zr-cene-MAO* catalysts

Catalyst	Zr wt %	$\dfrac{[Al]}{[Zr]}$	T, ^{0}C	A^a	$\overline{M}_W$ 10^{-3}	$\dfrac{\overline{M}_W}{\overline{M}_n}$	T_m ^{0}C	λ^b, %	I_{iso}^c %
ZSM-5(H$_2$O)/AlMe$_3$/ Et[Ind]$_2$ZrCl$_2$	0,012	5540	40	19800	40	13,4	124	55	82
ZSM-5(H$_2$O)/AlMe$_3$/ Et[Ind]$_2$ZrCl$_2$	0,015	4350	40	9170	43	8,8	123	50	82
ZSM-5(H$_2$O)/AlMe$_3$/ Et[Ind]$_2$ZrCl$_2$	0,018	5000	75	13160	1,5^d	-	wax	-	10
Et[Ind]$_2$ZrCl$_2$-MAO	0,026^e	1720	75	25000	2,0^d	-	wax	-	15

$^{a)}$ A=activity, kgPP/molZr·[C$_3$H$_6$]·h; $^{b)}$ polymer cristallinity; $^{c)}$ PP isotacticity by Luongo method; $^{d)}$ M$_n$; $^{e)}$ mmol/l.

Table 2. Ethylene polymerization with zirconocenes supported on MMT-H_2O/AlR$_3$

Catalyst	$Q_{Zr} \times 10^6$, mol/g MMT	[Al]/ [Zr]	Tp, ^0C	A^a
Cp$_2$ZrCl$_2$ - MAO	$5{,}6 \times 10^{-5}$ mol/l	1000	40	20800
MMT-H_2O/AlMe$_3$- Cp$_2$ZrCl$_2$	4,6	1100	40	10000
Et(Ind)$_2$ZrCl$_2$ - MAO	$4{,}0 \times 10^{-5}$ mol/l	1100	40	26000
MMT-H_2O/AlMe$_3$- Et(Ind)$_2$ZrCl$_2$	2,4	2000	40	14700
MMT-H_2O/AlMe$_3$- Et(Ind)$_2$ZrCl$_2$	2,4	2270	55	21450
MMT-H_2O/AlMe$_3$- Et(Ind)$_2$ZrCl$_2$	2,4	2000	65	54850
MMT-H_2O/AliBu$_3$- Et(Ind)$_2$ZrCl$_2$	2,4	1400	65	17800

[a] A = activity, kg PE/mol Zr [C$_2$H$_4$] h

the ethylene polymerization with catalysts based on montmorillonite as a support.

Polyethylene (PE) synthesized with the supported catalyst has a higher molecular weight, broader molecular weight distribution, higher melting point and degree of crystallinity than PE prepared with the suitable homogeneous catalyst (Table 3).

Table 3. Characteristics of PE prepared with the ZSM-5(H_2O)/AlMe$_3$/Cp$_2$ZrCl$_2$ catalyst and the Cp$_2$ZrCl$_2$-MAO homogeneous system (polymerization conditions: T=34^{0}C, [C$_2$H$_4$]=0,054 mol/l).

Catalyst	[Zr], wt.%	[Zr]/[Al]	M_w	M_w/M_n	Crystallinity, %	T_m ^0C
ZSM-5(H_2O)/AlMe$_3$/Cp$_2$ZrCl$_2$	0,07	4600	202750	22,7	85	133
ZSM-5(H_2O)/AlMe$_3$/Cp$_2$ZrCl$_2$	0,21	440	150500	18,7	80	136
ZSM-5(H_2O)/AlMe$_3$/Cp$_2$ZrCl$_2$ + MAO	0,16	470^a	52800	18	76	123
Cp$_2$ZrCl$_2$ - MAO	0,18^b	400	57538	4,6	72	121

[a] [Al]/[Zr]=[AlMe$_3$]/[Zr]+[MAO]/[Zr] =350 +50; [b] mmol/l.

It is interesting that addition of free MAO into the reaction zone during polymerization of ethylene with the prepared heterogenized catalyst is accompanied by the decrease of molecular weight and melting point. [8] As seen from gel chromatogram (Figure 1), two products with different molecular weights are present in the polymer in this case. PE contains a fraction with lower molecular weight which is absent in the polymer, synthesized without additional MAO. This means that the active sites in the prepared supported catalysts

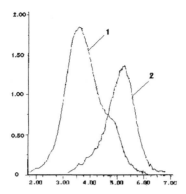

Figure 1. Gel-chromatograms of PE prepared with (1) ZSM-5(H_2O)/AlMe$_3$/Cp$_2$ZrCl$_2$ and (2) ZSM-5(H_2O)/AlMe$_3$/Cp$_2$ZrCl$_2$+MAO catalysts.

are located mainly on the surface during polymerization. Under the action of free MAO, the leaching of part of active sites from the catalyst surface into solution takes place.

The heterogenized products of partial hydrolysis of AliBu$_3$ activate metallocenes also. With the used Zr-cene, the supported catalyst exhibits in this case the lower activity in comparison with the catalyst containing the activator prepared with AlMe$_3$ (Table 2).

The specific activity of the catalyst depending on the content of Zr-cene in a supported catalyst has a maximum in the region of the low surface concentration of Zr-cene (Figure 2).[10] Similar dependence is typical for many immobilized catalysts of olefin polymerization, including the supported catalysts of Ziegler-Natta type. One of the possible reasons for the reduction in the activity of supported catalyst with an increase of Zr content is the decrease of Al/Zr ratio. With a decrease of Al/Zr ratio, the fraction of more active catalytic complexes, containing the several molecules of co-catalyst, decreases. Besides, on the surface of support-H_2O/AlR$_3$, there are the sites of different activity towards Zr-cene, and only a part of them is able to form with Zr-cene the centers of high activity in olefin polymerization.

It was observed that the order of polymerization rate with respect to the monomer concentration changes from 2 to 1 both in the ethylene and propylene polymerization (Figure 3).[10] This allows to suggest that two olefin molecules are involved in the monomer insertion into the active site occurring in the transition state, as in the olefin polymerization with homogeneous catalysts and the heterogeneous Ziegler-Natta catalysts.

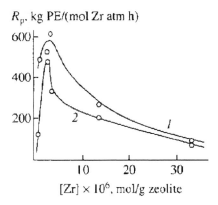

Figure 2. Effect of zirconocene concentration on the rate of ethylene polymerization with ZSM-5-H_2O/AlMe$_3$/Cp$_2$ZrCl$_2$; polymerization conditions: T = 34°C;. [C_2H_4] = 0.054 mol/l, [Al] = 4 mmol/g zeolite; 1- maximum rate; 2- stationary rate.

The structure of products of partial hydrolysis of AlMe$_3$ by the water of a support was compared with the structure of solid commercial MAO. For this we investigated the thermal destruction of both aluminumorganic compounds by the method of temperature programmed desorption with mass-spectrometric analysis of gaseous products (TPD-MS). [3,11]

The results showed that the composition of the desorption products arising by the thermal destruction of both compounds is identical and corresponding to the different fragments of

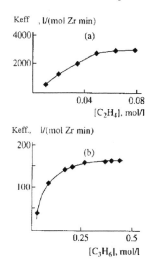

Figure 3. The effective rate constant of (a) ethylene and (b) propylene polymerization with the ZSM-5-(H_2O)/AlMe$_3$/Et(Ind)$_2$ZrCl$_2$ as a function of monomer concentration. Polymerization conditions: T= 40°C; [Zr] = 0.05 wt.%; Al/Zr=1400.

MAO molecule. Moreover, the TPD spectra for the same fragments, arising from both compounds, are in general similar (Figures 4-6). This allows to conclude that the structure of the prepared heterogenized co-catalyst is close to the commercial MAO structure. [12,13]

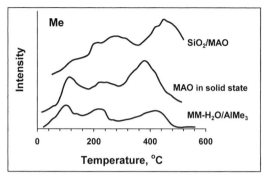

Figure 4. The curves of TPD for m/z=15.

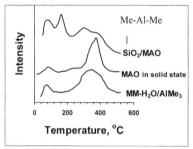

Figure 5.The curves of TPD for m/z=57.

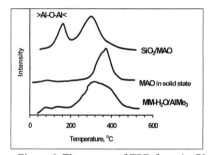

Figure 6. The curves of TPD for m/z=70.

At the same time we found out that after the supporting on the surface of dehydrated SiO_2, the commercial MAO looses part of the most active Me-groups. As seen from Figure 4, in TPD spectrum, which characterizes the detachment of methyl groups from the different fragments of MAO molecule, the maximum at low temperature is practically absent for SiO_2/MAO sample. It is important to note that the contributor of the recorded methyl groups is the thermal destructed MAO, and not $AlMe_3$. $AlMe_3$ is present in the tested samples, however the methyl groups, arising from the $AlMe_3$ destruction, are not registered by the method of mass-spectrometry.[14] The observed fact can be the reason why MAO, supported on the dehydrated SiO_2, does not activate the metallocene, and addition of free MAO or other aluminumalkyl is needed to produce the active heterogenized catalysts.

The initial stages of the ethylene polymerization with solid homogeneous $Cp_2ZrCl_2/$ MAO complex and with the heterogenized $MMT-H_2O/AlMe_3/Cp_2ZrCl_2$ catalyst were studied by the method of TPD-MS.[3,11] The method allows to record the evolution of short hydrocarbon molecules as a result of thermal destruction of Zr-C bonds in the living active sites depending on linear heating of sample. The results show the uniformity of active sites of homogeneous Cp_2ZrCl_2/MAO complex even in solid state (Figure 7a): there is only one narrow peak in the high temperature range on the TPD curve.

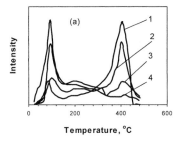 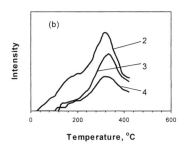

Figure 7. The TPD curves from solid Cp_2ZrCl_2-MAO complex (a) and the supported MM-$H_2O/AlMe_3/Cp_2ZrCl_2$ catalyst (b) after ethylene polymerization (ethylene pressure = 0,40kPa, temperature=20^oC) for m/z of: 70 (1), 98 (2), 154 (3), 182 (4).

Meanwhile in the heterogenized $MMT-H_2O/AlMe_3/Cp_2ZrCl_2$ catalyst, several active sites with different activation energies of thermal destruction of the active Zr-C bonds are generated, according to data of Figure 7b: the maximum on the TPD curve in the same temperature range is broad and has a complicated shape. On the base of TPD-MS data, the distribution of active sites over activation energy of thermal destruction of Zr-C bonds for both investigated catalysts was calculated (Figure 8a,b). The activation energy of thermal destruction of Zr-C bonds in active sites is about 32 kcal/mol for the solid homogeneous $Cp_2ZrCl_2/$ MAO catalyst and changes in the range from 25 to 32 kcal/mol for the heterogenized $MMT-H_2O/AlMe_3/Cp_2ZrCl_2$ catalyst.

154

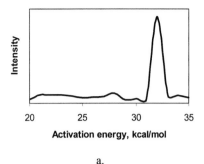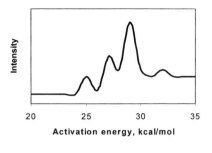

a.   b.

Figure 8. Calculated energy spectra of active sites for m/z=98 for solid Cp_2ZrCl_2-MAO complex (a) and the supported MMT-H_2O/AlMe$_3$/Cp_2ZrCl_2 catalyst (b).

Conclusion

The heterogenized activators – "support-H_2O/AlR$_3$", obtained directly on the surface, form with metallocene metal alkyl complexes which are active in olefin polymerization without use of commercial MAO. The aluminumorganic compound in the activator support-H_2O/AlR$_3$ is similar in general to the structure of commercial MAO. At the same time, it was found out that the commercial MAO, supported on the dehydrated SiO_2, looses part of the most active methyl-groups and obviously along with this, the ability to alkylate metallocenes. The heterogenization of Zr-cenes on support-H_2O/AlR$_3$ is accompanied by the appearance of the energy non-uniformity of active sites. The activation energy of thermal destruction of active Zr-C bonds in the active sites of the prepared heterogenized catalyst changes in the range from 25 to 32 kcal/mol.

[1] N.S.Enikolopov, F.S. D'yachkovskii, I.N.Meshkova, T.M.Ushakova USSR *Inventor's Certificate* 1066193 **1982**, *Byul.Izobr.* **1985**, *11*, 199.
[2] T.M.Ushakova, I.N.Meshkova, F.S.D'yachkovskii, *31st IUPAC Macromolecular Symposium,* Merseburg, GDR, 1987, 100.
[3] L.A. Novokshonova, I.N. Meshkova, N.Yu. Kovaleva, T.M. Ushakova, P.N. Brevnov, V.G. Krasheninnikov, T.A. Ladygina, I.O. Leipunskii, A.N. Zhigach, M.L. Kuskov Future in *"Technology for Polyolefin and Olefin Polymerization Catalysis"* Ed. Terano M., Shiono T. Tokyo, 2002. P. 162.
[4] W. Kaminsky *USA Pat.* 4431788, 1984; *Chem. Abstr.* **1981**, *95.* 151475.
[5] M. Chang *USA Pat.* 56292533, 1993; *Chem. Abstr.* **1997**, *127.* 34630.
[6] T. Tsutsui, T. Ueda. *USA Pat.* 5234878, 1993; *Chem. Abstr.* **1991**, *115.* 233130.

[7]   Y. Suga, Y. Maruyama, E. Isobe, T. Suzuki, F. Shimizu *USA Pat.* 5308811, 1994; *Chem. Abstr.* **1993**, *118*. 192487.

[8]   .N.Meshkova, T.M.Ushakova, T.A.Ladygina, N.Yu.Kovaleva, L.A.Novokshonova., *Polym. Bul.*, **2000**, *44*, 461.

[9]   I.N. Meshkova, T.A. Ladygina, T.M. Ushakova, N.Yu. Kovaleva, L.A. Novokshonova in *"Studies in Surface Science and Catalysis"* Ed. Aiello R., Giordano G., Testa F. Elsevier Science B.V., 2002. p. 691.

[10] I.N. Meshkova, T.A. Ladygina, T.M. Ushakova, L.A. Novokshonova Vysokomol. Soed. A, **2002**, *44*, 8, 1310.

[11] L.A.Novokshonova, N.Yu.Kovaleva, Yu.A.Gavrilov, V.G.Krasheninnikov, I.O. Leipunskii, A.N Zhigach, M.N.Larichev, M.V.Chebunin, *Polim. Bul.* **1997**, *39*, 59.

[12] H.Sinn, paper presented at Hamburg Makromoleculares Kolloquium, Hamburg, 22 – 23 September 1994.

[13] M.R.Mason, J.M.Smiths, S. G.Bott, A.R.Barron, *J. Am. Chem. Soc.,* **1993**, *115*, 4971.

[14] Cornu A., Massot R., in *"Compilation of Mass-spectral Data"* Heyden & Sons LTD, Press Universitaires de France, 1996.

Macromol. Symp. **2004**, *213*, 157-171

Fluorenyl Based Syndiotactic Specific Metallocene Catalysts Structural Features, Origin of Syndiospecificity

Abbas Razavi,[1] *Vincenzo Bellia,*[1] *Yves De Brauwer,*[1] *Kai Hortmann,*[1]
Liliane Peters,[1] *Sabine Sirole,*[1] *Stephan Van Belle,*[1] *Ulf Thewalt*[2]

[1] Atofina Research S.A., Centre de Recherche Du Groupe Total, Zone Industrielle C, B-7181 Seneffe (Feluy), Belgium
[2] Sektion für Röntgen und Elektronenbeugung Universität Ulm, Germany
E-mail: abbas.razavi@atofina.com

Summary: The stereochemistry of propylene insertion/propagation reactions with a variety of C_s symmetric fluorenyl- containing single site catalysts is discussed. Our recent results indicate that independent of the chemical composition of the ancillary ligand fragments, or nature of the transition metal, active sites with local C_s symmetry and enantiotopic coordination positions behave syndioselectively in the general context of chain migratory insertion mechanism. Perfect bilateral symmetry neither exists nor is required in these processes. In this context the mechanism of syndiospecific polymerization is revisited by taking into account the structural characteristics and catalytic behavior of the original metallocene based $(\eta^5\text{-}C_5H_4\text{-}CMe_2\text{-}\eta^5\text{-}C_{13}H_8)$ $MCl_2/$ MAO; M = Zr (**1**), Hf (**2**) catalyst systems and new syndiotactic specific systems including $(\eta^5\text{-}C_5H_4\text{-}CPh2\text{-}\eta5\text{-}3,6\text{-}di\text{-}tBut\text{-}C_{13}H_6)ZrCl_2$ (**3**), $\eta^1,\eta^5\text{-}(\mu Me_2Si)(3,6\text{-}di\text{-}tBut\text{-}Flu)(t\text{-}ButN)MCl_2/$ MAO; M =Ti (**4**) , Zr (**5**) and $\eta^1,\eta^5\text{-}(\mu Me_2Si)(2,7\text{-}di\text{-}tBut\text{-}Flu)(t\text{-}ButN)MCl_2$ / MAO ; M = Ti (**6**), Zr (**7**).

Keywords: catalysis; mechanism; metallocene; polypropylene; syndiospecificity

Introduction

Syndiotactic polypropylene was first isolated by Natta and his co-workers as a minor by-product of isotactic polypropylene produced with a $TiCl_3$ based catalyst.[1] The nature of the active site and the mechanism of formation of this polymer are still under debate. It is believed that it is formed on the sites with low chlorine coordination via a chain end controlled mechanism. Later Zambelli and coworkers produced syndiotactic polypropylene directly using a vanadium-based homogenous catalyst.[2] In this case, more is known about the nature of the active site and the mechanism of the polymerization is elucidated

 DOI: 10.1002/masy.200450916

satisfactorily. No X-ray structure of the catalyst precursor is, however, available due to its very temperature sensitive nature. After the discovery of the bridged cyclopentadienyl-fluorenyl metallocene based syndiospecific catalyst and the resulting syndiotactic polypropylene by us[3] one had for the first time the opportunity to make accurate statements about the nature of the active site and the mechanism of the polymerization. By studying the available X-ray structure data of the metallocenes and their stabilized alkylmetallocenium cation[3h] it has become possible to make reasonable deductions on the nature of the active site and its behavior during the polymerization. The concept of active site model and the mechanism of the polymerization have been refined gradually and continuously as new syndiotactic specific metallocenes were discovered and more elaborate calculation methods were applied. In this contribution we report the latest progress on catalysts development and mechanistic aspects of syndiospecific polymerization commencing with a review of the original model proposed for the isopropylidene (cyclopentadienyl-fluorenyl)MCl₂ / MAO; M = Hf, Zr catalyst systems.

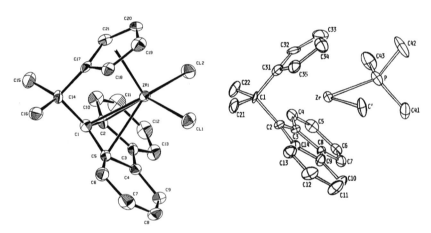

Figure 1. X-ray determined molecular structure of **1** (left) and its trimethyl phosphine stabilized methylmetallocenium cation (right).[3h]

Results and Discussion

Figure 1 presents the X-ray determined, molecular structure of the metallocenes isopropylidene(cyclopentadienyl-fluorenyl)MCl$_2$; M = Zr, Hf [3f] and the X-ray structure of the corresponding methylmetallocenium cations used to prepare the first ever highly crystalline syndiotactic polypropylene samples. Since their discoveries, we have been exhaustively discussing the structure, catalytic properties and origins of the formation of syndiotactic chains with …rrrrrrrrmrrrrrrrrrrrrrrrmmrrrrrrr… microstructure in numerous publications. [3,7,8,9,10,11,12] For the sake of continuity in the line of evidence, the highlights of these discussions will be reiterated in this paper and combined with the proposals made by other authors. 1) The stereorigid metallocene procatalyst (Figure 1) is prochiral and possesses a bilateral symmetry. 2) The chiral, cationic alkyl metallocenium species [3b] that are formed after the activation are composed of equal numbers of R and S monomer π-face selective enantiomers. 3) The re- and si–face selectivity is induced by the unique steric arrangement of the chelating ligand engulfing the resident transition metal center via a delicately balanced, cooperative, and non-bonded steric interaction contributed by different parts of the "living" catalytic species - ligand, polymer chain, and coordinating monomer (Figure 2). The non-bonded steric interactions govern the whole scenery of syndiospecific polymerization processes. 4) Since according to these assumptions each enantiomer, independently, would produce isotactic chains (yet syndiotactic polymer is exclusively formed), it must be concluded that the active enantiomeric species epimerize and are therefore interconvertable after each monomer insertion (Scheme 1). 5) The systematic interconversion of the two antipodes implies that the relative positions of at least two of the four ligands surrounding the transition metal are exchanged continuously. 6) Since the η^5 bonded aromatic ligands are tied together by a structural bridge and their rearrangement is not possible, such an epimerization can take place only when the alkyl group (or growing polymer chain) and the coordinating monomer exchange their positions uninterruptedly (provided no excessive steric restrictions are imposed). 7) A meso triad enantiomorphic site stereo chemical type error, mm, is formed whenever the said balanced, non-bonded steric interaction, is perturbed and the correct alignment of the substituents of all three main participants (ligand, polymer chain and the monomer) has not been realized. In such a case, a monomer with "wrong face" will be inserted and a unit with inverted configuration is enchained. This reverse-face selectivity emanates from inherent static structural factors and is independent of monomer concentration. Figure 3 depicts the model representing the transition state structure for propylene polymerization. This depiction is based on the relative importance of the non-bonded, steric interactions operating on different parts of the catalytic species and its - in the polymerization active participants - aromatic ligand, polymer chain and the coordinating monomer in the following order.

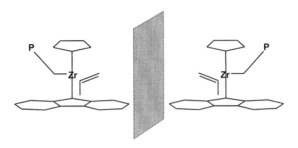

Figure 2. Interconverting, mirror image related active species produced with **1**.

The steric interaction between the flat and spatially extended fluorenyl ligand forces the growing polymer chain to be oriented towards the free space left (or right) to the unsubstituted cyclopentadienyl moiety of the ligand. To avoid excessive steric exposure, the incoming approaches with its methyl group trans to the growing polymer chain. The system reaches in this way a minimum energy state. In this orientation the coordinated monomer points with its methyl group head-down into the empty space in the central region of the fluorenyl ligand. The confirmation for the head-down coordination of the monomer was determined after extensive molecular mechanics and force field calculations performed by Corradini and coworkers.[4] The model underwent later additional refinement and took its current form after experiments conducted by several groups[5] supporting the idea of an α C-H agostic assistance in the transition state for the propylene polymerization.

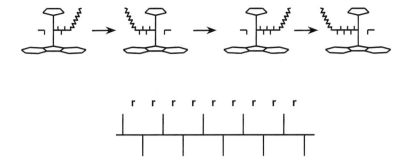

Scheme 1. Mechanism of syndiospecific polymerization (top). Fischer projection of an ideal syndiotactic chain (bottom).

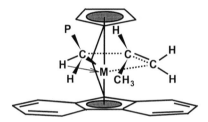

Figure 3. Transition state structure proposed based on the X-ray structures given in Figure 1.

The working hypotheses, active site model and the transition state structure discussed in the preceding paragraphs, account for syndiotactic specificity of the catalysts and formation of the syndiotactic polypropylene chains. Additionally, this justifies the formation of microstructural stereo errors - the meso triad (mm) and meso dyad (m) defects - encountered in the backbone of the polymers. The former – the so-called enantiomorphic site controlled defects - are well described and relate to reverse monomer enantioface insertion. The origin of the temperature and monomer concentration dependent m type stereo error, is more complex and more difficult to discern. Their formation has been explained according to the insertion-less site epimerization scheme shown in Figure 4.

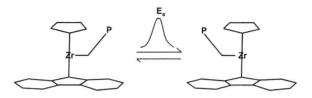

Figure 4. Coordination sites' equivalency and endothermic active site epimerization.

The R and S configured active sites are equi-energetic (Figure 4) and therefore may interconvert during the polymerization. The interconversion is particularly enhanced, in the absence of a coordinating monomers or solvent molecules. If the rate of this interconversion is faster than the actual rate of the monomer insertion, sites epimerize before the next insertion occurs. This increases the chance of two consecutive insertions taking place at the same enantiomorphic coordination position thus, adding two monomer units with the same prochiral face. This leads to the formation of meso dyad defects within predominantly syndio-regular polymer backbone. Even though the interconverting species are equi-energetic, the site epimerization requires activation energy for the chain's back or forth swing. The concept of site epimerization is, in principle, very close to the mechanism of chain back migration proposed by Cossee and Arlman in 1964. [6] The close proximity of a polar molecule, a counter ion, and/or temporarily blockage of one of the coordination sites (contact ion pairing) can provoke the site epimeriztion by favoring the polymer chain to move away and migrate back to its initial position. We have suspected this mechanism to be the cause for the formation of the short isotactic blocks that were detected in the backbone of the predominantly syndiotactic polymer chains formed with **2** (Figure 5).

It is conceivable that the MAO anion as counter ion, occupies the very same spot where the trimethyl phosphine is located in the structure shown in Figure 1 left. Under these conditions, its polarity and size would discourage the chain from moving to that position and provoke multiple insertions at the same coordination site, before the counter ion dissociates and regular chain migratory insertion process kicks in again. The occurrence of the site epimerization can be easily demonstrated and manipulated via polymerization temperature

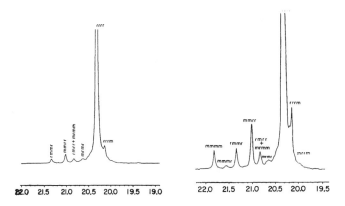

Figure 5. The methyl region of ^{13}CNMR of syndiotactic polypropylene produced by **1** (left) and **2** (right). [3f]

and monomer concentration modifications and also by distorting the coordination sites' equivalency. By playing with these parameters, the isotactic sequence lengths in predominantly syndiotactic polymers have been enlarged, to the extreme cases, where "pure" isotactic polymer chains were obtained. [7, 8]

To decrease the rmrr concentration (increase the overall stereoregularity of the syndiotactic polypropylene) we should lower the site epimerization rate to greatest possible extent. One way to accomplish this goal is to perform the polymerization in liquid propylene (high monomer concentration) at lowest possible temperatures. Of practical interest is, however, to accomplish this goal through structural modifications and/or by playing with the counter ion composition/interaction. The structure/counter ion modification which not only helps lower the frequency of rmrr stereodefects but also enhances the enantioselectivity of the catalysts (vide infra), can have an additional benefit in lowering the concentration of the rmmr pentads.[9b,3g] Figure 6 illustrates the X-ray determined molecular structure of (η^5-C$_5$H$_4$-CPh$_2$-η^5-3,6-di-tBut-C$_{13}$H$_6$)ZrCl$_2$ (**3**). After activation, complex **3** polymerizes propylene to highly stereo-regular polypropylene very efficiently, especially in the presence of hydrogen. Table 1 presents the polymerization results and polymer analysis of the syndiotactic polymers produced with catalyst system **3**/MAO, at different polymerization temperatures. Table 2 compares the meso dyad and triad concentrations found in syndiotactic polymer chains produced with metallocene **1** and **3**. A cursory glance at the data given in Tables 1 and 2 reveals that the t-butyl substitution in positions 3 and 6 of the fluorenyl moiety of the ligand in metallocene **1** causes substantial improvement in stereoselectivity of the final catalysts. It decreases the concentration of both meso dyad and triad related stereoerrors.

Table 1. Polymerization results and polymer analysis for **3**/MAO catalysts system

Temperature °C	Mw	rrrr %	Regiodefects %	T_m °C
40	766,000	91	nd	150
60	509,000	88.5	nd	143
80	443,000	79	nd	128

For example, the rmrr pentad concentrations in syndiotactic polymer chains produced with **3** are, by about 1/3, lower than that of the corresponding pentad concentrations in polymers made with **1**/MAO catalyst system. They double in size, in both systems, with every 20°C increase in polymerization temperature. On the other hand, the rmmr pentad concentrations in polymers produced with **3** are about half of the corresponding pentad concentrations observed in polymers made with **1**/MAO and vary little with temperature. The improved enantioselectivity (lower rmmr %) of catalyst **3**/MAO can be reasonably explained, by the enhanced substituent(s) effect in directing the polymer chain, to adopt the most preferred upward conformation - left or right to the cyclopentadienyl moiety. This provides a more effective guidance for the monomer's head down coordination mode (tighter "chiral pocket"). The explanation for lower rmrr pentads concentrations in polymers produced with **3**/MAO catalyst system, at the same polymerization temperatures and monomer concentrations, is however, less straightforward. It could be related to different cation/anion interaction mode for **3** and **1**.

Table 2. Stereo defects generated in the polymers produced with **3** (left) and **1** (right)

Temperature °C	rmmr %	rmrr %	rmmr %	rmrr %
40	0.73	0.93	1.55	1.15
60	0.85	1.99	1.65	2.7
80	1.06	3.79	2.20	4.82

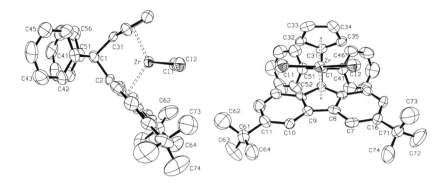

Figure 6. Side and front view perspectives of X-ray determined molecular structure of **3**.

A cursory check of the front view of the molecule **3** shown in Figure 6 and simple modeling experiments suggests that the spatial extension of the t-butyl groups can cause substantial steric interaction between the t-butyl substituents and MAO counter ion and the build up of strong repulsive forces against the formation of contact ion-pairs between metallocenium-methyl cations of **3** and anionic MAO molecules. Consequently, this could lower the site epimerization rate and the frequency of single m dyad formation.

We have recently reported two interesting syndiospecific cases for which similar substituent effects have been observed. [10] These new syndiospecific catalyst systems are composed of the amido complexes, η^1,η^5-(μMe$_2$Si)(3,6-di-tBut-Flu)(t-ButN)MCl$_2$; M =Ti (**4**) , Zr (**5**) and η^1,η^5-(μMe$_2$Si)(2,7-di-tBut-Flu)(t-ButN)MCl$_2$; M = Ti (**6**), Zr (**7**). After activation with MAO both Ti complexes (**4** and **6**) polymerize propylene to high molecular weight syndiotactic polypropylene, very efficiently. Conversely, the corresponding Zr homologues (**5** and **7**) give rise only to low molecular weight oligomers. [11] The polymerization results and some polymer properties of syndiotactic polypropylene produced with **4** and **6** are summarized in Table 3. The degree of stereoregularity of the polymers, produced at different polymerization temperatures, measured as the concentration of rmmr and rmrr pentads concentrations, is given is Table 4. The data presented in Tables 3 and 4 reveal that the stereoselectivity of both catalyst systems is higher than one would expect from catalyst systems with such flexible, low stereorigidity, precursor structures. Under more favorable conditions, in liquid monomer

Table 3. Polymerization results and polymer properties of **4**/MAO catalyst system

Temperature °C	Mw	rrrr %	Regiodefects %	T_m °C
40	>1000,000	81.6	nd	123
60	765,000	74.8	nd	105
80	703,000	69.4	nd	98

and comparatively low temperature (T< 30°C), the catalysts are even more selective. More recent investigations revealed that the polymers of **4** show quasi-perfect stereoregularity.[11b] The remaining few stereo defects are mostly related to mm-type errors arising from occasional reverse monomer enantio face selection. The stereoregularity, however, suffers with decreasing monomer concentration and/or increasing polymerization temperature, mainly as a result of an increased probability of consecutive monomer insertion at the same enantiomorphic coordination position (site epimerization) with the corresponding formation of m-type stereodefects. Since in the chain end control mechanism the formation of m defects is independent of monomer concentration, the monomer concentration dependency of rmrr pentads can be taken as the key clue for the correct interpretation of the polymerization mechanism.[11b,e,f]

Table 4. Stereodefects formed with **4** (left) and **6** (right) at different temperatures.

Temperature °C	rmmr %	rmrr %	rmmr %	rmrr %
40	1.49	3.37	1.55	4.80
60	1.73	6.45	3.03	8.55
80	1.75	8.34	8.00	11.20

The t-butyl substitution in positions 3 and 6 of the fluorenyl, brings about a much more dramatic improvement both with respect to stereo- and enantioselectivity than similar substitution in positions 2 and 7. The enantioselectivity of **4**, measured as the concentration of rmmr pentads, is higher than **6** and is close to the corresponding pentad values of **1** (Cf Table 2). Similarly, the rmmr pentad concentrations in polymers produced with **4** do not vary much with polymerization temperature. However, the overall stereoselectivity of **4** decreases much faster compared to **1** with increasing polymerization temperature due to a much sharper increase in rmrr pentad concentrations. For polymers produced with **6**, the temperature effect, with respect to both enantio- and stereoselectivity, is even more dramatic. At 40 °C the polymers of **6** exhibit almost the same level of rmmr pentads as is found in polymers produced with **4** (1.55 vs. 1.49). However, at 60°C they double in size (3.03 %) and reach the 8% level (1.75% for **4**) at 80°C. On the other hand, the site epimerization related pentads, rmrr, for **6** are about 30% higher than the corresponding rmrr pentad concentrations for polymers produced with **4** at 40 °C. They double in both cases at 60°C and reach the staggering value of over 11 % for **6**.

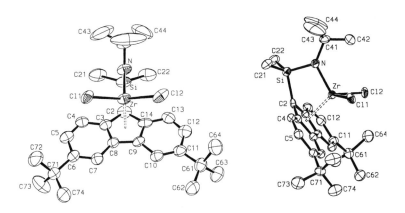

Figure 7. Front and side view perspectives of the X-ray determined structure of the **5** (**4**).[10a]

Figure 7 depicts two different views of the X-ray determined molecular structure of **5**. Similar structure has been recently determined for **4**. It reveals the most important structural features of the metallocene precatalysts, which is, for lower part of the molecules, very close to the structural properties of molecule **3** shown in Figure 6. Therefore similar improved stereoregulating behavior, like in the case of **3**, due to the substituents' steric interaction with the counter-ion(s), is expected and observed. Structures **4** and **6** are the only fully characterized examples of syndiotactic specific structures that have titanium as active transition metal and produce high molecular weight syndiotactic polymer chains. They are also in many other aspects different from the original syndiotactic specific systems. They consist of half sandwich molecules, contain amido type N-Ti (or Zr-N) bonds, and are 12 electron systems (14 electron systems at the most if one considers the participation of the lone pair electrons on the nitrogen to N-Ti bond). This leaves 10 (maximum 12) electrons only for the cationic active species. The most important difference between structure of **1** and **4** is, however, related to their stereo rigidity. Structure **4** (and also **6**), contains only one aromatic penta-hapto bonded ligand and possesses much lower stereo rigidity, yet it produces highly stereoregular syndiotactic polymers.

Concluding Remarks

The original model described in the "results and discussion" section including the chain migratory insertion mechanism explains the formation of syndiotactic polymer chains with a prochiral metallocene structure with bilateral symmetry. According to Scheme 1, the regularly alternating enantiofacial preference for monomer, at two enantiotopic coordination positions of the active site is the main characteristic of syndiospecific polymerization of propylene with C_s symmetric catalyst systems. Despite its simplicity and convenience the model has some handicaps. It is conceived of images coming from X-ray analysis performed on molecules contained in crystalline lattice in solid state and developed by force field calculations. It considers only frozen images of dynamic molecules and reflects only one aspect, the general shape and outline of the rigid molecules. These images reflect only a snapshot of the constantly vibrating, bending, "breathing" and freely floating molecules. The model disregards completely the dynamic nature of the active species and interactions of the counter-ion molecules and involved ion-pair association/dissociation equilibrium. Recent more flexible structures like **4** and **6**, structures with lower symmetry,[13] ^{13}C-NMR / ^{1}H-NMR data of the original syndiotactic specific structures, and their chemistry are all indicative of haptotropic and fluxional nature of the metallocene molecules during polymerization.[12d]

These "details" have often been neglected for the sake of simplicity and to prove their existence and impact has not been easy. However, one should be aware that at least two dynamic phenomena are actively involved, in one way or the other, in different steps of the polymerization and influence its mechanism, in one or the other direction. For example, the phenomenon of hapticity change or variation of bond order, between the transition metal and the aromatic ligands, should be seriously considered as being involved in certain metallocene catalysts as we have demonstrated.[12]

The haptotropy and ring slippage (Figure 8) can influence the instant electronic properties of the active site, the steric environment of its surrounding and impact the molecular weight and tacticity of their polymers. The second phenomenon is related to geometric change of the active species during the coordination and insertion steps. The pseudo tetrahedral geometry, which is assumed for the tetra-coordinated transition metal in the transition state, cannot be further extended to the geometry of the site in the step just after the insertion. At this stage the

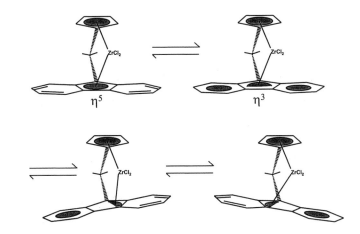

Figure 8. Fluorenyl bond haptotropy and ring slippage.

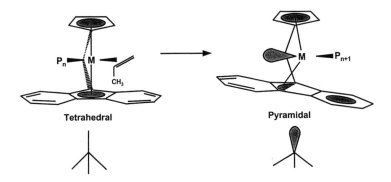

Figure 9. Systematic structural change and geometry variation during the insertion steps.

tetra-coordinated structure collapses, due to the disappearance of a ligands, leaving a tri-coordinated species behind in which the repulsive forces, acting upon the bonding electron pairs, are different and require a new geometry. The most logical structure that can be suggested for this step would be a trigonal pyramid (Figure 9 right). After the next monomer coordination, again the structure will adopt a tetrahedral geometry (Figure 9 left). This change in geometry, operating on all alkyl metallocenium cations, is probably of more importance for the syndiospecific case, where dynamic processes, such as chain migration and site epimerization (vide supra), are vital for its existence. Finally, another ligand/transition metal related dynamic behavior that can be envisaged to be acting on the transition structure is the lateral semi-rotational displacement of the whole ligand system around the transition metal (or vice versa) reported by Petersen[14] is also noteworthy in this context. This movement that can be described as a kind of wind shield wiper type oscillation of the metalacyclobutane moiety within the "fixed" ligand system might have a facilitating effect on the site epimerization and/or chain migration mechanisms. Especially in C_1 symmetric systems (Figure 10).

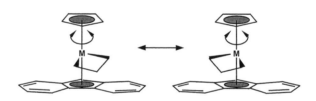

Figure 10. Lateral displacement of ligand system around the centroids-Zr bond axis (the bridge is omitted for the sake of clarity).

[1] [1a] G. Natta, I. Pasquon, P. Corradini, M. Peraldo, M. Pegoraro, A. Zambelli, *Rend Accad. Naz. Lincei,(8)*. **1960**, *28*, 539; [1b] G. Natta, I. Pasquon, P. Corradini, M. Peraldo, M. Pegoraro, A. Zambelli, *Chem. Abstr.* **1960**, 55, 9823i.
[2] [2a] G. Natta, I. Pasquon, A. Zambelli, *J. Am. Chem. Soc.* **1962**, *84*, 1488.
[3] [3a] A. Razavi, J. A. Ewen, *US patent* 6184326; [3b] A. Razavi, J. A. Ewen *US patent* 4892851; [3c] A. A. Razavi, J. A. Ewen, *US patent* 5,334,677; [3d] J. A. Ewen, A. Razavi, *US patent* 5,476,914; [3e] J. A. Ewen, L. R. Jones, A. Razavi, J. J. Ferrara, *J. Am. Chem. Soc.* **1988**, *110*, 6255. [3f] A. Razavi, J. J. Ferrara, *J. Organomet. Chem.* **1992**, *435*, 299; [3g] Razavi, J. L.Atwood, *J. Organomet. Chem.* **1993**, *459*, 117; [3h] A. Razavi, U. Thewalt, *J. Organomet. Chem.*, **1993**, *445*, 111.

[4]   L. Cavallo, G. Guerra, M. Vacatello, P. Corradini, *Macromolecules* **1991**, *24*, 1784.

[5]   [5a] W. E. Piers, J. E. Bercaw, *J. Am. Chem. Soc.* **1990**, *112*, 9406;   [5b] H. H. Brintzinger, H. Krauledat, *Angew. Chem., Int. Ed. Engl.* **1990**, *29*, 1412; [5c] H. H. Brintzinger, M. L. Leclerc, *J. Am. Chem. Soc.* **1995**, *117*, 1651; [5d] B. J. Burger, W. D. Cotter, E. B. Coughlin, S. T. Chascon, S. Hajela, T. A. Herzog, R. O. Koehn, J. P. Mitchell, W. E. Piers, P. J. Shapiro, J. E. Bercaw, J. E. in "Ziegler Catalyst"; G. Fink, R. H. Muelhaupt, H. H. Brintzinger, Eds; Springer verlag; Berlin, 1995. [5e] R. H. Grubbs, G. W. Goates, *Acc. Chem. Res.* **1996**, *29*, 85.

[6]   P. Cossee, E. J. Arlman, *J. Catal.* **1964**, *3*, 89. P. Cossee, The mechanism of Ziegler-Natta polymerization. II quantum chemical and crystal-chemical aspects. In: Ketly A. D, Editor. The stereochemistry of macromolecules, vol.1. New York: Marcel Dekker, 1967. p. 145 –175.

[7]   A. Razavi, L. Peters, L. Nafpliotis, D. Vereecke, K. Den Daw, *Makromol. Symp.* **1995**, *89*, 345.

[8]   A. Razavi, J. L. Atwood, *J. Organomet. Chem.* **1996**, *520*, 115.

[9]   [9a] A. Razavi, D. Vereecke, L. Peters, K. Den Daw, L. Nafpliotis, J. L Atwood, in "Ziegler Catalysts"; Fink, G.; Muelhaupt, R.; Brintzinger, H. H., Eds.; Springer Verlag; Berlin, 1993; [9b] R. Kleinschmidt, M. Reffke, G. Fink, Macromol, *Rapid Commun.* **1999**, *20*, 284; [9c] V. Busico, R. Cipullo, G. Talarico, *Macromolecules* **1997**, *30*, 4787; [9d] A. Razavi, *C.R. Aca. Sci. Paris, Serie IIc, chemie/chemistry* **2000**, *3*, 615 –625.

[10] [10a] A. Razavi, U. Thewalt, *J. Organomet. Chem.* **2001**, *621*, 267.   [10b] A. Razavi, European. Pat 96111127,5 and WO 98/02469.

[11] [11a] A. Razavi, U. Thewalt, *J. Organomet. Chem.* **2001**, 621, 267. [11b] V. Busico, R. Cupillo, F. Cutillo, G. Talarico, A. Razavi, *Macromol. Chem. Phys.* **2003**, 204, 1269; [11c] A. Razavi, V. Bellia, Y. De Brauwer, K. Hortmann, L. Peters, S. Sirole, S. Van Belle, V. Marine, M. Lopez, *J. Organomet. Chem.* **2003** in press; [11d] T. Shiomura, T. Asanuma, N. Inoue, *Macromol. Rapid Commun.* **1996**,17 ,9; [11e] H. Hagihara, T. Shiono, T. Ikeda, *Macromolecules*, **1997**, 30, 4783;   [11f] H. Hagihara, T. Shiono, T. Ikeda, *Macromolecules*, **1998**, 31, 84.

[12] [12a] A. Razavi, V. Bellia, Y. De Brauwer, K. Hortmann, M. Lambrecht, O. Miseque, L. Peters, S. Van Belle, In "Metalorganic catalysts for Synthesis and Polymerization. Ed. W. Kaminsky, Springer, Berlin (1999); [12b] A. Razavi, L. Peters, L. Nafpliotis, *J. Molecular Catalysis*, **1997**, *A. 115*, 129: [12c] H. G. Alt, M. Jung, G. Kehr, *J. Organomet. Chem.* 1998, *562*, 153-181; [12d] D. Drago, P. S. Pergosin, A. Razavi, *Organometallics* **2000**, *19*, 1802-1805; [12e] A. K. Dash, A. Razavi, A. Mortreux, C. Lehmann, J.F. Carpentier, *Organometallics* **2002**, *21*, 3238, [12f] E. Kirlov, L. Touqet, A. Razavi, S. Kahlal, J. Y. Saillard, J. F. Carpentier, *Organometallics* **2003** submitted.

[13] [13a] R. Leino, F. J. Gomez, A. P. Cole, R. M. Waymouth, *Macromolecules* **2001**, *34, 2082*; [13b] F. J. Gomez, R. M. Waymouth, *Macromolecules*, **2002**, 35, 3358.

[14] A. Kabi-satpathy, C. S. Bajgur, K. P. Reddy, J. L. Petersen, *J. Organomet. Chem.* **1989**, *364*, 105.

Zirconocene-Catalysed Propene Polymerisation: Kinetics, Mechanism, and the Role of the Anion

Fuquan Song, Mark D. Hannant, Roderick D. Cannon, Manfred Bochmann*

Wolfson Materials and Catalysis Centre, School of Chemical Sciences, University of East Anglia, Norwich NR4 7TJ, UK
E-mail: m.bochmann@uea.ac.uk

Summary: The olefin polymerisation activity of metallocene catalysts strongly depends on the counteranion provided by the activator system. The relative activities of a number of new diborate anions $[Z(BAr_3)_2]^-$ have been quantified (Z = CN, NH_2, $N(CN)_2$; Ar = C_6F_5 or o-$C_6F_4C_6F_5$). The kinetic parameters for the initiation, propagation and termination steps of propene polymerisations catalysed by $(SBI)ZrCl_2$ have been determined using quenched-flow kinetic and batch techniques [SBI = rac-$Me_2Si(1$-$Ind)_2$]. Comparison of two activator systems, (i) $CPh_3[B(C_6F_5)_4]$ / triisobutylaluminium (TIBA) and (ii) methylaluminoxane (MAO) shows, surprisingly, that the concentration of species actively involved in chain growth at any one time is comparable for both systems, although the MAO-activated catalyst is about 20 times less active than the borate system. It is concluded that the counteranion remains sufficiently strongly bound to the metal centre throughout the chain growth sequence to modulate the energetics of monomer insertion. A model suggesting that the monomer binding follows an associative interchange (I_a) mechanism is proposed.

Keywords: anion effects; kinetics; mechanism of polymerisation; metallocenes; propene polymerisation

Introduction

Although many aspects of metallocene catalysed 1-alkene polymerisations such as the relationship between ligand structure and polymer stereochemistry are well understood and quantifiable,[1-4] the relationship between catalyst structure and one of the most important catalyst properties, its activity, is much less certain. It is commonly accepted that the use of a variety of activator systems, such as methylaluminoxane (MAO), $B(C_6F_5)_3$, or trityl and ammonium salts of non-coordinating borates all lead to similar active species, generally formulated as ion pairs $[L_2M$-$R]^+[X]^-$, where the degree of ion interactions may range from molecular zwitterions as in $L_2ZrMe(\mu$-$Me)B(C_6F_5)_3$ to ion pairs of less distinct nature where the anion is thought to be sufficiently remote from the electron-deficient cationic metal alkyl

© 2004 WILEY-VCH Verlag GmbH & KGaA, Weinheim   DOI: 10.1002/masy.200450917

174

species not to impede the uptake of the 1-alkene monomer.[5,6] This view seems supported by the observation that catalytic activity increases as the nucleophilicity of the anion decreases.[7-9]

Our interest in the problem of quantitatively assessing the factors that control catalyst activity originated in the observation that the activity of the same metallocene precursor under comparable conditions could be increased by a factor of 30 – 40 by changing the activator from MAO to a trityl borate salt. Our initial assumption was that this was primarily the consequence of the differing concentrations of active species in these systems, as the result of a dissociation equilibrium:[5, 7-9]

$$L_2ZrR(X) \xrightleftharpoons{K} [L_2ZrR]^+ + X^- \qquad (1)$$

Results and Discussion

Significant research effort by a number of groups has concentrated on anion engineering, with the aim of maximising the potential of metallocene catalysts.[6,9-15] Some more recent examples of such anions prepared in our laboratory are shown in Chart I.[16-19]

Chart I

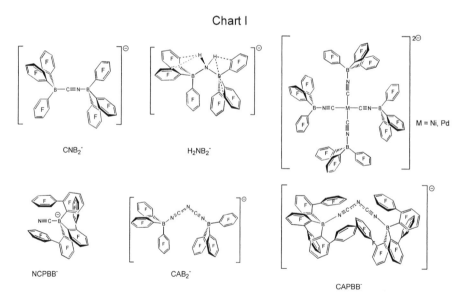

For ethene polymerisations, the activity sequence under normalised conditions (20 °C, 1 bar, toluene) is $CNB_2 \approx CAPBB > H_2NB_2 \approx B(C_6F_5)_4 > CAB_2 \approx NCPBB$. A similar sequence is found for propene: $CNB_2 \approx CAPBB > H_2NB_2 \approx B(C_6F_5)_4 >> CAB_2$. The trityl salts of all these anions, combined with $AlBu^i_3$ as scavenger, give significantly higher propene polymerisation activities than $B(C_6F_5)_3$; for example, while comparative tests with $(SBI)ZrMe_2 / AlBu^i_3 / CPh_3[X]$ had to be terminated after 30 – 60 s to prevent the effects of monomer depletion and system heterogenisation, reactions with $B(C_6F_5)_3$ required reaction times of 10 – 30 min. Typical activity profiles as a function of catalyst concentration are shown in Figure 1. Extrapolation to zero catalyst concentration enables concentration-independent activity data to be determined. From such data, the contribution of the counteranion to the activation barrier of propene polymerisation can be derived (Figure 2).[18]

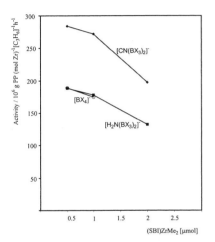

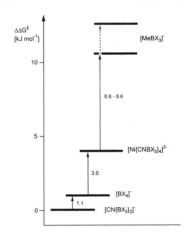

Figure 1. Anion dependence of propene polymerisation activity for selected anions (X = C_6F_5). Conditions: $(SBI)ZrMe_2 / AlBu^i_3$ (100 mmol) / toluene (100 mL), CPh_3:Zr = 1, 1 bar, 20 °C, 30 s.

Figure 2. Contribution of borate anions to the activation barrier for propene polymerisation in batch reactions, relative to CNB_2^- as arbitrary reference point.

While these studies support the assumption that more strongly coordinating anions, such as $[MeB(C_6F_5)_3]^-$, give the least active catalysts, the activities proved *independent* of total anion concentration. For example, the addition of $[PhCH_2NEt_3][B(C_6F_5)_4]$ up to a B/Zr ratio of 20:1 had no effect on the polymerisation rate.

Kinetics provide a more detailed insight into the role of anions during the chain growth process.[20] The (SBI)ZrMe$_2$ / AlBu$^i_3$ / CPh$_3$[CN{B(C$_6$F$_5$)$_3$}$_2$] system (1:100:1) was chosen for quenched-flow kinetic studies at 25.0±0.1 °C to explore the early phase ($t = 0.2 - 5$ s) of the polymerisation reaction.[21] Since at the initial stages the kinetics follows non-steady state conditions, the conversion of the precursor complex to the active species (initiation, k_i) must be taken into account, eq (2):

$$\frac{-\,d[M]}{dt} = k_i[C_0][M] + k_p[C][M] \qquad (2)$$

where [C$_0$] = initial concentration of the precursor and [C] = catalyst concentration. The polymer mass Y collected at flow rate F over time t_c is given by eq (3) (m = molar mass of monomer, [M]$_0$ = initial propene concentration = 0.59 mol L^{-1}).

$$Y = F\,m\,t_c\,[C]_0\,[(1 - e^{-k_i[M]_0 t}) + k_p[M]_0\,(t + \frac{1}{k_i[M]_0}\,e^{-k_i[M]_0 t} - \frac{1}{k_i[M]_0})] \qquad (3)$$

Figure 3 shows the time dependence of polymer yield for several zirconocene concentrations.

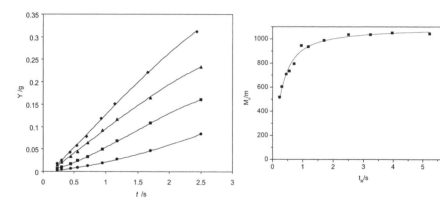

Figure 3. Time dependence of PP yield for the borate system. [Zr] = 2.38 × 10^{-5} (●); 4.76 × 10^{-5} (■); 7.14 × 10^{-5} (▲) and 9.52 × 10^{-5} mol L^{-1} (◆). The positive curvature at higher [Zr] and longer t indicates the effects of monomer depletion. [M]$_0$ = 0.59 mol L^{-1}, toluene, 25.0±01 °C.

Figure 4. Time dependence of $\bar{M}_n$; (SBI)ZrMe$_2$/AlBu$^i_3$/CPh$_3$[CN{B(C$_6$F$_5$)$_3$}$_2$, toluene, 25.0±01 °C.

In their pioneering kinetic studies on the Cp$_2$TiCl$_2$/AlEt$_2$Cl ethene polymerisation system, Fink et al. have suggested an "intermittent growth" model the polymerisation process, where the active species are in equilibrium with (observable) resting (dormant) states.[22] While over the duration of a polymerisation reaction all metal centres can become involved in the production of polymer chains (save those deactivated by impurities),[23] the percentage of total metallocene actively involved in polymer chain growth at any one time is not known for most systems but may vary widely. It was of obvious interest to quantify the distribution between active and dormant states, by measuring the time evolution of the number-average molecular weight $\bar{M}_n$ (Figure 4).

Very similar behaviour was found for the system (SBI)ZrCl$_2$ / MAO (Al:Zr = 2500:1), although the lower activity made it necessary to conduct the polymerisations at 40 °C for longer reaction times ([M] = 0.42 mol L^{-1}) (Figures 5 and 6).

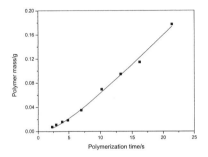 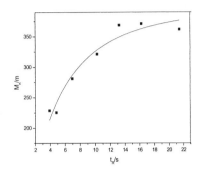

Figure 5. Time dependence of PP yield for the (SBI)ZrCl$_2$ / MAO system.

Figure 6. Time dependence of $\bar{M}_n$ for the (SBI)ZrCl$_2$ / MAO system.

We interpret these observations assuming an intermittent growth model where the precursor reacts with the monomer to give an active species at rate k_i, followed by chain propagation with rate constant k_p and involving active Zr species carrying polymeryl chains of varying chain length, P$_z$C*. The k_p values for different P$_z$C* are assumed to be identical. Chain termination (k_t) may take place from any of the species P$_z$C* (z = 1, 2, 3…), with $k_t \ll k_p$ (Scheme 1). The parameters k_i and k_p were estimated by curve fitting. Tests show that the data for k_p are rather insensitive to the value of k_i.

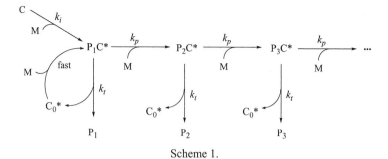

Scheme 1.

The number-average degree of polymerisation $<n>$ is defined by eq (4).

$$<n> = \frac{\overline{M}_n}{m_M} = \frac{\Sigma z P_z}{\Sigma P_z} \qquad (4)$$

It can be shown that the time dependence of $\overline{M}_n$ is given by eq (5).

$$\langle n \rangle = \frac{\dfrac{(k_p[M]_0 + k_t)t}{k_t} + \dfrac{k_p[M]_0 + k_t - k_i[M]_0}{k_i[M]_0(k_t - k_i[M]_0)}(e^{-k_i[M]_0 t} - 1) - \dfrac{k_p[M]_0 k_i[M]_0}{k_t^2(k_t - k_i[M]_0)}(e^{-k_t t} - 1)}{t + \dfrac{1}{k_i[M]_0}e^{-k_i[M]_0 t} - \dfrac{1}{k_i[M]_0}} \qquad (5)$$

Assuming that $k_i \ll k_p$ and $k_t \ll k_i[M]_0$, this equation simplifies to eq (6) which was used for modelling $\overline{M}_n$.

$$\langle n \rangle = \frac{k_p[M]_0}{k_t} \bullet \frac{t + \dfrac{k_t}{k_i[M]_0(k_t - k_i[M]_0)}(e^{-k_i[M]_0 t} - 1) - \dfrac{k_i[M]_0}{k_t(k_t - k_i[M]_0)}(e^{-k_t t} - 1)}{t + \dfrac{1}{k_i[M]_0}(e^{-k_i[M]_0 t} - 1)} \qquad (6)$$

Applying steady-state conditions with respect to initiation and termination, i.e. $\exp(-k_i[M]_0 t) \rightarrow 0$ and $\exp(-k_t t) \rightarrow 0$, eq (6) simplifies to eq (7) which has been ascribed to Natta and Pasquon[24] and has frequently been used to describe $<n>$ but is not applicable under the present reaction conditions.

$$\langle n \rangle = \frac{k_p[M]_0 t}{1 + k_t t} \qquad (7)$$

Curve fitting using eq (6) gives another set of data for k_p and k_i. Since the data in Figure 4 are derived from fewer measurements, the value of k_i is less certain but very comparable to that determined from $Y(t)$. On the other hand, the k_p value is an order of magnitude larger. To differentiate, we will refer to the propagation rate constant derived from polymer mass measurements as the *apparent* propagation rate k_p^{app}, since it assumes that 100% of the initially added zirconocene did indeed become catalytically active, while k_p will denote the rate constant derived from $\overline{M}_n$, eq (6). Pertinent kinetic parameters for the borate and the MAO-activated systems are given in Table 1.

The ratio k_p^{app}/k_p should then be a measure of the proportion of total zirconocene that was involved in chain growth at any one time. Surprisingly, this value was found to be almost identical in both the borate and the MAO-activated systems, ca. 0.08; i.e. under the given conditions less than 10% of the total zirconocene concentration may be regarded as "active species" P_zC^* in the sense of Scheme 1.

Table 1. Comparison of kinetic parameters for the (SBI)ZrMe$_2$ / borate and the (SBI)ZrCl$_2$ / MAO systems.

	L_2ZrR_2 / CPh$_3^+$		L_2ZrCl_2 / MAO	
	Yield vs Time	M_n *vs Time*	*Yield vs Time*	M_n *vs Time*
k_p^{app}, L mol^{-1}s^{-1}	1320 ± 20		48 ± 3	
k_p, L mol^{-1}s^{-1}		$17,200 \pm 1400$		600 ± 230
k_i, L mol^{-1}s^{-1}	5.2 ± 0.6	2.5 ± 3.2	0.25 ± 0.05	0.8 ± 2.6
k_t, s^{-1}		9.4 ± 0.7		0.6 ± 0.2
k_p^{app}/k_p	0.08		0.08	

Current thinking about the nature of active and dormant species is depicted in Scheme 2. Alkylation of the zirconocene precursor with AlBu$^i_3$ followed by activation with a trityl borate is thought to lead primarily to a zirconocene alkyl tight ion pair, a resting state **D1** that is observable and stable for CH$_2$R' = CH$_3$.[18] Removal of the anion from the equilibrium position provides entry into the catalytic cycle, consisting of an assembly of active species involved in chain growth, as depicted in box **A** of Scheme 2. Other dormant states are possible, such as that derived from 2,1-misinsertion of propene, to give a *sec*-alkyl (**D2**).[2] Further possibilities include the formation of a Zr-CMe$_2$P *tert*-alkyl formed by chain epimerisation.[25] Both *sec*- and *tert*-alkyls will be slow to insert further monomer.

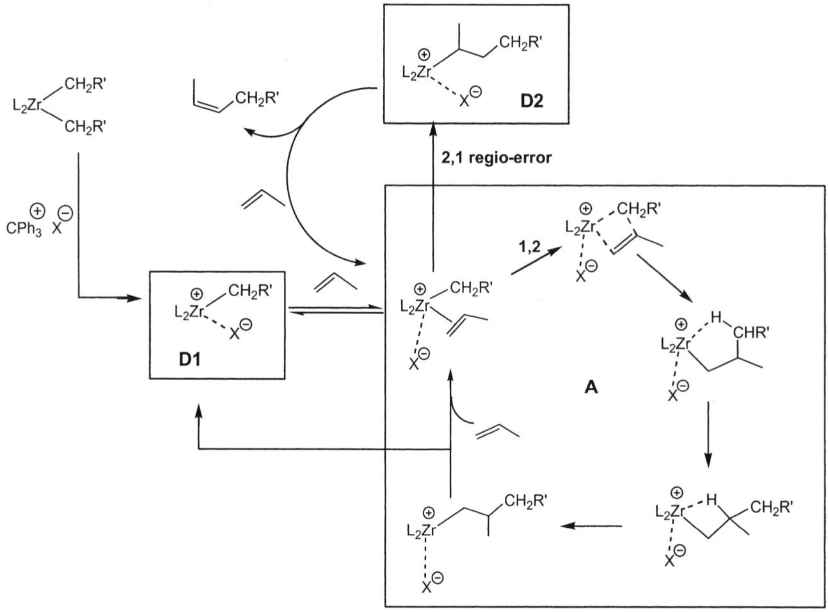

Scheme 2. Active and dormant states in zirconocene-catalysed alkene polymerisations.

Since the kinetic results suggest similar concentrations of active species for the borate and the MAO system, the differences in activity must relate to the rate with which the monomer insertion cycle is completed. In a low-dielectric medium such as toluene it is only to be expected that the anion remains associated with the metallocenium cation throughout the cycle, i.e. there is no dissociation equilibrium of the type shown in eq (1). This assumption is further supported by the absence of an anion concentration effect, which indicates that in a low-polarity hydrocarbon medium ions do not diffuse freely. The associated anion will influence the lifetime of one or more of the intermediates.

Two possible scenarios of monomer insertion are represented in Scheme 3. Path A involves anion dissociation, followed by solvent coordination and substitution by monomer. Path B involves the direct substitution of the anion by monomer, either by approach from the same side as the anion to lead, effectively, to an insertion of the monomer between the metal and the anion,[26] or by approach from the opposite side to the anion, which, in view of the considerable size of tetraarylborate anions, seems more likely. Path A may play a role in half-sandwich complexes where toluene coordination has been shown to be preferred to anion coordination, at least in the case of zirconium and hafnium.[27] For bis-Cp complexes of the type investigated here, however, the kinetic data rule out path A and favour path B.

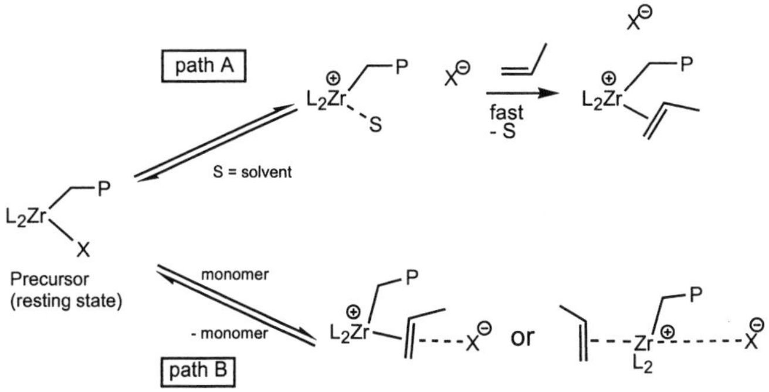

Scheme 3.

Brintzinger et al. have argued that, at least in the case of the relatively ionic zirconocenium [B(C$_6$F$_5$)$_4$]$^-$ salts, ion quadruples or higher aggregates may be present in hydrocarbon solutions.[28] Anion exchange rates determined by NMR methods for the borate system are of the same order of magnitude as the monomer insertion rates found here from the time dependence of $\bar{M}_n$. It is tempting to speculate, therefore, that the two processes may be related: in such an ion quadruple, the energy required to weaken the cation-anion interaction in the course of monomer coordination is balanced by the energy gained as a second anion approaches (Scheme 4). In such a case, monomer binding and anion displacement follows the path of an S$_N$2 reaction; it would be immediately followed by anion re-coordination. These mechanistic features are in fact those typical of an associative interchange, I_a, mechanism.

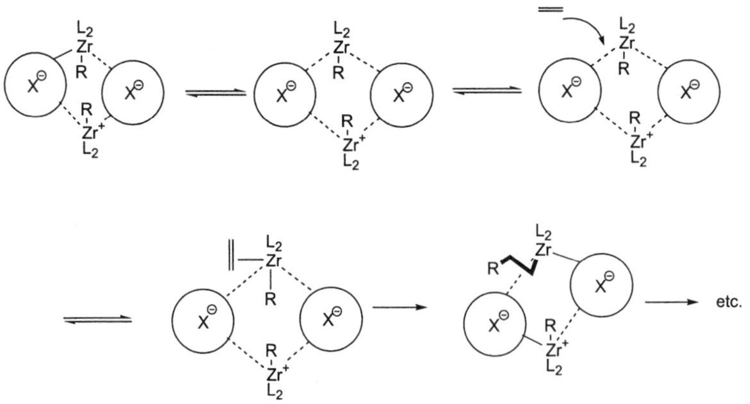

Scheme 4. Hypothetical monomer insertion mechanism coupled with anion exchange in ion quadruples.

Information about the nature of the main resting states is available by polymer end group analysis. Three different types of unsaturated polymers are found in this catalyst system, vinylidene, *cis*-butenyl, and internal C=C (Scheme 5).[29] The latter arise via C-H activation and formation of an η^3-allyl intermediate; in spite of the complex mechanism of their formation they are already apparent at reaction times of <1 s.

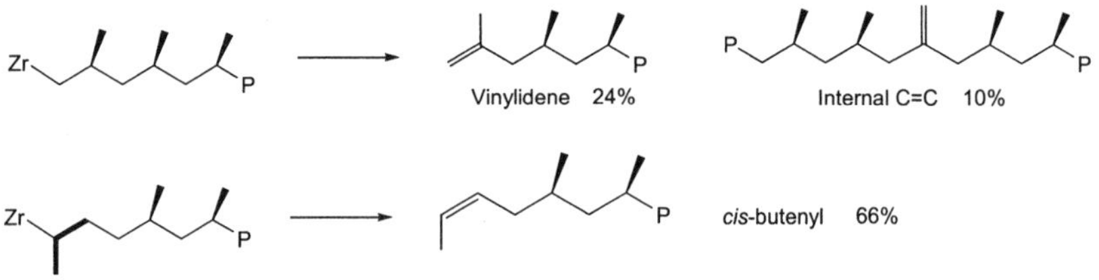

Vinylidene   24%   Internal C=C   10%

cis-butenyl   66%

Scheme 5. PP end groups.

The main type of unsaturated end group is *cis*-butenyl, formed by β-H elimination from *sec*-alkyl Zr species following 2,1-propene misinsertion. The polymer also contains about 0.2 mol-% enchained 2,1-misinsertions. Since 1,2-insertion following a 2,1 stereoerror is slow compared with the main propagation step, 2,1-misinsertions tie up a significant percentage of total zirconocene as dormant species, while the fact that 2,1-enchainments are present in the main polymer chain argues that termination following misinsertion is also comparatively slow. The mechanism shown in Scheme 1 should therefore be elaborated to take account of these findings, Scheme 6, where k_{21} and k_p' indicate, respectively, a 2,1-insertion into a Zr-*prim*-alkyl and 1,2-insertion into a Zr-*sec*-alkyl bond.

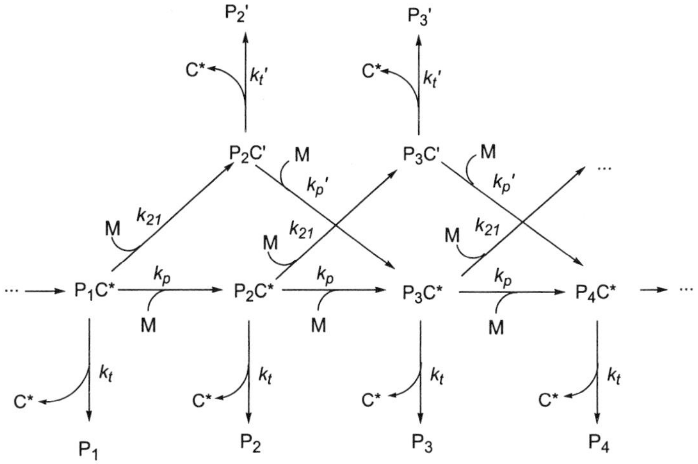

Scheme 6.

If one neglects differences in polymeryl chain lengths and considers all P_zC^* as one species, and all P_zC' as another, one can formulate a hypothetical equilibrium between a Zr-*sec*-alkyl and its product following propene 1,2-insertion: $K = \Sigma[P_zC']/\Sigma[P_zC^*] = k_{21}/k_p'$ (Scheme 7). The value for K is estimated to be about 12, which may be regarded as a quantitative measure for the steric repulsion of two adjacent methyl groups in the PP chain and the relative stabilities of primary and secondary alkyls.

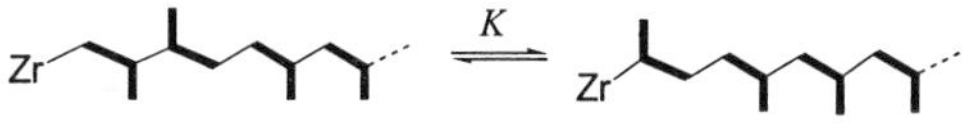

Scheme 7.

Conclusion

Quenched-flow kinetic techniques have allowed the quantification of the concentration of active species in metallocene-catalysed propene polymerisations. The rate constants for catalyst initiation, chain propagation and chain termination have been estimated. Surprisingly,

184

this value is not strongly dependent on the nature of the catalyst activator, at least for the catalyst system described here. Catalyst productivity, as a function of the time taken for the completion of a monomer insertion cycle, is strongly influenced by the counteranion. The data suggest that the anion remains closely associated during the transition state and influences the residence times of one or more catalytic intermediates in the assembly of active species that constitute the catalytic cycle. The non-active proportion of total [Zr] consists mainly of species produced by propene 2,1-misinsertions, with the rate of 2,1-misinsertion being about 500 times slower than that of 1,2-insertion. The overall polymerisation mechanism is consistent with an intermittent chain growth model with an equilibrium between dormant and active states; it is proposed that the anion re-coordinates after each monomer insertion step, in a manner reminiscent of an associative interchange mechanism.

Acknowledgement

This work was supported by the UK Engineering and Physical Sciences Research Council.

[1] Brintzinger, H.H.; Fischer, D.; Mülhaupt, R.; Rieger, B.; Waymouth, R. *Angew. Chem. Int. Ed. Engl.* **1995**, *34*, 1143.
[2] Resconi, L.; Cavallo, L.; Fait, A.; Piemontesi, F. *Chem. Rev.* **2000**, *100*, 1253.
[3] Coates, G. W. *Chem. Rev.* **2000**, *100*, 1253.
[4] Busico, V.; Cipullo, R. *Progr. Polym. Sci.* **2001**, *26*, 443.
[5] Bochmann, M. *J. Chem. Soc., Dalton Trans.* **1996**, 255.
[6] Chen, E. Y. X.; Marks, T. J. *Chem. Rev.* **2000**, *100*, 1391.
[7] Bochmann, M.; Lancaster, S. J. *Organometallics* **1993**, *12*, 633.
[8] Bochmann, M.; Wilson, L. M. *J. Chem. Soc., Chem. Commun.* **1986**, 1610. (b) Bochmann, M. *Angew. Chem. Int. Ed. Engl.* **1992**, *31*, 1181. (c) Bochmann, M.; Lancaster, S. J. *J. Organomet. Chem.* **1992**, *434*, C1.
[9] (a) Jordan, R. F. *Adv. Organomet. Chem.* **1991**, *32*, 325. (b) Yang, X.; Stern, C. L.; Marks, T. J. *Organometallics* **1991**, *10*, 840. (c) Metz, M. V.; Schwartz, D. J.; Stern, C. L.; Nickias, P. N.; Marks, T. J. *Angew. Chem. Int. Ed.* **2000**, *39*, 1312. (d) Metz, M. V.; Sun, Y.; Stern, C. L.; Marks, T. J. *Organometallics* **2002**, *21*, 3691. (e) Chen, M. C.; Marks, T. J. *J. Am. Chem. Soc.* **2001**, *123*, 11803. (f) Abramo, G. P.; Li, L.; Marks, T. J. *J. Am. Chem. Soc.* **2002**, *124*, 13966.
[10] (a) Strauss, S.H. *Chem. Rev.* **1993**, *93*, 927. (b) Lupinetti, A.J.; Strauss, S.H. *Chemtracts – Inorg. Chem.* **1998**, *11*, 565. (c) Ivanov, S. V.; Miller, S. M.; Anderson, O. P; Solntsev, K. A.; Strauss, S. H. *J. Am. Chem. Soc.* **2003**, *125*, 4694. (d) King, B. T.; Michl, J. *J. Am. Chem. Soc.* **2000**, *122*, 10255. (e) Tsang, C. W.; Yang, Q.; Sze, E. T.; Mak, T. C. W.; Chan, D. T.; Xie, Z. *Inorg. Chem.* **2000**, *39*, 5851.
[11] Reed, C. A. *Acc. Chem. Res.* **1998**, *31*, 133. (b) Stasko, D.; Reed, C. A. *J. Am. Chem. Soc.* **2002**, *124*, 1148.
[12] Rodriguez, G.; Brant, P. *Organometallics* **2001**, *20*, 2417.
[13] Kaul, F. A. R.; Puchta, G. T.; Schneider, H.; Grosche, M.; Mihalios, D.; Herrmann, W. A. *J. Organomet. Chem.* **2001**, *621*, 177. (b) Kaul, F. A. R.; Puchta, G. T.; Schneider, H.; Grosche, M.; Mihalios, D.; Herrmann, W. A. *J. Organomet. Chem.* **2001**, *621*, 184.
[14] (a) Williams, V. C.; Piers, W. E.; Clegg, W.; Elsegood, M. R. J.; Collins, S.; Marder, T. B. *J. Am. Chem. Soc.* **1999**, *121*, 3244. (b) Williams, V. C.; Irvine, G. J.; Piers, W. E.; Li, Z.; Collins, S.; Clegg, W.; Elsegood, M. R. J.; Marder, T. B. *Organometallics* **2000**, *19*, 1619. (c) Chase, P. A.; Piers, W. E.; Patrick, B. O. *J. Am. Chem. Soc.* **2000**, *122*, 12911. (d) Henderson, L. D.; Piers, W. E.; Irvine, G. J.; McDonald, R. *Organometallics*

2002, *21*, 340. (e) Roesler, R.; Has, B. J. N.; Piers, W. E. *Organometallics* **2002**, *21*, 4300.

[15] (a) LaPointe, R. E. WO 99/42467, 1999. (b) LaPointe, R. E.; Roof, G. R.; Abboud, K. A.; Klosin, J. *J. Am. Chem. Soc.* **2000**, *122*, 9560.

[16] Lancaster, S. J.; Walker, D. A.; Thornton-Pett, M.; Bochmann, M. *Chem. Commun.* **1999**, 1533.

[17] Lancaster, S. J.; Rodriguez, A.; Lara-Sanchez, A.; Hannant, M. D.; Walker, D. A.; Hughes, D. L.; Bochmann, M. *Organometallics* **2002**, *21*, 451.

[18] Zhou, J.; Lancaster, S. J.; Walker, D. A.; Beck, S.; Thornton-Pett, M.; Bochmann, M. *J. Am. Chem. Soc.* **2001**, *123*, 223.

[19] Hannant, M.; Bochmann, M., results to be published.

[20] See for example: (a) Busico, V.; Cipullo, R.; Esposito, V. *Macromol. Rapid Commun.* **1999**, *20*, 116. (b) Mori, H.; Tashimo, K.; Terano, M. *Macromol. Rapid Commun.* **1995**, *16*, 651. (c) Mori, H.; Tashino, K.; Terano, M. *Macromol. Chem. Phys.* **1996**, *197*, 895. (d) Mori, H.; Iguchi, H.; Hasebe, K.; Terano, M. *Macromol. Chem. Phys.* **1997**, *198*, 1249.

[21] Song, F.; Cannon, R. D.; Bochmann, M. *J. Am. Chem. Soc.* **2003**, *125*, 7641.

[22] (a) Schnell, D.; Fink, G. *Angew. Makromol. Chem.* **1974**, *39*, 131. (b) Fink, G.; Zoller, W. *Makromol. Chem.* **1981**, *182*, 3265. (c) Fink, G.; Schnell, D. *Angew. Makromol. Chem.* **1982**, *105*, 31. (d) Mynott, R.; Fink, G.; Fenzl, W. *Angew. Makromol. Chem.* **1987**, *154*, 1. (e) Fink, G.; Fenzl, W.; Mynott, R. *Z. Naturforsch. Teil B*, **1985**, *40b*, 158.

[23] (a) Liu, Z.; Somsook, E.; Landis, C. R. *J. Am. Chem. Soc.* **2001**, *123*, 2915. (b) Liu, Z.; Somsook, E.; White, C. B.; Rosaaen, K. A.; Landis, C. R. *J. Am. Chem. Soc.* **2001**, *123*, 11193.

[24] Natta, G.; Pasquon, I. *Adv. Catal.* **1959**, *11*, 1.

[25] Yoder, J. C.; Bercaw, J. E. *J. Am. Chem. Soc.* **2002**, *124*, 2548.

[26] (a) Lanza, G.; Fragalà, I. L. *Topics Catal.* **1999**, *7*, 45. (b) Chan, M. S. W.; Vanka, K.; Pye, C. C.; Ziegler, T. *Organometallics* **1999**, *18*, 4624. (c) Vanka, V.; Chan, M. S. W.; Pye, C. C.; Ziegler, T. *Organometallics* **2000**, *19*, 1841.

[27] (a) Beck, S.; Geyer, A.; Brintzinger, H. H. *Chem. Commun.* **1999**, 2477. (b) Beck, S.; Lieber, S.; Schaper, F.; Geyer, A.; Brintzinger, H. H. *J. Am. Chem. Soc.* **2001**, *123*, 1483.

[28] (a) Bochmann, M.; Robinson, O. B.; Lancaster, S. J.; Hursthouse, M. B.; Coles, S. J. *Organometallics*, **1995**, *14*, 2456. (b) Gillis, D. J.; Quyoum, R.; Tudoret, M. J.; Wang, Q. Y.; Jeremic, D.; Roszak, A. W.; Baird, M. C. *Organometallics*, **1996**, *15*, 3600. (c) Gillis, D. J.; Tudoret, M. J.; Baird, M. C. *J. Am. Chem. Soc.* **1993**, *115*, 2543.

[29] (a) Resconi, R.; Piemontesi, F.; Camurati, I.; Sudmeijer, O.; Nifant'ev, I. E.; Ivchenko, P. V. and Kuz'mina, L. G. *J. Am. Chem. Soc.* **1998**, *120*, 2308. (b) Moscardi, G.; Resconi, L.; Cavallo, L. *Organometallics* **2001**, *20*, 1918.

Highly Active and Stereoselective Olefin Polymerization Catalysts Generated by the Transfer-Epimetallation of Olefins or Acetylenes with Dialkyltitanium(IV) Complexes: Three-Membered Metallocycles as Active Catalyst Sites[‡]

John J. Eisch,* John N. Gitua, Peter O. Otieno, Adetenu A. Adeosun

Department of Chemistry, P.O. Box 6000, The State University of New York at Binghamton, Binghamton, New York 13902-6000, USA

E-mail: jjeisch@binghamton.edu; Fax: 607-777-4865

Summary: Efficient transfer-epimetallations of simple olefins and acetylenes by R_2TiL_2 reagents ($R = Bu^n$, Bu^t; $L = X$) are readily achieved in THF at $-78°C$ to generate titanacyclopropa(e)ne intermediates, readily capable of inserting various unsaturated addends (olefin, acetylene, nitrile). Analogous epimetallations conducted in hydrocarbons lead to the isotactic stereoselective polymerization of 1-alkenes and the cyclotrimerization of acetylenes. In place of the widely accepted Arlman-Cossee model for the active catalytic site, namely a Ti–C bond on the $TiCl_3$ crystal lattice, the 2-substituted-1-halotitanacyclopropyl cation formed in hydrocarbon media is proposed as the active site for stereoselective olefin polymerization.

Keywords: organotitanium catalytic sites; polyethylene; polyolefins; stereospecific polymers; Ziegler-Natta polymerization

Introduction

The essentially linear polymerization of ethylene achieved at low pressures by Karl Ziegler and coworkers in late 1953 and the stereoselective polymerization of *alpha*-olefins discovered by Giulio Natta and coworkers in early 1954 have transformed this period into an *annus mirabilis* of modern chemistry and industrial technology. These initial so-called Ziegler-Natta polymerization catalysts consisted of heterogeneous combinations of transition metal salts as the procatalyst with a main-group metal alkyl as the cocatalyst. In later modifications of titanium-based catalysts, titanium(III) chloride phases were produced on solid supports, such as $MgCl_2$, and the titanium(III) center then supposedly alkylated with R_nAlCl_{3-n} to produce the putative active chiral site for isotactic polymerization. More refined catalyst modifications thereafter

---

[‡] Part 28 of the series, Organic Chemistry of Subvalent Transition Metal Complexes; Part 27: J.J. Eisch, A.A. Adeosun, J.N. Gitua, *Eur. J. Org. Chem.* **2003**, in press.

employed a gamut of electron donors, which did in various degrees enhance the isotactic selectivity in *alpha*-olefin polymerization and whose action was explained in terms of selective coordination to and hence poisoning of the atactic sites by such donors.[1] Various theoretical models have been proposed for the nature of such chiral sites on the titanium(III) chloride crystal lattice. One of the most widely considered models is that proposed by Arlman and Cossee[2-5] and modified by subsequent workers, such as the variant put forward by Allegra.[6] A sobering, possibly prescient caveat about all such active-site models has been expressed by Pino and Mülhaupt in 1983: "The *weak point* of this approach (*i.e.*, theoretical calculations of models of the catalytic center) is the *choice of the model* on which calculations are made. As our knowledge of the structure of the catalytic center is *very scant*,... it only *seldom* happens that the calculations... *lead to a better understanding...*"[7] (italicized emphasis added).

That the occurrence of either ethylene polymerization or stereoselective olefin polymerization does not require a heterogeneous catalyst became evident from studies of metallocenes as procatalysts, which at least initially form hydrocarbon-soluble systems with aluminum alkyl cocatalysts. Already in 1957, two different groups, those of Natta[8] and of Breslow,[9] showed that combinations of titanocene dichloride with an aluminum alkyl can polymerize ethylene in solution into the largely linear high-density form. In 1984 Ewen demonstrated for the first time that combinations of *rac*-ethylene-bis(indenyl)TiCl$_2$ with MAO polymerize propylene in homogeneous solution to an isotactic polypropylene with an unusual stereoblock structure[10] and the very next year Kaminsky and Brintzinger were able to produce highly isotactic polypropylene by employing *rac*-ethylene-bis(tetrahydroindenyl)ZrCl$_2$ with MAO.[11] The final barrier to achieving stereoselective polymerization in solution fell in 1988 when Ewen reported that combinations of isopropylidene(cyclopentadienyl)(9-fluorenyl)ZrCl$_2$ with MAO cause the highly syndiospecific polymerization of propylene.[12]

Thus chemists now recognize that local molecular chirality at an active site in fluid solution can suffice to induce stereoselectivity in olefin polymerization. This insight removes the necessity of postulating a chiral active site on the crystal lattice of titanium(III) chloride, as in the Arlman-Cossee model for the isotactic polymerization of *alpha*-olefins with heterogeneous Ziegler-Natta catalysts. Although the Arlman-Cossee model with its TiCl$_3$ crystal lattice may remain a *possible mode* for stereoselective olefin polymerization, its solid-state features are no longer a *structural necessity*.

For Ziegler-Natta olefin polymerization catalysts not involving soluble metallocene- or nonmetallocene-ligated transition metal complexes, it has been difficult until now to envision models of the catalytic center that were other than those based on the face or edge of a transition metal salt crystal lattice. Quite recently, in the course of studies of the reactions of dialkyltitanium(IV) salts with olefins,[13] we have made the serendipitous discovery that the first-formed products of such reactions in THF are titanacyclopropanes (**1**, such as **12** and **27** in Schemes 2 and 4) and that when such intermediates as **1** are generated in heterogeneous hydrocarbon media, isotactic polymerization of the *alpha*-olefins present ensues. These findings persuade us to propose **1** as a chiral molecular active catalytic site, in place of the Arlman-Cossee site on the $TiCl_3$ crystal lattice.

Results

In studies extending over the last decade our group has been exploring the organic chemistry of early transition metal alkyls and subvalent metal salts.[14] From such research our particular attention has been drawn to the process of epimetallation, by which the oxidative addition of subvalent transition-metal reagents ($M_t^m L_n$, **2**) to various C=C, C≡C, C=O or C=N linkages (as in **3**) leads to the formation of three-membered metallocycles (**4**) (Eq. 1).[15] The bonding character

$$M_t^m L_n \ + \ -C{\equiv}E- \ \longrightarrow \ \overset{\diagdown}{C}\!{=}\!\overset{\diagup}{E}\underset{M_t^{m+2}L_n}{\diagdown\!\!\diagup} \ \longleftrightarrow \ \overset{\diagdown}{C}{\equiv}\overset{\diagup}{E}\underset{M_t^m L_n}{\big|} \ \ (1)$$

$$\underset{\textbf{2}}{} \qquad \underset{\textbf{3}:E=C,O,N}{} \qquad \underset{\textbf{4}}{} \qquad \underset{\textbf{5}}{}$$

of the metallocyclic adduct can range from a metallacycloprop(en)yl ring (**4**), having two *sigma*-like C–M_t bonds and a higher oxidation number for the metal center, M_t^{m+2}, to a *pi*-complex (**5**) having little change in the M_t^m oxidation number. Assigning the relative importance of resonance structures **4** and **5** for an epimetallated product requires careful evaluation of the structural parameters of the individual adduct if isolated or of its observed chemical reactions.[15-17] Recently, we have discovered that such epimetallations can be achieved more rapidly, cleanly and in higher yields by means of dialkyltitanium(IV) complexes (**6**) by a process of coordination-induced reductive elimination leading to titanium(II) carbenoid transfer (Eq. 2).[13] The adducts of such transfer-epimetallations (**7**) reflect their titanacyclopropa(e)ne-ring character by undergoing the insertion of various addends A=B (CO_2, R–C≡C–H, R–C≡N,

$R_2C=O$) and thereby expanding to five-membered titanocycles **8**. In donor solvents like THF such insertions are generally limited to single units of A=B.

$$R_2TiL_2 \; + \; -C\!\equiv\!\equiv\!E- \; \xrightarrow{-2\,R^{\bullet}} \; \overset{\displaystyle }{\underset{\underset{\textstyle 7}{\overset{\displaystyle L_2}{Ti}}}{\diagdown C\!\equiv\!\equiv\!E\diagup}} \; \xrightarrow{A=B} \; \overset{\diagdown C\!-\!A\diagup}{\underset{\underset{\textstyle 8}{\overset{\displaystyle L_2}{Ti}}}{\diagup E\diagdown B}} \qquad (2)$$

$$\underset{6}{} \qquad \underset{3}{}$$

$R = Bu^n, Bu^t, Pr^i$, etc.

$L = X, OR$

In an observation most pertinent to our quest for a defined chiral molecular catalytic site in Ziegler-Natta polymerizations, we now wish to report that such transfer-epimetallations by R_2TiL_2 reagents are readily achieved with simple olefins and acetylenes as well. More significantly, however, the resulting three-membered titanocycles show a dramatic difference in reactivity as a function of solvent. In donor solvents, for example as in THF, such adducts as **7** undergo a slow insertion of one or two olefin or acetylene units, as in Eq. 2, leading to the dimerization or trimerization of the monomer. In hydrocarbon medium, on the other hand, R_2TiL_2 (especially $L = X$) causes the immediate polymerization of ethylene and of *alpha*-olefins, as well as the cyclotrimerization of monosubstituted acetylenes, even at temperatures as low as $-78°C$. Moreover, the olefins, such as propylene, 1-hexene and styrene, were found to have undergone principally isotactic stereoselective polymerization in such hydrocarbons.

That the three-membered titanocycles (**12** or **13**) are formed in a donor solvent like THF at 25°C from the *transfer*-epimetallating action of $Bu^t_2TiCl_2$ (**9**) or $Bu^n_2TiCl_2$ with ethylene (**10**) or with diphenylacetylene (**11**), respectively, has been demonstrated by the chemical-trapping reactions depicted in Scheme 1. Titanacyclopropane **12** inserted benzonitrile to yield upon deuteriolysis β-deuteriopropiophenone (**15**), and titanacyclopropene **13** underwent deuteriolysis with acetic acid-d to produce *cis*-stilbene-$d_{1,2}$ (**16**). That both the three-membered titanocycles (**12** and **13**) underwent some insertion of a second monomer unit to form **17** and **19**, respectively, was revealed by the chemical derivatization with benzonitrile leading as well to ε-deuteriovalerophenone (**17** → **18**) and the isolation of some hexaphenylbenzene from the deuteriolytic workup (**13** → **19** → **20**). When, on the other hand, the interaction of $Bu^t_2TiCl_2$ (**9**)

Scheme 1

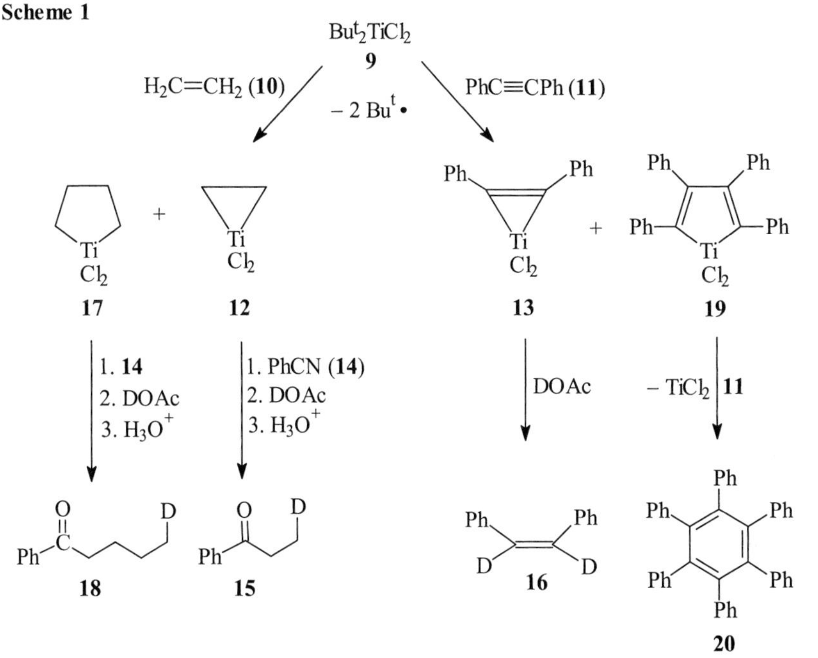

with ethylene (**10**) or with diphenylacetylene (**11**) is conducted starting at −78°C in either hexane or toluene, the ethylene (**10**) is immediately and exothermically polymerized to linear, high-density polyethylene (**21**) and the diphenylacetylene (**11**) undergoes rapid cyclotrimerization to hexaphenylbenzene (**20**) (Scheme 2). The absence of polymerization with ethylene (**10**) and the slow cyclotrimerization of diphenylacetylene (**11**) by Bu$^t_2$TiCl$_2$ in *tetrahydrofuran solution* may readily be attributed to the THF-solvation of the three-membered titanocycle active site (*e.g.*, **22** from **12**), which would reduce the electrophilicity of the titanium(IV) initiator and thus the ease of its attack upon the *pi*-electron system of **10** or **11**.[18] Similarly, the failure of Bu$^t_2$Ti(OPr^i)$_2$ (**23**) to initiate the polymerization of ethylene or the cyclotrimerization of diphenylacetylene at −78°C, even when generated in hexane or toluene (Scheme 3), most likely is due to the coordination complex **25**, which is formed from **23** and the hydrocarbon-soluble lithium salt **24**.[19] Such coordination would reduce the electrophilic character of the titanium(IV) center in **23**.

Scheme 2

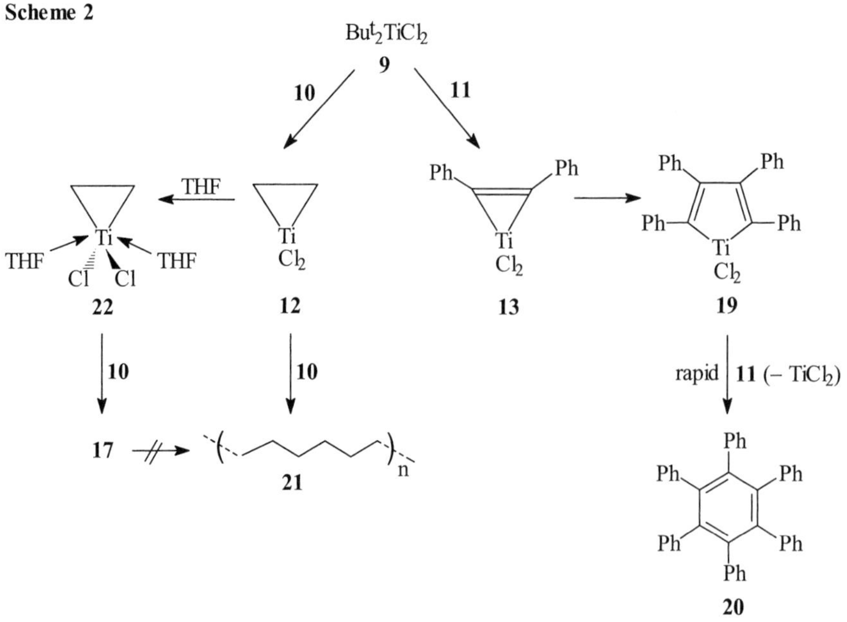

Scheme 3

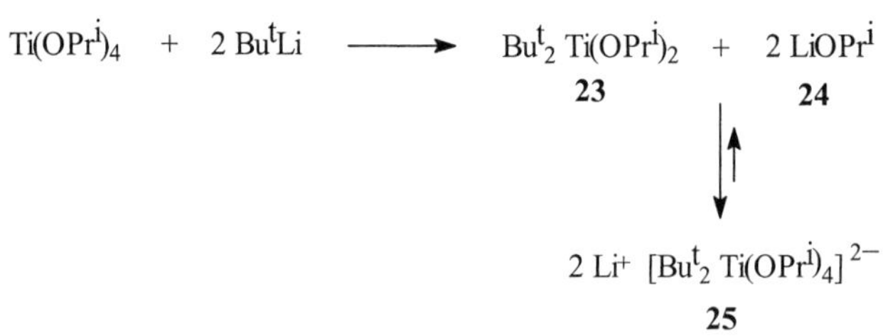

$$\text{Ti(OPr}^i)_4 \; + \; 2\,\text{Bu}^t\text{Li} \; \longrightarrow \; \underset{\textbf{23}}{\text{Bu}^t_2\,\text{Ti(OPr}^i)_2} \; + \; \underset{\textbf{24}}{2\,\text{LiOPr}^i}$$

$$\underset{\textbf{25}}{2\,\text{Li}^+ \; [\text{Bu}^t_2\,\text{Ti(OPr}^i)_4]^{2-}}$$

In a parallel fashion, passing propylene gas (**26**) into a THF solution of $\text{Bu}^t_2\text{TiCl}_2$ or $\text{Bu}^n_2\text{TiCl}_2$ starting at −78°C and bringing the solution to 25°C causes the formation of **27** by transfer-epimetallation. The presence of **27** was established by trapping with benzonitrile (**14**) and observing the generation of principally butyrophenone (**28**) (80%) (Scheme 4). Thus benzonitrile inserted selectively into the sterically more accessible C–Ti bond, **a**, of **27**; only 20% of isobutyrophenone (**29**), the insertion product of **14** into bond **b**, was detectable.

Scheme 4

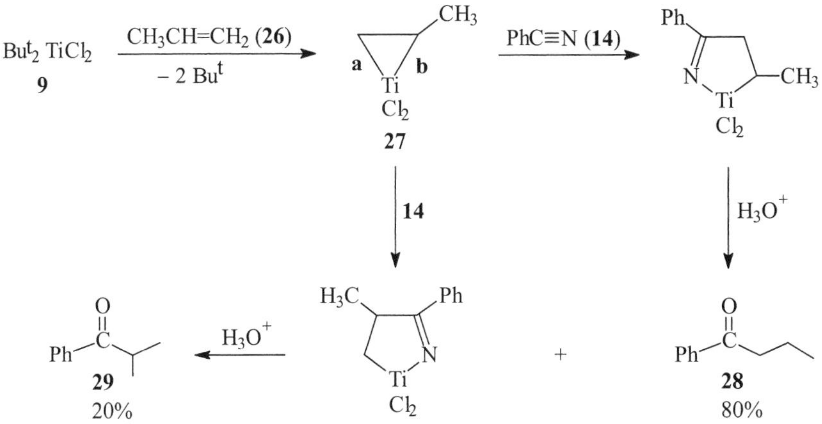

Discussion

The sterically controlled selectivity exhibited in the benzonitrile insertion into titanacyclopropane **27** forms the basis for an appealing model explaining the isotactic polymerization of propylene and other *alpha*-olefins by these R_2TiCl_2 catalysts in hydrocarbon medium.[1, 20] Up to the present, the most widely accepted model for the stereoselective polymerization of *alpha*-olefins in heterogeneous phase has been that of Arlman and Cossee cited above[2-5] or one of its modifications.[1] In the Arlman-Cossee view, an alkylated Ti center, Ti–R, located on the lateral face of the $TiCl_3$ crystal lattice, interacts with an *alpha*-olefin coordinated on an adjacent octahedral site (Figure 1). On steric grounds, the addition of the Ti–C bond to the coordinated olefin is greatly favored in the one conformation where the Ti–C bond and the C–C bond of the coordinated olefin are parallel to each other and as a result the ensuing polymer chain growth occurs preferentially with an isotactic configuration. In the Allegra counterproposal, the Ti–C bond and the C–C bond of the coordinated olefin are perpendicular to each other, thereby lessening steric repulsion (distances **a** and **b** in Figure 1).

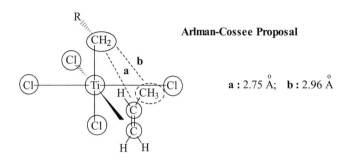

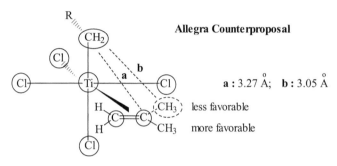

Figure 1. Arlman-Cossee proposal versus that of Allegra for the polymerization of *alpha*-olefins.

The present metallacyclopropane model hypothesizes that a titanacyclopropane, such as **27**, is formed in solution or in heterogeneous phase by the transfer-epimetallation of the *alpha*-olefin by a dialkyltitanium(IV) dichloride, as depicted in Schemes 4 and 5.[13b] Because of the high rate of polymerization, even at −78°C, **27** most likely is converted to the titanacyclopropyl cation **30** by coordinative abstraction of the chloride anion into the LiCl lattice.[21] For steric reasons, the next propylene unit would insert into bond **a** of **30**, approaching the ring from the side opposite (underneath) to that of the first projecting methyl group and with its own methyl group distal to the Ti–Cl cationic center and oriented exocyclic to the ring (**31**) (Scheme 5). The resulting five-membered titanocyclic cation **32** would have established a pattern for the head-to-tail, isotactic union of propylene units, which union could be propagated by further, similar insertions of propylene units at the sterically more accessible C–Ti bond, namely the underside of bond **c** in **32**.

Scheme 5

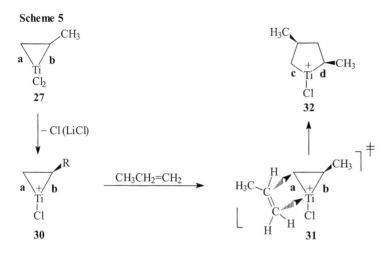

Although the present experimental results give ample support for the generation of titanacyclopropanes **12** and **27** and titanacyclopropene **13** through the transfer epititanation of ethylene, propylene or diphenylacetylene *in THF solution*, respectively, one might ask what convincing evidence can be marshaled in favor of such intermediates being formed as well in hydrocarbon solvents, *i.e.* under polymerization conditions. It is therefore pertinent to point out that $Bu^n_2TiCl_2$ or $Bu^n_2Ti(OPr^i)_2$ can also effect the stoichiometric transfer-epititanation in hexane or toluene solution of both diphenylacetylene and 1,2-disubstituted ethylenes, such as *cis*-stilbene and acenaphthylene, at low temperatures without ensuing polymerization.[13a] The 1,2-substituents at the olefinic carbons appear to block further monomer insertions into the C–Ti bonds of the titanacyclopropane intermediates. But such clean stoichiometric transfer-epititanations of these hindered olefins in hydrocarbons argue for similar epititanation of ethylene and α-olefins as the crucial first step in their polymerization.

A further objection to this present proposal of titanacyclopropane intermediates as the crucial first step in stereoselective α-olefin polymerization is to argue that some form of Ti(II) or Ti(III) may rather be involved in the present $Bu^i_2TiCl_2$ catalyst system. To counter any such suggestion, we offer these further observations as well as related commentary on previous work. Previous research on the polymerization of ethylene has involved passing the gas into mixtures of *n*-butyllithium and $TiCl_4$ combined in ratios of 0.75:1.0 to 6.0:1.0 in hydrocarbon media <u>under a</u>

nitrogen atmosphere at room temperature.[22, 23] Also by use of *n*-butyllithium and $TiCl_4$ propylene has been polymerized under similar conditions and the poly(propylene) found to be about 60% isotactic.[24]

With reference to our present work, however, we wish to point out that $Bu^n_2TiCl_2$ and $Bu^t_2TiCl_2$ intermediates, generated from two equivalents of RLi and one equivalent of $TiCl_4$, are kinetically stable at $-78°C$ to decomposition into $TiCl_2$. But at 25°C such intermediates in THF decompose rapidly to $TiCl_2$, and their solutions in hydrocarbons decompose slowly as well. The hydrocarbon suspensions of the resulting $TiCl_2$ and LiCl at 25°C can polymerize either ethylene or propylene, albeit at a slower rate (5g PE/g Ti-atm-h).[25] Noteworthy also is that $TiCl_2$ is very reactive toward the "inert" nitrogen atmosphere via a redox reaction, such that ammonia is evolved upon hydrolysis.[26] Thus although previous workers may have generated $Bu^n_2TiCl_2$ in their experiments, it is not clear that this alkyl survived under their conditions or was the specific catalyst for the observed polymerizations. It equally follows that under our conditions (under argon and at $-78°C$) our observed polymerizations cannot be attributed to $TiCl_2$-catalysts or, *a fortiori*, Ti(III)-catalysis.

Conclusion

In summary, we have demonstrated that dialkyltitanium(IV) dichlorides in donor solvents are able to effect the transfer epimetallation of olefins or acetylenes with the formation of titanacyclopropane or titanacyclopropene intermediates. With 1,2-disubstituted olefins or acetylenes as substrates, the resulting epimetallated adducts are stable as formed, even in hydrocarbon media, and can be shown to be present by chemical trapping with D_2O (cf. **13** in Scheme 1). Such intermediates should no doubt also be formed either in THF or hydrocarbon solution, as we have shown, or in a heterogeneous phase, as we propose. These three-membered titanocycles in donor solvents are able to insert one or more unsaturated monomers with the generation of 5- or 7-membered titanocycles. These findings support the proposal that such titanocycles are also formed from R_2TiCl_2 and the olefins or acetylenes in hydrocarbon media (*cf. supra*). In such noncoordinating media, however, such titanocycles are able to promote rapid and stereoselective olefin polymerization and acetylene cyclotrimerization, most likely via insertion of monomers into titanacycloalk(en)yl cations. The postulated 2-substituted-1-titanacyclopropyl cation intermediate, such as **30** in Scheme 5, thus constitutes an excellent model for rationalizing the observed isotactic stereoselectivity in the formation of isotactic poly(*alpha*-olefins) from terminal alkenes. Finally, the experimental evidence supporting the Arlman-Cossee model has been the XRD studies of the titanium(III) chloride salts that have been used in Ziegler-Natta catalysts. But to our knowledge, there has been no direct observation of the Ti–C bond postulated in their model (Figure 1). On the other hand, reliable chemical-trapping experiments have provided concrete evidence for the formation of titanacyclopropa(e)ne intermediates, which are formed from R_2TiCl_2 and olefins or acetylenes, both in donor solvents such as THF and in hydrocarbon dispersing media. Accordingly, the titanacyclopropyl-cation model for the active catalytic site may be judged at this point to have the better experimental corroboration.

Acknowledgments. Our investigations of Group 4 metal alkyls over the last decade have been supported at various times by Akzo Corporate Research America, The Boulder Scientific Company, the U.S. National Science Foundation and Solvay, S.A., Brussels, Belgium. The present research has been conducted under the auspices of the Alexander von Humboldt Stiftung, Bonn, Germany, which has provided the principal investigator with a Senior Scientist Award.

[1] The masterly reference work for research on Ziegler-Natta olefin polymerization prior to 1979: J. Boor, Jr., "*Ziegler-Natta Catalysts and Polymerization*", Academic Press, New York, 1979, 670 pp.
[2] E.J. Arlman, P. Cossee, *J. Catal.* **1964**, *3*, 99.
[3] E.J. Arlman, *J. Catal.* **1964**, 3, 89.
[4] E.J. Arlman, *J. Catal.* **1966**, *5*, 178.
[5] E.J. Arlman, *Recl. Trav. Chim. Pays-Bas.* **1968**, *87*, 1217.
[6] G. Allegra, *Makromol. Chem.* **1971**, *145*, 235.
[7] P. Pino, R. Mülhaupt, in: "*Transition Metal Catalyzed Polymerizations. Alkenes and Dienes*", Part A, R.P. Quirk, Ed., MMI Press, Harwood Academic Publishers, 1983, p. 14.
[8] G. Natta, P. Pino, G. Mazzanti, U. Giannini, E. Mantica, M. Peraldo, *Chem. Ind. (Milan)* **1957**, *39*, 19.
[9] D.S. Breslow, N.R. Newburg, *J. Am. Chem. Soc.* **1957**, *79*, 5072.
[10] J.A. Ewen, *J. Am. Chem. Soc.* **1984**, *106*, 6355.
[11] W. Kaminsky, K. Külper, H.H. Brintzinger, F.R.W.P. Wild, *Angew. Chem., Int. Ed. Engl.* **1985**, *24*, 507.
[12] J.A. Ewen, L. Jones, A. Razavi, *J. Am. Chem. Soc.* **1988**, *110*, 6255.
[13] a) J.J. Eisch, J.N. Gitua, *Organometallics* **2003**, *22*, 24; b) *ibid.* **2003**, *22*, 4172.
[14] Initial publication: J.J. Eisch, X. Shi, J. Lasota, *Z. Naturforschung* **1995**, *50b*, 342.
[15] J.J. Eisch, *J. Organomet. Chem.* **2001**, *617-618*, 148.
[16] J.J. Eisch, X. Ma, K. I. Han, J.N. Gitua, C. Krüger, *Eur. J. Inorg. Chem.* **2001**, 77.
[17] J.J. Eisch, J.N. Gitua, P.O. Otieno, X. Shi, *J. Organomet. Chem.* **2001**, *624*, 229.
[18] Inhibiting action of Lewis bases on Ziegler-Natta olefin polymerizations: reference 1, pp. 112-115.
[19] J.J. Eisch, J.N. Gitua, *Eur. J. Inorg. Chem.* **2002**, 3091.
[20] More recent and authoritative surveys of Ziegler-Natta polymerization include: H.H. Brintzinger, D. Fischer,

R. Mülhaupt, B. Rieger, R.M. Waymouth, *Angew. Chem., Int. Ed. Engl.* **1995**, *34*, 1143; and L. Resconi, L. Cavallo, A. Fait, F. Piemontesi, *Chem. Rev.* **2000**, *100*, 1253.

[21] The Lewis acididty of unsolvated ion-pairs of lithium chloride for chloride anions is evident in the experimental detection of diamond shaped, bridged lithium chloride dimers in the vapor phase at over 800°C by means of electron diffraction ($\angle$ Cl–Li–Cl = 108° ± 4): S.H. Bauer, T. Ino, R.F. Porter, *J. Chem. Phys.* **1958**, *33*, 685.

[22] H.N. Friedlander; K. Oita, *Ind. Eng. Chem.* **1957**, *49*, 1885.

[23] M. Frankel; J. Rabani; A. Zilkha, *J. Poly. Sci.* **1958**, *28*, 387.

[24] A. Zilkha; N. Calderon; A. Ottolenghi; M. Frankel, *J. Poly. Sci.* **1959**, *40*, 149.

[25] J.J. Eisch; S.I. Pombrik, X. Shi; S.C. Wu, *Macromol. Symp.* **1995**, *89*, 221.

[26] J.J. Eisch, T. Chan, unpublished studies.

Polymerization of Acrylate Monomers by Iron(II) Complexes Bearing Bis(imido)pyridyl or Phosphine Ligands

Pascal M. Castro, Mikko P. Lankinen, Anne-Marja Uusitalo, Markku Leskelä, Timo Repo*

Department of Chemistry, Laboratory of Inorganic Chemistry, PO Box 55, FIN-00014, University of Helsinki, Finland
E-mail: timo.repo@helsinki.fi

Summary: Iron(II) complexes bearing bis(imido)pyridyl or phosphine ligand were used for the polymerization of *tert*-butylacrylate and other (meth)acrylates after activation with methylaluminoxane. Phosphine based catalysts turned to be more active than the bis(imido)pyridyl ones. *tert*-Butylacrylate was the best polymerized monomer, while methylmethacrylate was slowly converted to poly(methylmethacrylate) and no activity was observed in the polymerization of styrene. Influence of the MAO to complex ratio was investigated and kinetics studies were conducted. Based on these results, a schematic representation of the polymerization mechanism was proposed.

Keywords: acrylate monomers; bis(imido)pyridyl ligand; methylaluminoxane; organometallic catalysts; phosphine ligand; transition metal chemistry

Introduction

(Meth)Acrylate polymers are easily formed via anionic or radical polymerization. Nevertheless, control of the polymerization (*e.g.* molar mass or polydispersity) requires specific conditions, for instance low temperature (anionic)[1] or dioxygen-free atmosphere (controlled radical).[2] Therefore, an interesting challenge in acrylate polymerization is to introduce catalytic control as it is known for transition metal catalyzed olefin polymerization. This would facilitate the use of mild reaction conditions and the control of the polymerization in terms of molar mass, polydispersity or stereostructure.

The recent development of late transition metal catalysts[3] opens interesting possibilities for this purpose, since late transition metals are less oxophilic than early transition metals, and are therefore supposed to be more tolerant towards Lewis bases. A wide variety of late

© 2004 WILEY-VCH Verlag GmbH & KGaA, Weinheim   DOI: 10.1002/masy.200450919

transition metal complexes are currently used for radical polymerization through atom transfer radical polymerization (ATRP),[4,5] and palladium(II) complexes[6] or organonickel/methylaluminoxane[7-9] systems are reported to homopolymerize methyl(meth)acrylate. Recently, iron(II) complexes bearing tridentate nitrogen ligands proved to be active for the polymerization of *tert*-butylacrylate after activation with methylaluminoxane (MAO). The activity of the catalyst and the molar mass of the polymer are clearly influenced by the ligand structure and the polymerization conditions.[10]

In the present work, polymerization kinetics of *tert*-butylacrylate and other (meth)acrylate monomers were investigated with the above mentioned iron(II) catalysts bearing bis(imido)pyridyl ligands, and a new polymerization system based on iron(II) complexes bearing phophine ligands (Figure 1). Phosphine based catalysts turned out to be more active for the polymerization of acrylates than the bis(imido)pyridyl ones, despite that the polymerization behaves similarly in both cases.

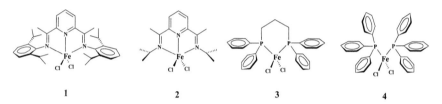

Figure 1. Complex chart.

Experimental

All the solvents were dried over sodium and purified by distillation prior to use. All other chemicals were used as received. MAO (30% in toluene) was received from Borealis Polymers Ltd. Polymerizations of *tert*-butylacrylate were carried out under argon atmosphere at room temperature in a Schlenk tube with standard Schlenk techniques, as described earlier in the literature.[10] Monomer conversion was determined by gas chromatography using *n*-decane as the internal standard. Molecular weights and molecular weight distributions were measured at 40°C in THF, by gel permeation chromatography relative to polystyrene standards using a Waters 515 HPLC pump GPC fitted with Styragel columns HR 2, HR 4, and HR 6, a UV-detector Waters 2487 and a RI-detector Waters 2410.

Four different complexes were used for this study. 2,6-bis[1-(2,6-bis(isopropyl)phenyl)imido)ethyl]pyridine iron dichloride (**1**)[10] and 2,6-bis[1-(isopropylimido)ethyl]pyridine iron dichloride (**2**)[11] complexes were synthesized according to published procedures. Bis(triphenylphosphine) iron dichloride (**3**) and 1,3-bis(diphenylphosphino)propane iron dichloride (**4**) were prepared via a straightforward synthesis.[12]

Results and Discussion

Activity

Batch polymerizations of *tert*-butylacrylate (tBA) were carried out with complexes **1-4** after activation with MAO, and monomer consumption was monitored by gas chromatography (Figure 2). The difference in activity observed between the bis(imido)pyridyl iron(II) catalysts **1**/MAO and **2**/MAO was attributed to the variation of steric hindrance around the metal, as bulky 2,6-bis(isopropyl)anilino substituents present on **1** prevent the access of tBA monomers to the reactive center.[10] Similarly, the difference between **3**/MAO and **4**/MAO can be explained by the distinct bridged (**3**) and unbridged (**4**) phosphine ligands: the 1,3-bis(diphenylphosphino)propane ligand markedly displays a more constrained coordination sphere than the unbridged bis(triphenylphosphine).[13] Anyhow, the definitive reason why **4**/MAO gives a higher activity than **3**/MAO remains open.

The phosphine based iron(II) catalysts **3**/MAO and **4**/MAO turned out to be more active than the bis(imido)pyridyl based ones, **1**/MAO and **2**/MAO. Firstly, one can ascribe these observations to the steric hindrance around the active center and to the rigidity of the ligand backbone, as discussed above. Secondly, the different influence from the bis(imido)pyridyl and the phosphine ligands on the electron density of the iron center have to be taken in account. Lastly, the distinct coordination sphere of complexes **1-2** and **3-4** can eventually affect the reactivity of iron towards tBA, as the four-coordinated complexes **3-4** would have a stronger tendency to expand their coordination number than the five-coordinated complexes **1-2**.

The time-conversion logarithmic plots presented in Figure 2/left show an apparent first-order kinetic for all the catalysts. Nevertheless, a deviation from linearity after a certain period of time indicates that the amount of active species does not remain constant during the polymerization. The same data plotted as catalyst activity (kg of polymer / mol of catalyst / h) vs. time gives more information (Figure 2, right). It illustrates that the activation of the complexes is very fast without any induction period. High initial polymerization activities are measured, but then decline with time, reaching lower steady state values.

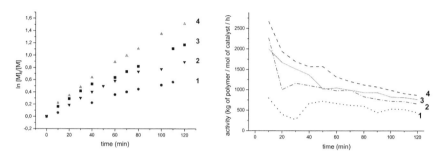

Figure 2. Time-conversion logarithmic plot (left) and activity dependence on time (right) for the polymerization of tBA with 1-4/MAO. Conditions: [complex] = 6,33·10^{-5} mol/L, [tBA] = 1,1 mol/L, MAO : complex = 250:1, in THF at room temperature.

We assume that the reason for this reduction of activity is twofold. As reported earlier, the activity of iron catalysts in tBA polymerization is depending on the monomer concentration,[10] and in batch polymerization, this monomer concentration gradually decreases. Since both the rate of declination and steady state value depend on the catalyst precursor (Figure 2, right), the observed changes can be related to the activation process itself. After fast activation the active species undergo a series of reversible and irreversible reactions involving e.g. condensation reactions with MAO and, at the steady state, the polymerization process is dominated by a sensitive equilibrium between active and dormant metal centers (i.e. catalytically inactive iron species), as described earlier for metallocene catalysts.[14, 15] A simple decay of the catalysts would rather show a slow death of the polymerization activities instead of the observed steady state polymerization kinetics.

Molar Mass

In order to verify the nature of the polymerization, the time dependence of molar mass was investigated (Figure 3). Surprisingly, the higher M_n value was attained at the beginning of the polymerization (after only two minutes with **2**/MAO), and then it is gradually decreased. This is in agreement with previously published results where M_n dependence on monomer concentration was demonstrated.[10] Hence, the decrease of molar mass with time is due to the consumption of tBA, leading to a decrease of its concentration. Accordingly, it means that each polymer chain is formed instantaneously, and that every catalyst molecule produces several ones. This is corroborated by the calculation of the catalyst efficiency after two hours ($f = M_{n(obs)}/M_{n(calc)}$) which gives $f_{(2/MAO)}= 0,0126$ and $f_{(4/MAO)}= 0,0085$. In other words, it signifies that every single molecule of **2**/MAO and **4**/MAO produces 79 and 117 poly(tBA) chains, respectively. In addition, the polydispersity index (PDI) remained constant in each case. These facts definitely rule out any sort of living character for the studied system.

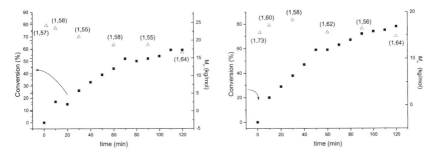

Figure 3. M_n and Conversion (%) *vs.* time for the polymerization of tBA with **2**/MAO (left) and **4**/MAO (right). The polydispersity index is given between the brackets. Conditions: [complex] = $6,33 \cdot 10^{-5}$ mol/L, [tBA] = 1,1 mol/L, MAO : complex = 250:1, in THF at room temperature.

Influence of MAO

The cocatalyst MAO is an important component of the catalytic system as it converts the iron dichloride precursor into a cationic species presumably responsible for the catalytic activity.[16, 17] The concentration of cocatalyst, and furthermore its ratio to the complex, will necessarily have an influence on the polymerization behavior like any other component of the polymerization system. The polymerization of tBA with **2** and different amount of MAO revealed a linear relationship between the MAO to complex ratio and both the molar mass of the obtained poly(tBA) and the activity of the catalyst (Figure 4). The rise of activity is to be related to the augmentation of the presence of alkylaluminum species in the media, which is able to re-activate the dormant centers and shift the equilibrium towards active iron species, thus accelerating the polymerization of tBA. The decrease of the molar mass indicates the presence of chain transfer to aluminum as a chain termination mechanism. Similar results were reported for ethylene polymerization with 1/MAO.[11]

It is important to notice that, due to its Lewis acidic nature, MAO has a strong tendency to coordinate to polar monomers like acrylates. In the polymerizations presented here, a high monomer to MAO ratio (typically 70 to 1) guarantees the presence of free monomers in solution, and that MAO does not have a significant role as a protecting group.

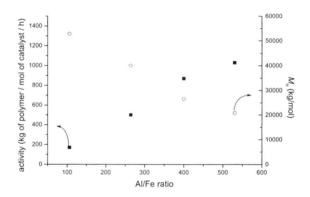

Figure 4. Activity and M_n dependence on the MAO to complex ratio for the polymerization of tBA with 2/MAO. Conditions: [complex] = $6,33 \cdot 10^{-5}$ mol/L, [tBA] = 2,3 mol/L, in THF at room temperature.

Polymerization of different monomers with complex 4

The polymerization of methylacrylate (MA), n-butylacrylate (nBA), methyl methacrylate (MMA) and styrene (St) was carried out using 4/MAO and compared with tBA polymerization (Figure 5). The highest activity is achieved with the bulky tBA, while MA and nBA are polymerized both with a similar conversion rate, and MMA bearing a methyl group at its α –position, is slowly converted to poly(MMA) by 4/MAO. Nevertheless, this latter result has to be examined in relation to a recently published study by Carlini *et al.* about the polymerization of MMA with MAO activated nickel(II) complexes.[9] The best activity obtained was 155 kg of PMMA / (mol of catalyst x h) with 32% conversion, while **4**/MAO gives 86,8 kg of PMMA / (mol of catalyst x h) after two hours. In the case of tBA, 29% of monomer consumption corresponds to an activity of 1937 kg of tBA / (mol of catalyst x h) (Figure 2, right). No activity was observed when styrene was employed as the monomer. Similar results were obtained with **2**/MAO.[18]

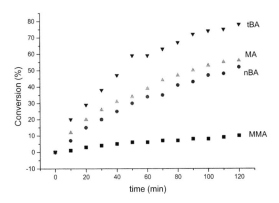

Figure 5. Conversion (%) *vs.* time for the polymerization of various (meth)acrylates with **2**/MAO. Conditions: [complex] = $6,33 \cdot 10^{-5}$ mol/L, [monomer] = 1,1 mol/L, MAO : complex = 250:1, in THF at room temperature.

Schematic Representation of the Mechanism

The presence of activated iron complexes is required for an efficient polymerization of acrylates, as both iron(II) complexes and MAO alone do not promote the polymerization of tBA.[10] Similar observations were also reported about nickel(II) complexes for MMA polymerization.[7, 9] MAO reacts with bis(imido)pyridyl iron(II) complexes to form a cationic Fe^{+}-Me species,[16, 17] and consequently, these cationic species are responsible for the polymerization (Scheme 1, step 1). This assertion is supported by Figure 3 which shows that the activity increases with the amount of aluminum present in the reaction mixture.

The propagation mechanism is not clarified yet. Radical polymerization was ruled out for both iron(II)[10] and nickel(II)[9] systems. After activation of the iron complex, two different

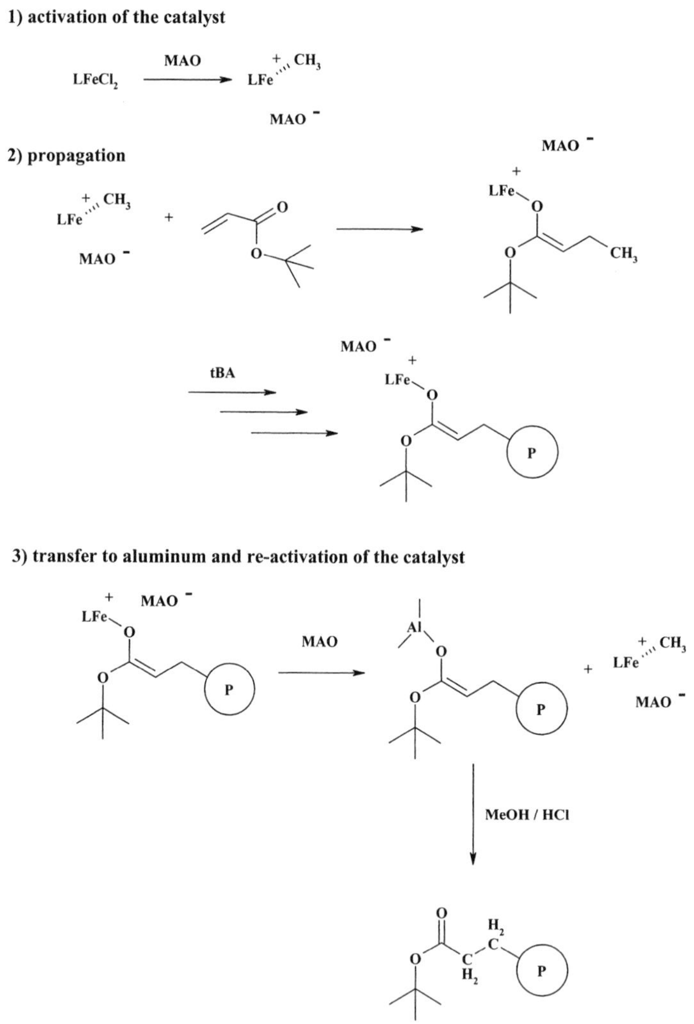

206

1) activation of the catalyst

2) propagation

3) transfer to aluminum and re-activation of the catalyst

Scheme 1. Schematic representation of the polymerization mechanism. L represents the different ligands for complexes **1-4**.

phenomena can occur: methyl transfer from iron to monomer at the β –position (anionic like polymerization), or coordination of tBA to the iron followed by 2,1-insertion into the iron-methyl bond (coordination-insertion mechanism) (Scheme 1, step 2). Nevertheless, the moderate activity observed in the case of MMA polymerization and the absence of activity for styrene polymerization, which is easily polymerized by both radical or ionic pathway, strengthens the idea of a coordination-insertion mechanism. Molar mass is clearly influenced by the MAO to catalyst ratio (Figure 3), which suggests chain transfer to aluminum as the dominating termination process (Scheme 1, step 3). Furthermore, Figure 4 strongly suggests the instantaneous formation of polymer chains, followed by chain transfer to aluminum and reactivation of the iron catalyst into the cationic Fe^+-Me species able to start a new chain.

The mechanism proposed in Scheme 1 is only a schematic representation of how the polymerization occurs, as the intimate propagation mechanism (Scheme 1, step 2) remains unclear.

Conclusion

The polymerization of *tert*-butylacrylate (tBA) was carried out using different MAO activated iron(II) complexes and consumption of the monomer was followed as a function of time. After high initial activity the polymerization capability of the catalysts declines to lower steady state values. The phosphine based iron(II) catalysts turned out to be more active than the bis(imido)pyridyl iron(II) ones. Polymerization is more efficient with acrylates, among them tBA, than with methylmethacrylate, while styrene is not polymerized.

The fact that the molar mass does not increase with time indicates that each polymer chain is formed instantaneously, and that every catalyst molecule produces several chains. The decrease of the molar mass with increasing MAO concentration indicates the presence of chain transfer to aluminum as a chain termination mechanism. According to these results, a schematic representation of the polymerization pathway was proposed.

Acknowledgement is made to Academy of Finland for financial support (Center of Excellence: Bio- and Nanopolymer Research Group, code number 77317). We are grateful to Marjut Wallner for GPC measurements made in the Polymer Laboratory of the Chemistry Department, University of Helsinki.

208

[1]G.Odian, "Principles of Polymerization", 3rd ed., J. Wiley and Sons, New York 1991, p.357.

[2] G.Odian, "Principles of Polymerization", 3rd ed., J. Wiley and Sons, New York 1991, p.264.

[3] For a review about non-metallocene olefin polymerization catalysis in general and late-transition metal olefin polymerization catalysts in particular, see V. C. Gibson, S. K. Spitzmesser, *Chem. Rev.* **2003**, *103*, 283.

[4] K. Matyjaszewski, J. Xia, *Chem. Rev.* **2001**, *101*, 2921.

[5] M. Kamigaito, T. Ando, M. Sawamoto, *Chem. Rev.* **2001**, *101*, 3689.

[6] G. Tian, H. W. Boone, B. M. Novak, *Macromolecules* **2001**, *34*, 7656.

[7] K. Endo, A. Inukai, *Polym. Int.* **2000**, *49*, 110.

[8] E. Ihara, T. Fujimura, H. Yasuda, T. Maruo, N. Kanehisa, Y. Kai, *J. Polym. Sci. Part A: Polym. Chem.* **2000**, *38*, 4764.

[9] C. Carlini, M. Martinelli, A. M. Raspolli Galletti, G. Sbrana, *J. Polym. Sci. Part A: Polym. Chem.* **2003**, *41*, 2117.

[10] P. M. Castro, K. Lappalainen, M. Ahlgrén, M. Leskelä, T. Repo, *J. Polym. Sci. Part A: Polym. Chem.* **2003**, *41*, 1380.

[11] G. J. P. Britovsek, M. Bruce, V. C. Gibson, B. S. Kimberley, P. J. Maddox, S. Mastroianni, S. J. McTavish, C. Redshaw, G. A. Solan, S. Strömberg, A. J. P. White, D. J. Williams, *J. Am. Chem. Soc.* **1999**, *121*, 8728.

[12] M. P. Lankinen, P. M. Castro, M. Leskelä, T. Repo, *manuscript in preparation.*

[13] P. W. N. M. van Leeuwen, P. C. J. Kamer, J. N. H. Reek, P. Dierkes, *Chem. Rev.* **2000**, *100*, 2741.

[14] E. You-Xian Chen, T. J. Marks, *Chem. Rev.* **2000**, *100*, 1391.

[15] E. Zurek, T. Ziegler, *Organometallics* **2002**, *21*, 83.

[16] E. P.Talsi, D. E. Babushkin, N. V. Semikolenova, V. N. Zudin, V. N. Panchenko, V. A. Zakharov, *Macromol. Chem. Phys.* **2001**, *202*, 2046.

[17] G. J. P. Britovsek, V. C. Gibson, S. K. Spitzmesser, K. P. Tellmann, A. J. P. White, D. J. J. Williams, *J. Chem. Soc., Dalton Trans.* **2002**, *6*, 1159.

[18] P. M. Castro, M. P. Lankinen, A. M. Uusitalo, M. Leskelä, T. Repo, *manuscript in preparation.*

Macromol. Symp. **2004**, *213*, 209-220

Styrene Polymerization by Ziegler-Natta Catalysts Based on Bis(salicylaldiminate)nickel(II) Complexes and Methyl Aluminoxane

Carlo Carlini,* Alessandro Macinai, Anna Maria Raspolli Galletti, Glauco Sbrana

Dipartimento di Chimica e Chimica Industriale, Università di Pisa, Via Risorgimento 35, 56126 Pisa, Italy
E-mail: carlini@dcci.unipi.it

Summary: The homopolymerization of styrene by using different catalytic systems based on bis(salicylaldiminate)nickel(II) and methylaluminoxane was investigated. In particular, the effect on catalyst activity and polymer characteristics by electron withdrawing groups located on the phenolic moiety was studied. The influence of the bulkiness of the substituents on the N-aldimine ligand was also ascertained. Finally the catalytic performances were investigated as a function of the main reaction parameters, such as temperature, Al/Ni molar ratio and duration.

Keywords: addition polymerization; bis(salicylaldiminate)nickel complexes; methylaluminoxane; organometallic catalysts; styrene

Introduction

It is well established that isotactic poly(styrene) (i-PS) was obtained since the discovery of conventional Ziegler-Natta catalysts based on titanium chlorides and organoaluminium derivatives.[1-3] Quite recently, mono-cyclopentadienyl- and other titanium compounds, when combined with methylaluminoxane (MAO) or with tris(pentafluorophenyl)borane and corresponding borates were found to give in homogeneous phase syndiotactic poly(styrene) (s-PS).[4-9] Organometallic nickel complexes were also reported to be active in the polymerization of styrene. In particular, the cationic $[(\eta^3$-methallyl)$(\eta^4$-cycloocta-1,5-dienyl)Ni(II)]$ hexafluorophosphate complex was found to afford very low molecular weight i-PS without any addition of Lewis acids as co-catalysts.[10] When solvents with different polarity and phosphines as ancillary ligands were used in combination with the above nickel complex the catalytic activity and polymer characteristics were markedly affected.[10,11]

© 2004 WILEY-VCH Verlag GmbH & KGaA, Weinheim

DOI: 10.1002/masy.200450920

Neutral σ-acetylide nickel(II) complexes were found to be good initiators in chloroform solution for styrene polymerization, but a radical mechanism was proposed.[12] Recently, several nickel complexes, when combined with MAO, resulted active in the styrene polymerization.[13-16] In the presence of the bis(acetylacetonate)Ni(II) [Ni(acac)$_2$]/MAO catalyst, contradictory results concerning the PS microstructure were reported. Indeed, subsequent studies allowed to ascertain that the stereoregularity is strongly dependent on the content of trimethylaluminium (AlMe$_3$) present in MAO.[17,18] On the other hand, MAO alone behaves as ionic initiator for some vinyl monomers.[19] When Al/Ni molar ratios in the 60-90 range were adopted, catalyst activity was improved, but a detrimental effect on the PS stereoregularity was observed.[17] A detailed study of styrene polymerization with nickel derivatives/MAO catalysts allowed to conclude that Ni(acac)$_2$ and NiCl$_2$ resulted the best precursors, although rather low activities (~ 220 mol of converted styrene/mol of Ni × h) and molecular weights of the resulting polymers were obtained.[20] Moreover, the heterogenization of the aforementioned catalyst on SiO$_2$ allowed to improve the activity.[21] Novel nickel-based/MAO catalysts, prepared from bis(α-nitroacetophenonate)nickel(II) precursor were recently proposed by us for the polymerization of styrene.[22, 23] They displayed higher activity as compared with the Ni(acac)$_2$/MAO catalyst. However, the molecular weight of the obtained PS resulted rather low (< 20.000 Da) and significant amounts of oligomeric products were present. When AlMe$_3$ free MAO was used an increase of polymer isotacticity degree (up to 70%) was achieved. GC/MS analysis of oligomers allowed us to conclude that styrene insertion is mainly of secondary type. In this context it appeared very interesting to study the polymerization of styrene in the presence of novel catalytic systems based on bis(salicylaldiminate)nickel(II) complexes (Chart 1) which, in combination with MAO, have been found to be active in the polymerization of methyl methacrylate to a syndiotactic-rich polymer.[23,24]

In particular, the influence on the catalyst activity and polymer characteristics will be investigated by the insertion of electron withdrawing nitro groups on the phenol moiety as well as of bulky substituents on the N-aldimine moiety of the salicylaldiminate ligand.

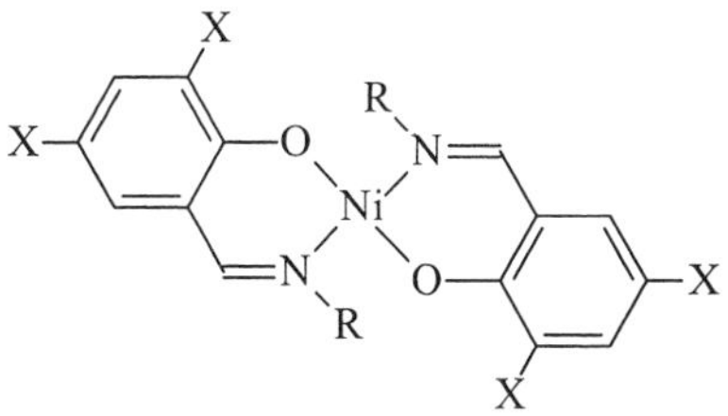

A (X = H, R = isopropyl)
B (X = H, R = 2,6-diisopropylphenyl)
C (X = NO$_2$, R = 2,6-diisopropylphenyl)
D (X = NO$_2$, R = phenyl)

Chart 1

Experimental Part

Materials

Anhydrous toluene (Baker) and *n*-hexane (Carlo Erba) were obtained by distillation on K/Na alloy under dry argon and stored on molecular sieves (4Å).

Methanol (Carlo Erba), Ni(OAc)$_2$·4 H$_2$O (OAc = acetate) (Carlo Erba), methyl ethyl ketone (MEK) and 3,5-dinitro-salicylaldehyde (Aldrich) were used as received.

Methylaluminoxane (MAO) (Witco), containing about 30 wt-% of AlMe$_3$, was stored under dry argon and used in toluene solution (4.5 M) as received.

Styrene (Aldrich) was purified by distillation under vacuum after stirring for 4 h on CaH$_2$. It was stored under dry argon in a refrigerator.

2,6-diisopropylaniline (Aldrich), aniline (Aldrich), diisopropylamine (Aldrich) and salicylaldehyde (Carlo Erba) were distilled *in vacuo* (ca. 1 mm Hg) and stored under dry argon in the refrigerator.

Bis(salicyladiminate)nickel(II) complexes were prepared according to a general method.[25]

As an example the procedure is described for the synthesis of bis[(3,5-dinitro-N(2,6-diisopropylphenyl)]nickel(II) complex (**C**): 3,5-dinitro-salicylaldehyde was allowed to react with an equimolar amount of Ni(OAc)$_2$·4H$_2$O and 3 equivalents diisopropylamine in 100 ml of absolute ethanol at the refluxing temperature for 2 h. After cooling, the solid was filtered and crystallized from *n*-heptane.

C was thus obtained as yellow-green crystals.[24] ^{1}H-NMR (CDCl$_3$): δ: 9.5 (d, 2H, aromatic);

8.4 (d, 1H, HC=N); 7.5 (m, 6H, aromatic); 7.2 (t, 2H, aromatic); 3.8 [hept, 4H, C$\underline{H}$(CH$_3$)$_2$]; 1.44 [d, 24H, CH(C$\underline{H}_3$)$_2$] ppm.

FT-IR: 3089 (v_{C-H} aromatic); 2965 (v_{C-H} aliphatic); 1621 ($v_{s_{CH=N}}$) cm^{-1}. mp = > 270 °C.

Bis[3,5-dinitro-N(phenyl)salicylaldiminate]nickel(II) (**D**) was obtained as yellow crystals.[24]

^{1}H-NMR (CDCl$_3$): δ: 9.5 (s, 2H, aromatic); 8.4 (d, 1H, HC=N); 7.5 (m, 8H, aromatic); 7.2 (t, 2H, aromatic) ppm;

FT-IR: 3083 (v_{C-H} aromatic); 1621 ($v_{s_{CH=N}}$) cm^{-1}. mp = > 270°C.

Bis(N-isopropylsalicylaldiminate)nickel(II) (**A**) was obtained as green crystals.[26]

FT-IR: 3053 (v_{C-H} aromatic); 2970 (v_{C-H} aliphatic); 1605 ($v_{s_{CH=N}}$) cm^{-1}. mp = 203°C.

Bis[(N-2,6-diisopropylphenyl)salicylaldiminate)]nickel(II) (**B**) was obtained as green crystals.[27]

FT-IR: 3059 (v_{C-H} aromatic); 2960 (v_{C-H} aliphatic); 1606 ($v_{s_{CH=N}}$) cm^{-1}. mp = >320°C.

Styrene Homopolymerization Experiments

Polymerization experiments were usually carried out at 25 °C in a 50 ml Carius vessel under dry argon and magnetic stirring. In a typical procedure, the desired amount of nickel precursor, dissolved at room temperature in dry toluene (15 ml), was transferred under dry argon into the reaction vessel. Then 5 ml of styrene (44 mmol) were introduced and the desired amount of MAO (Al/Ni = 30-100 mol/mol) was added. The reaction mixture was maintained, through a thermostated oil-bath, at the desired temperature for the chosen duration. At the end of the polymerization the reaction mixture was analyzed by gas-chromatography (GC) in order to determine styrene conversion and to check the presence of oligomeric products. After pouring the mixture in a large excess of 5% aq. HCl acidified methanol the coagulated polymer was filtered, dried under vacuum, weighed and finally characterized.

Analytical Procedures

Oligomers analysis was performed by GC on a Hewlett Packard 5890 chromatograph equipped with a flame ionization detector, a SE-52 capillary column (50 m × 0.2 mm × 0.33

μm) and a Hewlett-Packard (HP) 3396 integrator. The following temperature program of the oven was adopted: 100 °C for 5 min, then the temperature was increased by a 15 °C/min heating until 250 °C was reached; this value was maintained constant for further 60 min. The oligomeric products were finally analyzed by GC/MS employing a HP 5973 Mass Selective Detector an a HP 6890 series GC System.[22] In this way it was possible to separate and identify in the oligomeric mixtures 2,3-diphenyl butene (DPB) (m/z values: 208, 193, 115, 105, 91 and 77) and 2,3,5-triphenyl-1-hexene (TPH) (m/z values: 312, 207, 193, 115, 105, 91, and 77).

Physico-chemical Measurements

The molecular weights of PS samples were determined by size exclusion chromatography (SEC) measurements, performed on a Perkin-Elmer Series 10 instrument, equipped with a Jasco 830-RI refractive index detector and an UV-VIS Perkin-Elmer LC-25 model working at 270 nm wavelength, by using a HPLC PL gel 5 μm MIXED-C column and chloroform as eluent. Calibration curve was obtained by using several monodisperse PS samples having different molecular weights.

^{1}H- and ^{13}C-NMR spectra were performed on a FT Varian Gemini 200, working at 200 and 50 MHz, respectively, by using tetramethylsilane (TMS) as internal standard.

FT-IR spectra were performed on powdered samples by using a Perkin Elmer Spectrum One spectrophotometer, equipped with an Attenuated Total Reflectance (ATR) apparatus.

Results and Discussion

Polymerization of styrene by bis[(N-substituted)salicylaldiminate]nickel(II)/MAO catalytic systems

The experiments were preliminarily carried out by using as catalyst precursors bis(salicylaldiminate)nickel(II) characterized by the absence of substituents on the phenoxy moiety and by the presence of groups with different steric hindrance directly bound to the aldimine nitrogen.

Therefore, complexes **A** and **B** have been tested in combination with MAO for the polymerization of styrene. Both complexes were previously employed by us in the activation of ethylene but they substantially resulted oligomerization catalysts.[26] However, when used in the polymerization of styrene in the presence of MAO (Table 1), they resulted appreciably active, although rather low molecular weight PS samples were formed.

Table 1. Styrene polymerization by **A** or **B**/MAO catalytic systems. Polymerization conditions: Nickel complex, 0.04 mmol; solvent (toluene), 15 ml; styrene (5 ml), 44 mmol; $T = 25°C$; time, 1 h.

Entry	Nickel complex	Al/Ni mol/mol	Crude polymeric product				$A^{a)}$
			g	Yield (%)	$M_w^{b)}$ $(x\ 10^{-3})$	$PDI^{b,c)}$	
1	A	30	0.90	20	10	2.2	22.5
2	A	50	1.00	22	10	2.3	25.0
3	A	100	0.70	15	14	2.4	17.5
4	B	30	0.50	11	22	2.0	12.5
5	B	50	0.70	15	22	1.9	17.5
6	B	100	1.55	34	27	1.8	38.8

[a)] Activity expressed as: Kg of PS /(mol of Ni x h).
[b)] Determined by SEC.
[c)] Polydispersity index, M_w/M_n.

Analogously to what was proposed in the case of ethylene activation with the same systems,[26] bis(salicylaldiminate)nickel(II) precursors can react with MAO giving rise to a pristine monochelate methyl nickel species characterized by the presence of a vacancy able to coordinate styrene and initiate its polymerization (Scheme). These active sites are supposed to evolve into intermediate hydride species due to the growing chain termination step; in any case, they are characterized by the presence of a residual salicylaldiminate ligand whose electronic and steric requirements may significantly influence catalyst performances.[28]

Independently from the type of catalyst precursor adopted, the polydispersity index (PDI), of the resulting PS samples, as determined by the M_w/M_n ratio, was found to be near to 2. The relatively higher activity of **B** as compared with **A** and the different behaviour of the two systems in terms of activity as a function of Al/Ni molar ratio may be addressed not only to the different steric constraints around the metal sites in **A** and **B** precursors[29,30] but also to the different basic character of the aldimine nitrogen when either an alkyl group or a

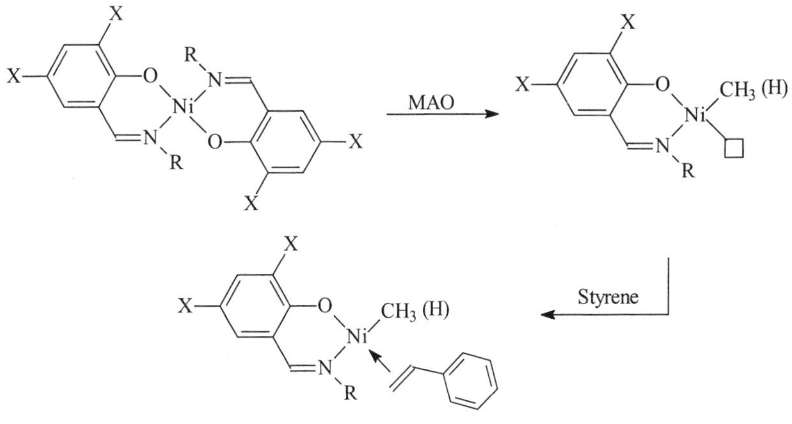

Scheme

substituted phenyl ring is directly bound to it. It is noteworthy that when **A** complex was used as catalyst precursor a very low amount (yield: up to 2 %) of oligomeric products containing ca. 85 % of dimers, was formed in addition to PS. When **A** was replaced by **B** only traces of oligomeric products were obtained. However, in both cases the GC/MS analysis allowed to establish that the dimeric product was essentially constituted by 2,3-diphenyl-1-butene (DPB) and the trimeric one by 2,3,5-triphenyl-1-hexene (TPH) (Chart 2). No detectable amount of tetramers and higher oligomers was ascertained even by SEC analysis.

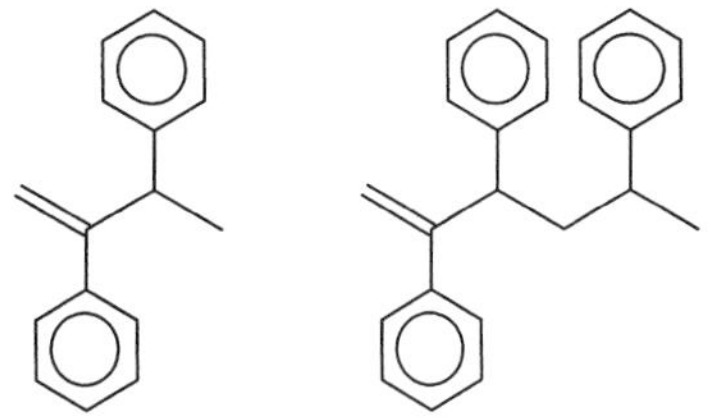

DPB   TPH

Chart 2

This occurrence is in agreement with a mechanism of oligomers formation via styrene secondary insertion which stabilizes the resulting nickel species through an equilibrium between η_1 and η_3 species, analogously to what was previously proposed.[10,22,23,31-33] In fact, the formation of these oligomers is readily explained by assuming that either one or two secondary styrene insertions on a pristine nickel hydride species occur, followed by a final primary insertion. The crude polymers were also fractionated by extraction with boiling solvents using n-hexane, MEK and toluene in that order. The first solvent was adopted for extracting a very low molecular weight polymer fraction, whereas the other solvents were used for dissolving a-PS and i-PS, respectively. When the crude PS samples were subjected to fractionation (Table 2), the **A**/MAO catalyst gave a higher percentage of very low molecular weight polymer in agreement with the larger oligomerization capability of this system. In all cases, the polymers were completely soluble in MEK, thus indicating that a-PS was formed, as confirmed by ^{13}C-NMR analysis.

Table 2. Characterization of the polymer fractions obtained by extraction with boiling solvents of the crude PS obtained by **A**/MAO (entry 2) or **B**/MAO (entry 6) catalysts.

Polymer fraction[a]	Entry 2			Entry 6		
	wt.%	M_w[b] $(\times 10^{-3})$	PDI[b,c]	wt.%	M_w[b] $(\times 10^{-3})$	PDI[b,c]
1	23.0	3.7	1.5	4.9	5.2	1.8
2	77.0	10.0	2.2	95.1	23.8	1.9
3	0.0	-	-	0.0	-	-

[a] 1 n-Hexane soluble fraction; 2 n-Hexane insoluble, MEK soluble fraction; 3 MEK insoluble, toluene soluble fraction.
[b] Determined by SEC analysis.
[c] Polydispersity index.

Polymerization of Styrene by bis[3,5-dinitro-N(substituted)salicylaldiminate]Nickel(II)/ MAO Catalytic Systems

With the aim to improve the catalytic activity two electron withdrawing substituents, such as nitro groups, were introduced on the phenoxy moiety of the salicylaldiminate ligands bound to the nickel(II). Indeed, this structural situation should reduce the electron density on the metal center and hence increase its propensity to the coordination by an electron-rich monomer, such as styrene. Moreover, the influence on catalytic performances by the bulkiness of the substituents on the N-aryl moiety of the ligand was checked. Therefore, **C** and **D** complexes were tested as catalyst precursors in combination with MAO (Table 3). On the basis of the obtained results we may conclude that the presence of the two electron withdrawing nitro groups on the ligands caused a large improvement of catalyst activity (up to 255 Kg of PS/mol Ni x h). Moreover, also the molecular weight of the resulting PS samples was significantly increased. However, when **C** was replaced by the less hindered **D** (entries 12-15, Table 3) despite a substantially similar activity, a decrease of the molecular weights was achieved, thus confirming that also steric factors around the metal play an important role in determining the competition between chain growth and termination steps.[28]

Table 3. Styrene polymerization by **C** or **D**/MAO catalytic systems. Polymerization conditions: Nickel complex, 0.04 mmol; solvent (toluene), 15 ml; styrene (5 ml), 44 mmol.

Entry	Nickel complex	Al/Ni mol/mol	Temp. °C	Time h	Crude polymeric product				$A^{a)}$
					g	Yield (%)	$M_w^{b)}$ $(x\ 10^{-3})$	$PDI^{b,c)}$	
7	C	30	25	1	1.90	43	121	10.7	47
8	C	100	25	1	3.45	78	103	10.8	86
9	C	100	25	4	4.00	89	99	6.0	25
10	C	100	25	0.33	3.40	77	37	3.5	255
11	C	100	70	4	3.30	74	31	2.3	18
12	D	30	25	1	3.31	72	51	2.8	83
13	D	50	25	1	3.34	74	52	3.0	85
14	D	100	25	1	3.65	80	28	1.8	91
15	D	100	25	0.33	3.40	77	23	2.2	255

[a] Activity expressed as: Kg of PS /(mol of Ni x h).
[b] Determined by SEC.
[c] Polydispersity index, M_w/M_n.

The polymers obtained in entries 9 and 13 were also fractionated with the same series of boiling solvents employed in the preceding section. As shown in Table 4, the crude polymers prepared with **C** or **D**/MAO catalysts, differently from those obtained with nickel precursors without nitro substituents, were characterized by the presence of a small amount (8-10%) of a largely polydispersed (PDI ca. 10) high molecular weight (M_w = ca. 400000 Da) MEK insoluble fraction which should be constituted by i-PS. Indeed, the ^{13}C-NMR analysis of these last fractions allowed to determine a content of mmmm pentads in the 60-70% range.

In order to elucidate the polymerization mechanism, the polymer end-chain structure of low molecular weight n-hexane soluble fractions of PS obtained from entry 2 and 9 was analyzed by ^{13}C-NMR.

Table 4. Characterization of the polymer fractions obtained by extraction with boiling solvents of the crude PSt obtained by **C**/MAO (entry 9) or **D**/MAO (entry 13) catalysts.

Polymer fraction[a]	Entry 9			Entry 13		
	wt.%	M_w^b (x 10^{-3})	$PDI^{b,c}$	wt.%	M_w^b (x 10^{-3})	$PDI^{b,c}$
1	8.1	3.7	2.0	2.5	5.2	1.8
2	82.1	44.4	3.7	89.5	29.3	2.0
3	9.8	359	10.1	8.0	413	10.3

[a] 1 n-Hexane soluble fraction; 2 n-Hexane insoluble, MEK soluble fraction; 3 MEK insoluble, toluene soluble fraction.
[b] Determined by SEC analysis.
[c] Polydispersity index.

In both cases signals at 37.8 and 20.9, as well as at 36.7 and 21.5 ppm, assignable respectively to erythro and threo 2,4-diphenylbutyl end groups,[22] were detected, suggesting a secondary styrene insertion on a nickel hydride bond in the initial chain growth step. Moreover, signals at 128.4 and 133.2 ppm were present and assignable to 1,3-diphenyl-1-butenyl end-groups[14] deriving by β-hydrogen elimination from the ultimate inserted styrene secondary unit. Indeed, no detectable regioirregular arrangement of styrene units due to primary insertions was observed in the PS samples. The above data support a polymerization mechanism different from that proposed for the oligomerization, this last probably occurring at a very low extent on different catalytic species.

Conclusions

On the basis of the obtained results the following concluding remarks can be drawn:

1. Bis(salicylaldiminate)nickel(II)/MAO catalysts resulted active in the styrene

polymerization. The catalytic performances, in terms of activity and polymer characteristics, are significantly affected by the structure of the nickel precursor and by the reaction parameters.

2. The presence of electron withdrawing nitro groups on the phenoxy moiety of the ligand markedly enhanced the activity of the catalytic system. The average molecular weight of the polymer was also influenced by the bulkiness of the substituents on the N-aryl moiety.

3. The analysis of oligomeric and polymeric products suggested that they are probably derived from different catalytic species.

Acknowledgement

The financial support from MIUR through the Research Project of National Interest (PRIN) 2002: "New strategies for the control of reactions: interactions of molecular fragments with metallic sites in unconventional species" is gratefully acknowledged.

[1] G. Natta, P. Pino, E. Mantica, F. Danusso, G. Mazzanti, M. Peraldo, *Chimica Industria* **1956**, *38*, 124.
[2] F. Danusso, D. Sianesi, *Chimica Industria* **1958**, *40*, 450.
[3] G. Natta, F. Danusso, D. Sianesi, *Makromol. Chem.* **1958**, *28*, 253.
[4] N. Ishihara, T. Seimiya, M. Kuramoto, M. Uoi, *Macromolecules* **1986**, *19*, 2464.
[5] A. Zambelli, P. Longo, C. Pellecchia, A. Grassi, *Macromolecules* **1987**, *20*, 2035.
[6] P. F. Foster, J. C. W. Chien, M. D. Rausch, *Organometallics* **1996**, *15*, 2404.
[7] W. Kaminsky, S. Lenk, V. Scholz, H. W. Roesky, A. Herzog, *Macromolecules* **1997**, *30*, 7647.
[8] C. Pellecchia, P. Longo, A. Proto, A. Zambelli, *Makromol. Chem. Rapid Commun.* **1992**, *13*, 265.
[9] G. Xu, *Macromolecules* **1998**, *31*, 586.
[10] J. Ascenso, A. R. Dias, P. T. Gomes, C. C. Romão, D. Neibecker, I. Tkatchenko, A. Revillon, *Makromol. Chem.* **1989**, *190*, 2773.
[11] J. Ascenso, A. R. Dias, P. T. Gomes, C. C. Romão, Q.-T. Pham, D. Neibecker, I. Tkatchenko, A. Revillon, *Macromolecules* **1989**, *22*, 1000.
[12] H. Sun, Q. Shen, M. Yang, *Eur. Polym. J.* **2002**, *38*, 2045.
[13] N. Ishihara, M. Kuramoto, M. Uoi, *Macromolecules* **1988**, *21*, 3356.
[14] P. Longo, A. Grassi, L. Oliva, P.Ammendola, *Makromol. Chem.* **1990**, *191*, 237.
[15] K. Endo, K. Masaki, *Macromol. Rapid Commun.* **1995**, *16*, 779.
[16] K. Endo, K. Masaki, Kobunshi Rombushu **1994**, *51*, 652; *Chem Abstr* **1994**, *121*, 281314 j.
[17] G. L. Crossetti, C. Bormioli, A. Ripa, A. Giarrusso, L. Porri. *Macromol. Rapid Commun.* **1997**, *18*, 801.
[18] J. Bliemeister, W. Hagendorf, A. Harder, B. Heitmann, I. Schimmel, E. Schmedt, W. Schnuchel, H. Sinn, L. Tikwe, N. von Thienen, K. Urlass, H. Winter, O.Zarncke, " The Role of MAO-Activators", in: *Ziegler Catalysts*, G. Fink, R. Mülhaupt, H. H. Brintzinger, Eds., Springer-Verlag, Berlin 1995, p 58.
[19] N. Cardi, R. Fusco, L. Longo, R. Po, S. Spera, C. Bacchilega. *Proc. Internat. Symp. on Ionic Polymerization*, Paris, 1997.
[20] R. Po, N. Cardi, R. Santi, A. M. Romano, C. Zannoni, S. Spera, *J. Polym. Sci., Polym. Chem. Ed.* **1998**, *36*, 2119.
[21] L. C. Ferreira, M. A. S. Costa, P. I. C. Guimaraes, L. C. de Santa Maria, *Polymer*, **2002**, *43*, 3857.C.
[22] Carlini, A.M. Raspolli Galletti, G. Sbrana, D. Caretti, *Polymer* **2001**, *42*, 5078.
[23] C. Carlini, M. Martinelli, E. Passaglia, A.M. Raspolli Galletti, G. Sbrana, *Macromol. Rapid Commun.* **2001**, *22*, 664.

[24] C.Carlini, M. Martinelli, A.M. Raspolli Galletti, G. Sbrana, *J. Polym. Sci., Part A: Polym. Chem.*, **2003**, *41*, 2117.

[25] R. H. Holm, K. Swaminathan, *Inorg. Chem.* **1963**, *2*, 181.

[26] C. Carlini, M. Isola, V. Liuzzo, A.M. Raspolli Galletti, G. Sbrana, *Appl. Catal. A: General* **2002**, *231*, 307.

[27] S. Yamada, K. Yamanouchi, *Bull. Chem. Soc. Jpn.* **1982**, *55*, 1083.

[28] M. S. W. Chan, L. Deng, T. Ziegler, Organometallics, **2000**, *19* 2741.

[29] M. Schumann, A. von Holtum, K.J. Wannowius, H. Elias, *Inorg. Chem.* **1982**, *21*, 606.

[30] S. Yamada, K.Yamanouchi, *Bull. Chem. Soc., Jpn.* **1982**, *55*, 1083.

[31] L. Oliva, P. Longo, A. Grassi, P. Ammendola, C. Pellecchia, *Makromol. Chem. Rapid Commun.* **1990**, *11*, 519.

[32] L. Porri, A. Giarrusso, G. L. Crossetti, A. Ripa, *ACS Polym Prep.* **1996**, *37*, 470.

[33] J. Ascenso, A. R. Dias, P. T. Gomes, C. C. Romão, I. Tkatchenko, A. Revillon, Q.-T. Pham, *Macromolecules* **1996**, *29*, 4172.

Macromol. Symp. **2004**, *213*, 221-233

Observation and Identification of the Catalytically Active Species of Bis(phenoxy-imine) Group 4 Transition Metal Complexes for Olefin Polymerization Using ¹H NMR Spectroscopy

Haruyuki Makio,* Terunori Fujita*

Mitsui Chemicals, Inc., 580-32 Nagaura, Sodegaura-City, Chiba 299-0265, Japan
E-mail: Haruyuki.Makio@mitsui-chem.co.jp
E-mail: Terunori.Fujita@mitsui-chem.co.jp

Summary: Solution structures of bis(phenoxy-imine) group 4 transition metal complexes (FI Catalysts) were investigated using ¹H NMR spectroscopy. At least two isomers exist in equilibrium for FI Catalysts precursors, bis[*N*-(3-*tert*-butylsalicylidene)anilinato]zirconium(IV) dichloride (**1**), and bis[*N*-(3,5-dicumylsalicylidene)anilinato]zirconium(IV) dichloride (**2**), while bis[*N*-(3-*tert*-butylsalicylidene)-2,3,4,5,6-pentafluoroanilinato]titanium(IV) dichloride (**3**) exhibits only one isomer under the conditions examined. Upon activation with MAO, all FI Catalysts (**1-3**) generate two species at ambient temperature judging from some key signals in the ¹H NMR. When temperature is raised (up to 75 °C), one species (**1a-3a**) converts irreversibly to the other species (**1b-3b**). The resulting species, **1b-3b**, are stereochemically rigid, in contrast to precursors **1** and **2**. Species **3b**, derived from a living FI Catalyst, exhibited virtually no reactivity toward olefin insertion. The imine protons of species **1b-3b** are temperature and solvent polarity sensitive. Two possibilities are proposed for the assignment of species **1b-3b**, i) heterobinuclear complexes of group 4 metal and alkylaluminum with methyl and/or chlorine as bridging groups and ii) phenoxy-imine ligated aluminum complexes whose ligands are transferred from the group 4 metal. The latter is more probable from the separate synthesis of LAlMe₂ (L: phenoxy-imine ligand). When **3** was activated with MAO in the presence of olefins, a new imine signal was observed. This species (**3c** for ethylene and **3d** for propylene) is thermally more robust than **3a** toward transformation to **3b** and assignable to the living propagating species.

Keywords: dynamic behavior; FI Catalysts; living polymerization; NMR; solution structure

Introduction

One important development in the field of single-site olefin polymerization catalysis has been the elucidation of a relationship between the dynamic structure of the active catalyst species and the

 DOI: 10.1002/masy.200450921

structure of resulting polyolefins. Taking advantage of this dynamic nature, unique polymerization processes based on the reversible transformation of catalytically active species have been developed.[1] Transformation of the catalyst species proceeds at a rate either much slower than chain-propagation, where the catalytic system will take on a multi-site character, or significantly faster than chain-propagation, where the system will apparently follow single-site polymerization mechanisms. When the transformation takes place on a time scale between catalyst initiation and polymer chain termination, the structural changes of the active species are reflected in the microstructures of individual polymer chains, which allows us unusual control over each monomer enchainment and syntheses of unique polymers with unprecedented structures. Fluxional processes of transition metal complexes that are observable on the NMR time scale (10^{-2} to 10^{-5} s) are similar in frequency to a typical propagation reaction in olefin polymerization. Therefore, such fluxionality can have an affect on polymerization reactivity or selectivity.

Significant time and effort have gone into exploring the potential of bis(phenoxy-imine) group 4 transition metal complexes as olefin polymerization catalysts (FI Catalysts).[2] By choosing the appropriate metal center, cocatalysts, and diversified ligand structures, these catalysts mediate ethylene polymerization with exceptionally high activity and desired molecular weight (molecular weight distribution),[3] living ethylene polymerization at ambient temperature or higher,[4] and living propylene polymerization with high syndiotacticity.[5,6] Schematically, FI Catalysts can have at least five isomers depending on the ligand coordination modes in an octahedral configuration.[2] In most cases, a stable *cis-N*, *trans-O*, *cis-Cl* isomer could be isolated in the solid state. However, we sometimes encountered multi-site characteristics with the FI Catalysts, which can probably be attributed to the presence of isomers in solution through intramolecular ligand rearrangement. One example includes the synthesis of polyethylene with controllable unimodal, bimodal, and trimodal molecular weight distributions by a Zr-FI catalyst, bis[*N*-(3,5-dicumylsalicylidene)anilinato]zirconium(IV) dichloride (2) / MAO.[3] Recently, a theoretical study involving the sydiospecific polymerization of propylene by bis[*N*-(3-*tert*-butylsalicylidene)-2,3,4,5,6-pentafluoroanilinato]titanium(IV) dichloride (3) and MAO suggested that the specificity was a result of rapid isomerization between active catalyst diastereomers.[7] Qualitatively, the theoretical results are in good agreement with experimental observations obtained from '*ligand-directed chain-end controlled*' syndiospecific propylene polymerizations.[2,5c]

In general, octahedral complexes are considered to be fairly stereochemically rigid unless metal-ligand bond cleavage is involved.[8] However, there is growing evidence of stereochemical nonrigidity with FI Catalysts[3] and other related octahedral complexes[9] with $[N_2O_2]^{2-}$ donor ligands. These complexes isomerize under the safe assumption that dissociation of a neutral coordinating atom is relatively easy. Even a non-dissociation mechanism may be allowed in the case of Ziegler-type polymerization catalysis where one ligand of the catalytically active cationic species is a loosely bound counter anion.

Investigation into the solution structures of FI Catalysts is of fundamental importance to understanding polymerization kinetics, selectivity, and polymer microstructures. Herein, we describe our attempts to observe the solution structures of some FI Catalysts, bis[N-(3-*tert*-butylsalicylidene)anilinato]zirconium(IV) dichloride (**1**), bis[N-(3,5-dicumylsalicylidene)anilinato]zirconium(IV) dichloride (**2**), and bis[N-(3-*tert*-butylsalicylidene)-2,3,4,5,6-pentafluoroanilinato]titanium(IV) dichloride (**3**), where fluxional processes between configurational isomers might lead to multimodal molecular weight distributions or syndiotactic polymers.

Experimental

Materials and Methods. All manipulations of air-sensitive materials were performed in Schlenk-type glassware on a dual-manifold Schlenk line or in a nitrogen-filled glove box with a high capacity recirculator (< 1 ppm O_2). Deuterated solvents were obtained from Cambridge Isotope Laboratories (all $\geq$ 99 atom %D), freeze-pump-thaw-degassed, dried over activated molecular sieves, and stored in resealable flasks. Precatalysts **1-3** were prepared as described in our previous papers.[3,4,10] Commercially available MAO (Albemarle, 1.2 M in toluene) was dried under vacuum to obtain white powder, dissolved in deuterated toluene, and dried again to remove a trace amount of residual non-deuterated toluene. This dried MAO (DMAO) was added as a powder whose molecular weight is 58 based on -Al(CH_3)O- unit.

Physical and Analytical Measurements. NMR spectra were recorded on JEOL GSX 270 spectrometer (FT, 270 MHz, ^{1}H). Chemical shifts were referenced using internal solvent resonances and are reported relative to tetramethylsilane. NMR experiments on air-sensitive samples were conducted in Teflon valve-sealed sample tubes (J. Young). In VT-NMR experiments, each temperature transition was conducted at intervals around 5–10 min including temperature equilibration and data collection.

Results and Discussion

In our previous paper,[3] fluxional processes for the FI Catalyst precursors, bis[N-(3-*tert*-butylsalicylidene)anilinato]zirconium(IV) dichloride (**1**), and bis[N-(3,5-dicumylsalicylidene)-anilinato]zirconium(IV) dichloride (**2**) were observed by ^{1}H and ^{15}N NMR spectroscopy. The isomers detected in solution were assigned to a *cis-N, trans-O, cis-Cl* isomer with C_2 symmetry (I) and a *cis-N, cis-O, cis-Cl* isomer with C_1 symmetry (II) that has a chemically nonequivalent phenoxy-imine ligand pair (Scheme 1). In general, with early transition metal complexes, shorter metal-O bonds are trans to one another to alleviate repulsion between valence shell electron pairs.[9] This is observed with the major solution-state isomer for **1**. However, the major isomer for **2** was found to have a *cis-N, cis-O, cis-Cl* configuration.[3] On the other hand, at room temperature the titanium complex, bis[N-(3-*tert*-butylsalicylidene)-2,3,4,5,6-pentafluoro-anilinato]titanium(IV) dichloride (**3**), exhibits a set of sharp signals for one isomer in solution (toluene-d_8 or CDCl$_3$), which is in sharp contrast to the broad, fluxional signals observed for the

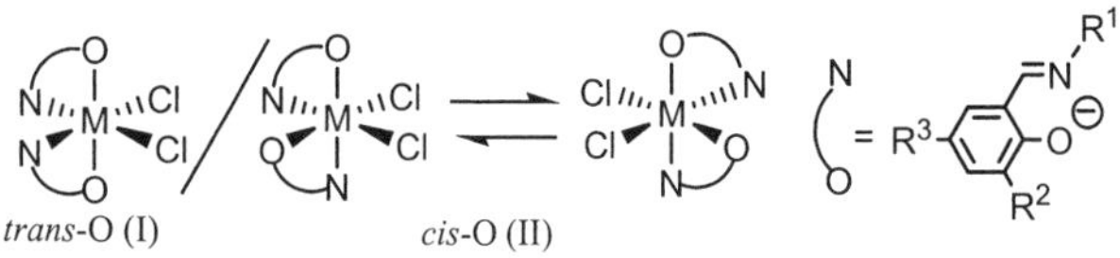

trans-O (I) *cis*-O (II)

1: R^1 = C$_6$H$_5$; R^2 = C(CH$_3$)$_3$; R^3 = H; M = Zr; I / II = 84/16[3]
2: R^1 = C$_6$H$_5$; R^2, R^3 = C(CH$_3$)$_2$Ph; M = Zr; I / II = 12/88[3]
3: R^1 = C$_6$F$_5$; R^2 = C(CH$_3$)$_3$; R^3 = H; M = Ti; I / II = 100/0

Scheme 1

Zr complexes **1** and **2** under the same conditions. Considering the crystallographically characterized structure as well as the relative formation energy calculated by DFT for **3**,[4b] the dominant species of **3** in solution is most likely the *cis-N, trans-O, cis-Cl* isomer.

Upon activation of **1** with DMAO (Al/Zr = 110, [Zr] = 7.0 μmol) at ambient temperature in toluene-d_8, two new imine signals appear at 7.81 and 7.51 ppm distinct from the peak at 7.54 ppm for **1** (Figure 1A). With increased temperature, the imine peak at 7.81 ppm decreases in intensity and shifts downfield slightly (7.85 ppm at 50 °C). The imine peak around 7.81 ppm disappears at about 50 °C, while the peak at 7.51 ppm increases in intensity and shifts downfield significantly (7.60 ppm at 50 °C and 7.65 ppm at 75 °C). This indicates that two species are generated and that one species (**1a**, Figure 1) is transformed into the other (**1b**, Figure 1). After cooling to room temperature, the signals for **1b** remain but those for **1a** do not reappear, whichsuggests that the transformation of **1a** to **1b** is irreversible. At all temperatures examined, judging by the single imine in the ¹H NMR spectra, species **1a** and **1b** seem to have one set of ligand signals and are stereochemically rigid except for the temperature-sensitive imine proton of **1b**, which may be involved in dynamic processes. Another feature is the appearance and growth of a sharp singlet at 0.18 ppm during the transformation of **1a** to **1b** (Figure 1B), which turns out to be assignable to methane by comparison with an authentic sample. This methane formation is observed in all experiments below.

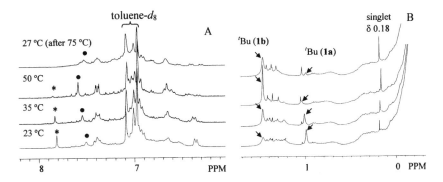

Figure 1. Variable-temperature ¹H NMR spectra of aromatic (A) and aliphatic region (B) of 1/DMAO. The imine signals for species **1a** and **1b** are marked * and •, respectively.

Precatalyst **2** (Al/Zr = 130, [Zr] = 6.0 µmol) was similarly activated with DMAO at ambient temperature in toluene-d_8. The ¹H NMR spectra involving DMAO activated precatalyst **2** are more complicated than those of activated precatalyst **1** and rather uninformative. However, as the temperature is increased, broad signals at 7.95 ppm decrease in intensity and disappear around 75 °C. A new imine signal similar in chemical shift to that of **1b** as well as another signal at 7.7 ppm appears. Irreversible increase or decrease in intensity of the other signals, which are poorly resolved and unassignable, suggests that a transformation from **2a** to **2b** proceeds in a similar manner to the 1/DMAO system. After complete formation of species **2b** the spectra are less complicated. The temperature-sensitive imine signal of **2b** (• in Figure 2) is clearly seen as a singlet. In addition, a doublet around 7.7 ppm (J = 2.31 Hz, denoted as × in Figure 2), whose chemical shift is rather temperature independent, can be observed (Figure 2A). In order to investigate solvent effects, all volatiles were then removed from the NMR tube *in vacuo* and deuterated tetrachloroethane ($C_2D_2Cl_4$) was added (Figure 2B). Using $C_2D_2Cl_4$ as a solvent, resolution was significantly improved. Notable is the dramatically downfield shifted singlet imine proton (~0.7 ppm at 25 °C, ● in Figure 2B), which demonstrates that the structure of species **2b** is sensitive to solvent polarity as well as the temperature. In contrast, the chemical shift of the doublet around 7.6 ppm (× in Figure 2B) does not shift significantly in $C_2D_2Cl_4$. These two signals are almost equal in intensity at all temperatures examined and the integration approximates one proton for each signal. The imine protons of FI Catalysts or phenoxy-imine ligands usually appear as singlets but sometimes as poorly resolved doublets or multiplets. In the latter cases, the coupling constants observed with the imine proton usually do not match those on the phenol ring, indicating that the multiplicity on imine protons is not caused by a long-range coupling between the imine proton and 6-H of 3-alkylsalicylidene moiety.[11] Since two other doublets having equal coupling constants of 2.31 Hz are observed at 6.86 ppm and 7.36 ppm (75 °C), the doublet at 7.62 ppm is assignable to an aromatic proton on a phenol-ring. Since neither peak broadening nor coalescence are observed (Figure 2), a configurational isomerization does not take place with **2b,** but some dynamic processes must be considered for the temperature and solvent sensitive imine chemical shifts. A sharp singlet at 0.18 ppm (methane) appeared in a similar manner to the reaction of **1/DMAO**.

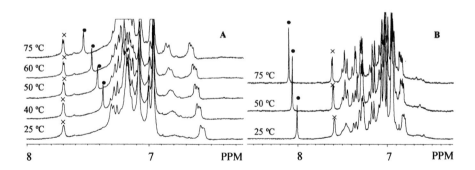

Figure 2. VT-^{1}H NMR spectra of **2/DMAO** in toluene-d_8 (A) and in $C_2D_2Cl_4$ (B), respectively (Al/Zr = 224, [Zr] = 6.3 mmol).

To further examine the generality of the observations made with precursors **1** and **2**, bis[*N*-(3-*tert*-butylsalicylidene)-2,3,4,5,6-pentafluoroanilinato]titanium(IV) dichloride (**3**) was chosen as a precursor in the next experiment. There are several advantages of investigating this complex: i) the perfluorinated phenyl ring simplifies the aromatic region of spectra; ii) greater solubility improves resolution even at room temperature; and iii) the living nature of this catalyst allows in-situ observation of subsequent reactions with olefins. In CDCl$_3$ or toluene-d_8 at room temperature, **3** shows only a single set of ligand signals in contrast to **1** and **2** under the same conditions. This indicates that **3** exists as a single isomer possessing C_2 symmetry (Scheme 1). Activation of **3** with DMAO at room temperature leads to the formation of two species (Figure 3). With the perfluorinated ligand, each species is fairly well separated from each other and the solvent signals, which allowed us to determine that both species have only one set of ligand signals. Thus, as shown in Figure 3, each species gives rise to five signals (*i.e.*, imine (s, 1H), phenol (dd, dd, t, 3H) and *tert*-butyl (s, 9H) signals) though one triplet overlaps with the solvent signals. In addition to these signals assignable to the ligands, another broad singlet appears at 1.97 ppm. Upon increasing the temperature, the signals of the dominant species at room temperature (**3a**) decreases in intensity and the signals for the other minor species (**3b**) increases. The singlet at 1.97 ppm decreases in intensity in accordance with the decrease in intensity of the signals derived from **3a**. The singlet integrates to 3H when the imine proton of **3a** at 8.20 ppm is integrated to 2H.

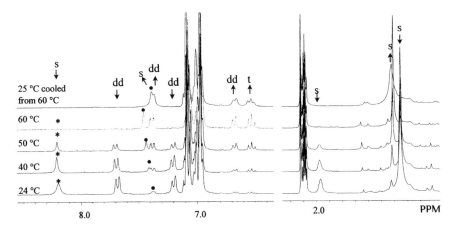

Figure 3. VT-^{1}H NMR spectra of **3**/DMAO in toluene-d_8 (Al/Ti = 108, [Ti] = 14.0 mmol). Symbols denote the imine protons of **3a** (*) and **3b** (•).

Thus, the singlet can be assigned to the methyl group attached to titanium; therefore, **3a** may be identified as the cationic methyl species, $L_2TiCH_3^+$ (L: phenoxy-imine ligand). The conversion of **3a** to **3b** is irreversible and the imine signal for **3b** shifts downfield with increased temperature in contrast to the other temperature independent signals. In addition, the formation of methane is again observed. These observations are similar to those seen in the previously described reactions involving precursors **1** and **2**, demonstrating that these characteristics have generality in the activation processes of FI Catalysts with MAO. Since MAO does not produce methane upon heating, the cationic species would be involved in the methane formation probably via C–H bond activation.

To identify species **3b**, various olefins were added to investigate whether or not olefin insertion will occur. Introduction of ethylene to the in-situ generated **3b** (Al/Ti = 53; ethylene/Ti ≈ 2) leads to formation of solid polyethylene. The ^{1}H NMR spectra of the reaction mixture shows large signals of **3b** remaining in the mixture and additional small broad signals in conjunction with broad polyethylene signals (spectrum iii in Figure 4). Unfortunately, it is almost impossible to ensure uniform reaction conditions in the NMR tube because the high MAO concentration tends to solidify the reaction mixture especially at the low temperatures preferable for reaction NMR experiments. To prevent this from occurring slower reactions with bulkier olefins were examined. Styrene did not react with **3b** even at 75 °C. With the reaction of **3b** with 1-hexene, only isomerization to internal olefins was observed. The low reactivity toward α-olefin insertion leads to the postulation that species **3b** (and presumably **1b** and **2b**) is not the active species during olefin polymerization.

In the real polymerization experiments, activation of FI Catalysts with MAO is usually conducted in the presence of a monomer. Polymerization starts without an induction period even below room temperature indicating extremely rapid formation of active species. Therefore, it is reasonable to assume that an immediately formed active species, **3a**, turns gradually into an inactive species, **3b**, in the absence of a reacting monomer. Activation in the presence of ethylene was conducted by injecting a toluene solution of **3** into DMAO (~70 equiv to Ti) and ethylene (1.4 or 7 equiv to Ti) dissolved in toluene at 0 °C. After removing all volatiles *in vacuo*, the residue was dissolved in toluene-d_8. A new set of signals (species **3c**) was observed in addition to those of **3b** and growing polyethylene signals (spectra iv and v in Figure 4). The peaks corresponding to **3c** are similar to those of **3a** except for the imine chemical shift, which appears slightly upfield (7.9 – 8.0 ppm) in comparison to the corresponding signal of **3a** (8.19 ppm) but much further downfield than the signal of **3b** (7.37 ppm). There are some signals in the olefinic region (4.8 – 6.6 ppm) which are unassignable at this time.

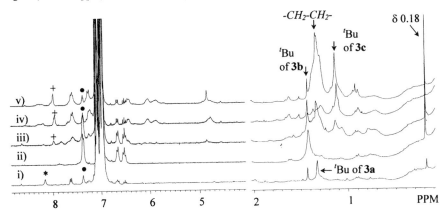

Figure 4. ^{1}H NMR spectra of **3**/DMAO prepared under various conditions. i) **3**/DMAO (1/108) at 24 °C; ii) **3**/DMAO (1/108) after reaction at 60 °C; iii) **3**/DMAO (1/53) after reaction at 60 °C for 5 min, ethylene (2 equiv to Ti) was added; iv) activation in the presence of ethylene, DMAO/ethylene/**3** (67/1.4/1); v) activation in the presence of ethylene, DMAO/ethylene/**3** (71/7/1). Characteristic imine signals are marked with * for **3a**, • for **3b**, and + for **3c**.

Olefin polymerization mediated by FI Catalysts is conceived to be catalyzed by a cationic metal–alkyl species like other well-defined group 4 transition metal catalysts. Thus, downfield shift of imine protons is most likely attributed to an increase in cationic nature. This assumption rationalizes the chemical shifts of imine protons observed in ^{1}H NMR, *e.g.*, two reactive species which appear downfield are assignable to the cationic methyl species, L_2Ti^+-CH_3 (**3a**) and the cationic propagating species, L_2Ti^+-$(CH_2CH_2)_n$-CH_3 (**3c**), respectively. The less reactive species **3b** with a diminished cationic character may be assigned to i) a heterobinuclear complex with alkylaluminum compounds, $[L_2Ti-\mu_2-(CH_3)(X)-Al(CH_3)(X)]^+$ (X: -CH_3 or -Cl) or ii) a L_nAlMe_{3-n} (n = 1 or 2) complex where L^- is transferred from the Ti center (Scheme 2). There are examples of the formation of similar complexes in metallocenes and other early transition metal complexes.[9a,12] The reaction between L–H and $AlMe_3$ (~0.5 equiv) in toluene-d_8 at room temperature monitored by ^{1}H NMR indicates the formation of a compound similar to **3b** as a major product, of which imine proton shows similar temperature dependence to that of **3b**. From the integration, the major product is assignable to LAlMe$_2$ (^{1}H NMR (δ, C$_7$D$_8$): 7.39-7.36 (m, 3H), 6.60 (dd, 1H), 6.51 (t, 1H), 1.43 (s, 9H), -0.35 (t, 6H)) which coexists with unreacted L–H, suggesting that **3b** is most likely to be LAlMe$_2$.

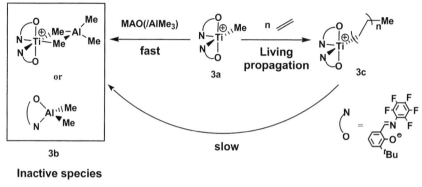

Scheme 2

Similar species were observed through the reaction of **3**, AlMe$_3$, and Ph$_3$CB(C$_6$F$_5$)$_4$. Two species very similar to **3a** and almost identical to **3b** are observed in the ^{1}H NMR spectrum (Figure 5). The signal of Ph$_3$CC*H$_3$* is clearly seen at 2.0 ppm, indicating methide abstraction from titanium,[13] which in turn suggests the generation of a cationic titanium species. These observations are in agreement with the generation of [L$_2$Ti-CH$_3$]$^+$[B(C$_6$F$_5$)$_4$]$^-$ similar to **3a** with concomitant formation of **3b**.

The activation of **3** was attempted in the presence of propylene by the same method used to activate **3** in the presence of ethylene. As observed with the ethylene system, species **3b** (the

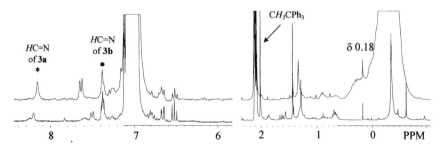

Figure 5. ^{1}H NMR spectra of **3** activated with DMAO (top; Al/Ti = 108) and AlMe$_3$ / Ph$_3$CB(C$_6$F$_5$)$_4$ (bottom; Al/Ti/B = 2.8/1/1.04).

imine protons are marked with • in Figure 6) and **3d** (supposedly with a propagating chain of propylene, the imine proton are marked with + in Figure 5) were observed. As temperature increases, **3d** is slowly transformed into **3b** but a significant amount of **3d** remains even at 75 °C for more than 15 min. Considering the fact that species **3a** turned to **3b** completely at 60 °C in the absence of propylene, the thermal resistance of **3d** toward formation of **3b** probably stems from a bulkier growing chain of propylene, which will discourage the formation of **3b**. This thermal stability is important in establishing the living nature of this species (Scheme 2).

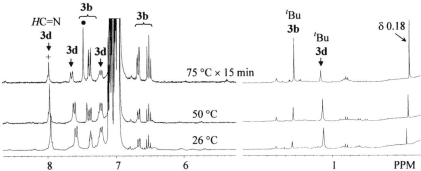

Figure 6. VT-^{1}H NMR spectra of **3** activated with DMAO in the presence of propylene (Al/Ti/propylene = 72/1/1.4). The imine proton of **3b** and **3d** is marked with • and +, respectively.

These interpretations qualitatively explain many general characteristics of olefin polymerizations mediated by FI Catalysts, *e.g.*, i) a bulky substituent next to the phenoxy-oxygen increase catalytic activity;[2,10] ii) DMAO exhibits higher activity than commercially available MAO (regular MAO) that contains a larger amount of AlMe$_3$ than DMAO;[14] and iii) reaction of FI Catalysts with a cocatalyst in the absence of a monomer lowers catalytic activity compared to activation in the presence of a monomer. These observations can be rationalized in terms of discouraging the formation of inactive species with AlMe$_3$ by i) steric hindrance of substituents close to reaction center, ii) lower AlMe$_3$ content in DMAO, and iii) steric protection by a bulkier growing polymer chain. To spectroscopically examine the phenomenon ii) mentioned above, trimethylaluminum was added to DMAO (DMAO: 1.4 mmol; AlMe$_3$: 0.35 mmol) to create similar conditions to those of regular MAO. Treatment of precursor **2** with the prepared MAO (**2**: 5.6 μmol, Al/Zr = 315) results in immediate formation of **2b** without heat, which verifies the hypothesis. The differences between DMAO and regular MAO have a significant effect on syndiotacticity in propylene polymerization with non-fluorinated Ti-FI Catalysts, that is, regular MAO affords higher syndiotacticity than DMAO.[14] This might not be explained in terms of the formation of inactive LAlMe$_2$ and therefore, direct interaction of AlMe$_3$ with active species, e.g., [L$_2$Ti-μ_2-(CH$_3$)$_2$Al(CH$_3$)$_2$]$^+$, cannot be excluded.

Conclusion

Reaction NMR experiments were conducted for the activation of three FI Catalyst precursors with DMAO. The chemical shifts of imine protons are well-separated at lower field and are sensitive to the structure and cationic character of the generated species to serve as a good probe for the identification of the catalytically active species. The results suggest that with all three precursors the active cationic methyl species are generated immediately and slowly transformed into inactive species, which are most likely to be aluminum complexes bearing phenoxy-imine ligands. Activation of the living precatalyst 3 in the presence of ethylene or propylene leads to the formation of a cationic propagating species, which is thermally more robust than the corresponding methyl species toward transformation to the aluminum species. These interpretations qualitatively explain general polymerization characteristics observed with FI Catalysts. Expected fluxionality between configurational isomers could not be examined in detail because the systems suffer from heterogeneity due to solidified MAO at lower temperatures and the formation of the aluminum complex at higher temperatures.

Acknowledgement

We thank Amber Kawaoka and Yasuhiko Suzuki for helpful discussions.

[1] For example, a) G. W.Coates, R. M. Waymouth, *Science* **1995**, *267* (5195), 217; b) C. Muller, D. Lilge, M. O. Kristen, P. Jutzi, *Angew. Chem. Int. Ed.* **2000**, *39*, 789.
[2] As review articles, a) H. Makio, N. Kashiwa, T. Fujita, *Adv. Synth. Catal.* **2002**, *344*, 477; b) N. Matsukawa, S. Ishii, R. Furuyama, J. Saito, M. Mitani, H. Makio, H. Tanaka, T. Fujita, *e-Polymers* **2003**, no. 021; c) M. Mitani, T. Nakano, T. Fujita, *Chem.Eur.J.* **2003**, *9*, 2396; d) Y. Suzuki, H. Terao, T. Fujita, *Bull. Chem. Soc. Jpn.* **2003**, *76*, 1493.
[3] Y. Tohi, H. Makio, S. Matsui, M. Onda, T. Fujita, *Macromolecules* **2003**, 36, 523.
[4] a) J. Saito, M. Mitani, J. Mohri, Y. Yoshida, S. Matsui, S. Ishii, S. Kojoh, N. Kashiwa, T. Fujita, *Angew. Chem. Int. Ed.* **2001**, *40*, 2918; b) M. Mitani, J. Mohri, Y. Yoshida, J. Saito, S. Ishii, K. Tsuru, S. Matsui, R. Furuyama, T. Nakano, H. Tanaka, S. Kojoh, T. Matsugi, N. Kashiwa, T. Fujita, *J. Am. Chem. Soc.* **2002**, *124*, 3327.
[5] a) J. Saito, M. Mitani, J. Mohri, S. Ishii, Y. Yoshida, T. Matsugi, S. Kojoh, N. Kashiwa, T. Fujita, *Chem. Lett.*

2001, 576; b) M. Mitani, R. Furuyama, J. Mohri, J. Saito, S. Ishii, H. Terao, N. Kashiwa, T. Fujita, *J. Am. Chem. Soc.* **2002,** *124*, 7888; c) M. Mitani, R. Furuyama, J. Mohri, J. Saito, S. Ishii, H. Terao, T. Nakano, H. Tanaka, T. Fujita, *J. Am. Chem. Soc.* **2003,** *125*, 4293.

[6] J. Tian, P. D. Hustad, G. W. Coates, *J. Am. Chem. Soc.* **2001,** 123, 5134.

[7] G. Milano, L. Cavallo, G. Guerra, *J. Am. Chem. Soc.* **2002,** *124*, 13368.

[8] F. A. Cotton, G. Wilkinson, C. A. Murillo, M. Bochmann, "*Advanced Inorganic Chemistry*", 6th ed., A Wiley-Interscience Publication, New York, 1999, p. 13.

[9] a) X. Bei, D. C. Swenson, R. F. Jordan, *Organometallics* **1997,** *16*, 3282; b) T. Tsukahara, D. C. Swenson, Richard F. Jordan, *Organometallics,* **1997,** *16*, 3303; c) I. Kim, Y.Nishihara, R. F. Jordan, R. D. Rogers, A. L. Rheingold, G. P. A. Yap, *Organometallics,* **1997,** *16*, 3314; d) T. Toupance, S. R. Dubberley, N. H. Rees, B. R. Tyrrell, P. Mountford, *Organometallics,* **2002,** *21*, 1367.

[10] a) S. Matsui, M. Mitani, J. Saito, Y. Tohi, H. Makio, H. Tanaka, T. Fujita, *Chem.Lett.* **1999,** 1263; b) S. Matsui, M. Mitani, J. Saito, Y. Tohi, H. Makio, N. Matsukawa, Y. Takagi, K. Tsuru, M. Nitabaru, T. Nakano, H. Tanaka, N. Kashiwa, T. Fujita, *J. Am. Chem. Soc.* **2001,** *123*, 6847.

[11] For example, at a lower temperature of –25 °C in $C_2D_2Cl_4$, the imine protons for **1** appears in doublet for both isomers shown in Scheme 1 (major: δ 8.12; 2.3 Hz, minor: δ 8.07 and 7.86; 2.3 Hz). The coupling on the imine is clearly larger than those on the double doublets of the phenol ring of **1** (7.8 Hz, 1.4 Hz).

[12] For example, M. Bochmann, S. J. Lancaster, *Angew. Chem. Int. Ed. Engl.,* **1994,** *33*, 1634.

[13] No reactions were observed between $AlMe_3$ and $Ph_3CB(C_6F_5)_4$.

[14] R. Furuyama, J. Saito, S. Ishii, M. Mitani, S. Matsui, Y. Tohi, H. Makio, N. Matsukawa, H. Tanaka, T. Fujita, *Journal of Molecular Catalysis A: Chemical,* **2003,** *200*, 31.

Polymerization of Propene by Post-Metallocene Catalysts

Marina Lamberti,[1] *Mina Mazzeo,*[1] *Daniela Pappalardo,*[2] *Adolfo Zambelli,*[1]
Claudio Pellecchia*[1]

[1] Dipartimento di Chimica, Università di Salerno, I-84081 Baronissi (SA), Italy
E-mail: cpellecchia@unisa.it
[2] Dipartimento di Scienze Geologiche ed Ambientali, Università del Sannio, via
Port'Arsa 11, I-82100 Benevento, Italy

Summary: In the last few years several non-metallocene catalysts have been
disclosed as efficient catalysts for the stereospecific polymerization of propene.
In this paper we summarize some recent literature data and some new results
concerning the stereochemical mechanism of propene polymerization promoted
by late transition metal systems and group 4 metal bis(phenoxyimine) systems.
NMR analysis of the fine structure of the polymers obtained, in some cases using
isotopically enriched reagents, provides valuable information on the
regiochemistry and stereochemistry of the polymerization.

Keywords: catalysis; microstructure; NMR; poly(propylene); stereochemical
mechanism

Introduction

The renaissance era of olefin polymerization catalysis has been mainly driven by the

successful design and application of Group 4 metallocenes, some of which are now used in

industrial processes.[1] While research in the metallocene area appears far from over, in the

last years increasing interest has been directed toward the search for new class of catalysts,

involving complexes of transition metals throughout the Periodic Table.[2] Examples of

successful new classes of highly active non-metallocene catalysts include late-transition metal

systems such as the Brookhart's Ni and Pd diimine catalysts,[3] and the Fe bis(imino)pyridyl

catalysts, and group 4 metal bis(phenoxyimine) catalysts,[4] all of them showing a number of

interesting peculiar features. One of the outcomes from these developments has been the

expansion of stereospecific polymerizations.

In this paper we review some recent literature data and some new results concerning the

© 2004 WILEY-VCH Verlag GmbH & KGaA, Weinheim   DOI: 10.1002/masy.200450922

stereochemical mechanism of propene polymerization promoted by "post-metallocene" catalysts, namely the above mentioned late-transition metal and bis(phenoxymine) systems. Most of the results discussed were achieved by using techniques, developed by Zambelli in the 70's, based on the NMR analysis of suitably labelled polymers obtained by employing isotopically enriched catalysts or monomers. In particular, ^{13}C can be introduced in different points of the polymer chain as a sort of magnifying lens to observe fine aspects of the microstructure, and to establish a chronology in the sequence of the monomer insertions. Thus, NMR analysis of ^{13}C enriched end groups provides information about the regiochemistry and the stereochemistry of the monomer insertion in the initiation steps. A useful method to investigate the chain propagation involves the synthesis of copolymers of propene containing isolated units of 1-^{13}C ethylene: NMR analysis provides information on the regiochemistry and the stereochemistry of propene insertion before and after ethylene entrance. When the ethylene unit bridges regioregular, either primary or secondary, polypropylene blocks, a three-methylene sequence is formed (see Scheme 1 a, b). An enriched two-methylene sequence can derive only from ethylene spanning propene units of opposite regiochemistry, according to scheme 1c. Finally, a four-methylene sequence can derive from the succession of insertions displayed in scheme 1d.

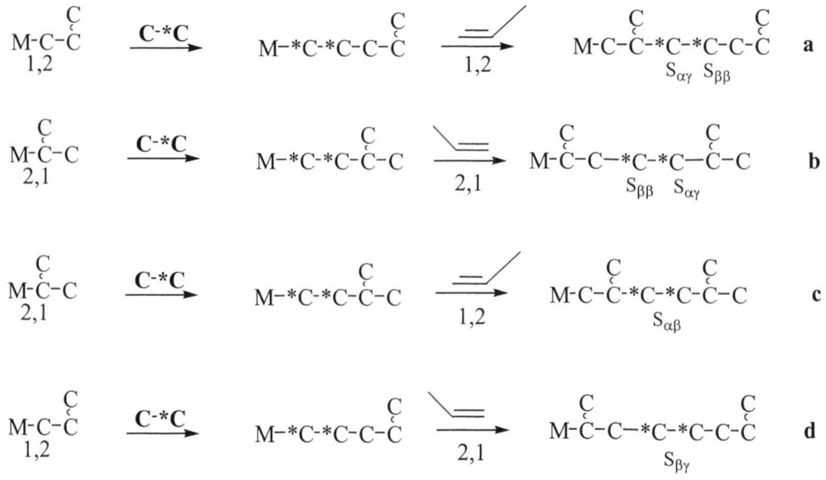

Scheme 1

Since the chemical shifts of the enriched methylenes[5] are affected by the stereochemical arrangements of the neighbouring propene units, valuable information on the mechanism of steric control can be obtained. In particular, if the steric control crosses intervening ethylene units, the stereochemistry of the insertion is controlled by the chirality of the active sites, while if it does not cross the intervening ethylene units, the stereospecificity of the insertion arises from the asymmetric induction by the last (propene) unit of the growing chain. The whole matter has been recently revisited analyzing, by high field NMR, several propene-enriched ethylene copolymers prepared with a variety of catalysts.[6]

1. Iron bis(imino) Pyridine Catalysts

Iron bis(imino)pyridine complexes, activated by MAO, were independently reported by Brookhart[7] and Gibson[8] as very active ethylene polymerization catalysts producing strictly linear, high-molecular weight polymers. These catalysts are moderately active in the polymerization of propene.[9-10] The bis(imino)pyridine complex with bulky isopropyl *ortho* substituents on the arylimino moieties (**Fe-1**, Scheme 2) is able to produce regioregular isotactic polypropylene with a 69% *mm* triad content at 0 °C.

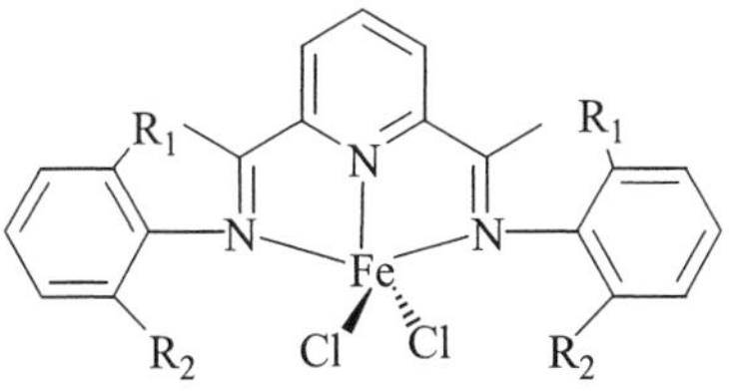

Fe-1: $R_1 = R_2 = CH(CH_3)_2$

Fe-2: $R_1 = CH_3$; $R_2 = CH(CH_3)_2$

Scheme 2

NMR analysis of the polypropylene microstructure suggested that a chain end mechanism of steric control is operative. Interestingly, this steric control is still effective above room temperature, e.g. at 50 °C (*mm*=60%).[11]

The regiochemistry of the monomer insertion in the initiation step was investigated by preparing a polymer sample bearing ^{13}C enriched end groups.[11] The major ^{13}C enriched end groups were *sec*-butyls, deriving from 2,1 insertions of propene into Fe-[13]CH₃ bonds (Scheme 3).

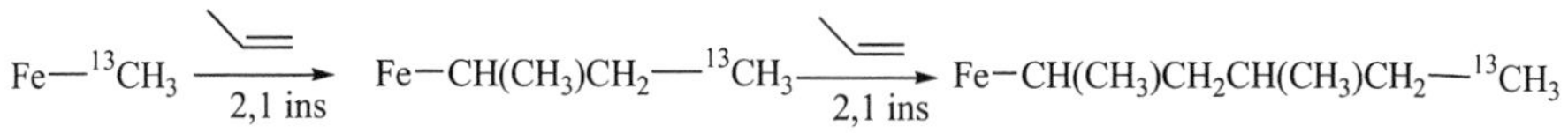

$Fe-^{13}CH_3 \xrightarrow[2,1 \text{ ins}]{} Fe-CH(CH_3)CH_2-^{13}CH_3 \xrightarrow[2,1 \text{ ins}]{} Fe-CH(CH_3)CH_2CH(CH_3)CH_2-^{13}CH_3$

Scheme 3

Isopentyl end groups were also detected, deriving from initiation via 1,2 insertion followed by 2,1 insertion (Scheme 4).

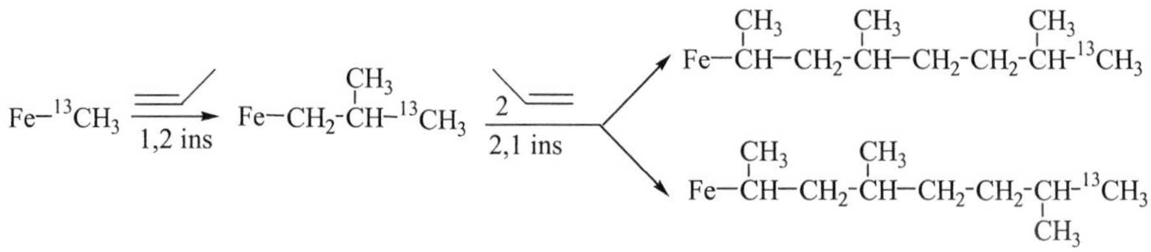

$$Fe-^{13}CH_3 \xrightarrow[1,2 \text{ ins}]{} Fe-CH_2-\overset{CH_3}{\underset{|}{CH}}-^{13}CH_3 \xrightarrow[2,1 \text{ ins}]{2}$$

$Fe-\overset{CH_3}{\underset{|}{CH}}-CH_2-\overset{CH_3}{\underset{|}{CH}}-CH_2-CH_2-\overset{CH_3}{\underset{|}{CH}}-^{13}CH_3$

$Fe-\overset{CH_3}{\underset{|}{CH}}-CH_2-\overset{CH_3}{\underset{|}{CH}}-CH_2-CH_2-\underset{\underset{CH_3}{|}}{\overset{}{CH}}-^{13}CH_3$

Scheme 4

Analysis of the natural abundance end groups showed the presence of unsaturated allyl groups deriving from β-H-elimination at the methyl of a secondary growing chain (Scheme 5) and *n*-butyl end groups, originating from 1,2 insertion of propene into Fe-H bonds followed by a 2,1 insertion (Scheme 6).

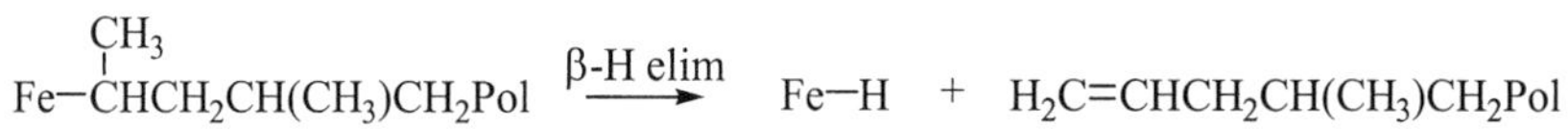

$Fe-\overset{CH_3}{\underset{|}{CH}}CH_2CH(CH_3)CH_2Pol \xrightarrow{\beta\text{-H elim}} Fe-H + H_2C=CHCH_2CH(CH_3)CH_2Pol$

Scheme 5

$Fe-H \xrightarrow[1,2 \text{ ins}]{} Fe-CH_2CH_2CH_3 \xrightarrow[2,1 \text{ ins}]{} Fe-CH(CH_3)CH_2CH_2CH_2CH_3$

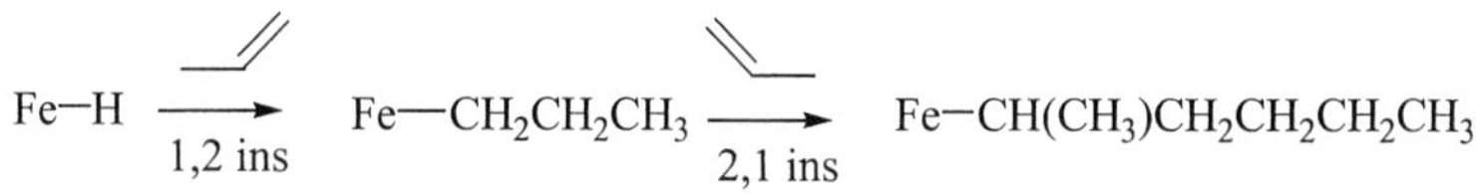

Scheme 6

Since the polymer backbone is highly regioregular, the above end group analysis indicated that primary insertion is favoured into Fe-H bond, primary and secondary insertion have comparable probabilities into Fe-CH$_3$ bonds, while secondary insertion is preferred into Fe-alkyl bonds (for alkyl > CH$_3$). In conclusion secondary insertion is the most probable event during the propagation.

Complex **Fe-2** affords a polypropylene with a similar content of *mm* triads, although the polymer microstructure slightly deviates from the composition expected for a simple chain-end mechanism. Thus, a copolymer of propene containing about 1% of 1-^{13}C enriched ethylene was prepared with complex **Fe-2**.[6] NMR analysis of the copolymer showed only three-methylene sequences, in agreement with a highly regioregular secondary insertion of the monomer into any Fe-R bond (R > CH$_3$). Further information came from close examination of the fine structure of the enriched carbons (Figure 1). The relative areas of the resonances of the m-S$_{\alpha\gamma}$ and r-S$_{\alpha\gamma}$ enriched methylenes are the same as the molar fraction of m and r dyads inside the polypropylene blocks as determined from the methyl pentad resonances. This is again in agreement with the above mentioned mechanism since, for a secondary mode of propagation, the S$_{\alpha\gamma}$ carbon resonance is diagnostic of the stereochemical arrangement of the two propene units incorporated *before* the insertion of ethylene unit.

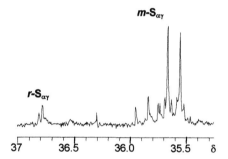

Figure 1. S$_{\alpha\gamma}$ region of the 100.6 MHz ^{13}C NMR spectrum (C$_2$D$_2$Cl$_4$, 100 C) of a copolymer of propene with a little amount of 1-^{13}C-ethylene prepared with **Fe-2**-MAO. δ in ppm from hexamethyldisiloxane.

In the S$_{\beta\beta}$ region of the spectrum (Figure 2) two major resonances attributable to *mmMmm* and *mmRmm* pseudohexads were detected, [6] confirming a chain end mechanism of steric control, *i.e.* the steric control does not cross the intervening ethylene unit (see Scheme 7).

$$mmMmm \quad C-\overset{\overset{\textstyle C}{|}}{C}-C-\overset{\overset{\textstyle C}{|}}{C}-C-\overset{\overset{\textstyle C}{|}}{C}-C-C-C-\overset{\overset{\textstyle C}{|}}{C}-C-\overset{\overset{\textstyle C}{|}}{C}-C-\overset{\overset{\textstyle C}{|}}{C}-C$$

$$mmRmm \quad C-\overset{\overset{\textstyle C}{|}}{C}-C-\overset{\overset{\textstyle C}{|}}{C}-C-\overset{\overset{\textstyle C}{|}}{C}-C-C-C-\underset{\underset{\textstyle C}{|}}{C}-C-\underset{\underset{\textstyle C}{|}}{C}-C-\underset{\underset{\textstyle C}{|}}{C}-C$$

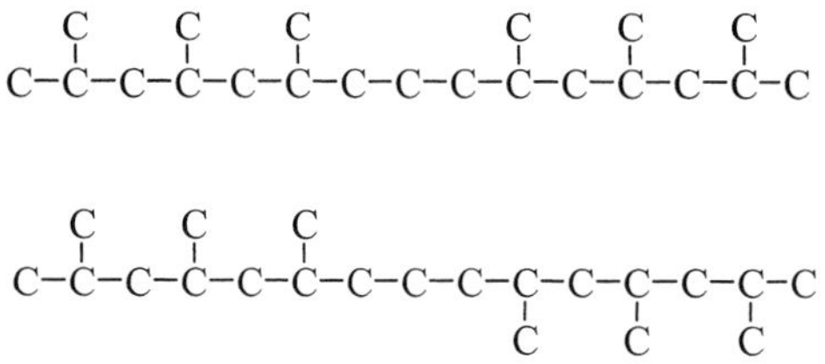

Scheme 7

A similar spectrum was observed for a copolymer prepared with the Ewen's achiral Cp_2TiPh_2-MAO catalyst which affords isotactic polypropylene via chain-end control at low temperature.[12]

On the contrary, only the *mmMmm* pseudohexad was observed for an isospecific C_2-symmetric *ansa*-zirconocene such as *rac*-$Me_2Si(2$-Me-4-Ph-Ind$)_2ZrCl_2$[13] in agreement with an enanthiomorfic-site mechanism of steric control.[6]

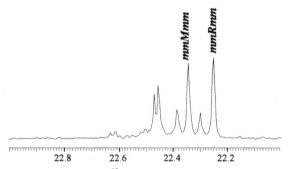

Figure 2. $S_{\beta\beta}$ region of the 100.6 MHz ^{13}C NMR spectrum ($C_2D_2Cl_4$, 100 °C) of a copolymer of propene with a little amount of 1-^{13}C-ethylene prepared with **Fe-2**-MAO. δ in ppm from hexamethyldisiloxane.

2. Nickel α-diimine Catalysts

Nickel α–diimine catalysts disclosed by Brookhart polymerize ethylene and α-olefins to high molecular mass polymers with activities comparable to early transition metal systems.[14] The polymer microstructure and properties are greatly dependent on the reaction conditions and on the ligand structure.[3] These catalysts are also moderately active in the polymerization of propene. The nickel complex displayed in Scheme 8 produces syndiotactic polypropylene at sub-ambient temperature, with a *rr* triad content up to 80% at -78°C.[15]

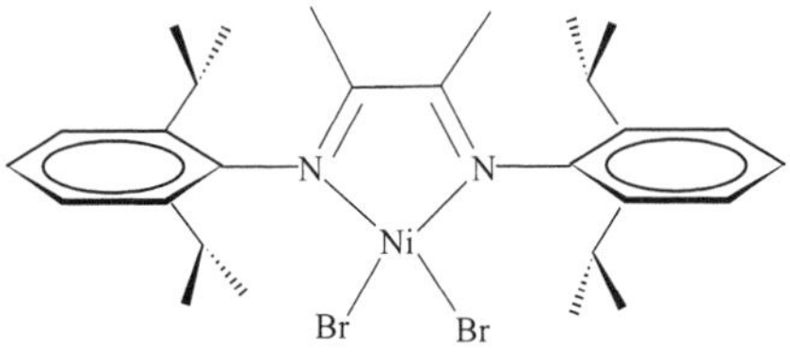

Scheme 8

Analysis of pentad region of the ^{13}C NMR spectrum suggested a chain end mechanism of steric control, since isolated *m* diads are the major stereodefects. A significant amount of regioirregular monomer units was also detected (about 12%). The whole picture is similar to the case of the classical V-based sydiospecific catalyst.[16-17] A significant difference emerged from investigation of the regiochemistry of the monomer insertion. The insertion in the initiation steps was determined using a ^{13}C enriched catalyst, producing a polypropylene with enriched isobutyl end groups indicating a primary (1,2) monomer insertion (Scheme 9). Also, only two of four possible diasteroisomers were observed indicating that steric control becomes active when the first chiral carbon is generated in the growing chain,[18] while in the case of vanadium the steric control is effective only when monomer insertion turns from primary to secondary.[19]

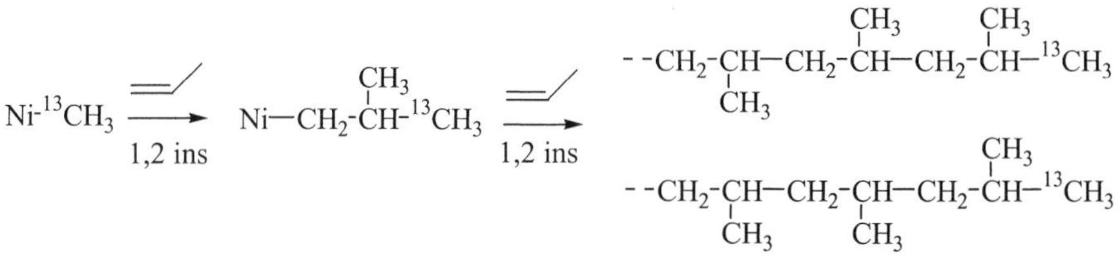

$$Ni\text{-}^{13}CH_3 \xrightarrow[\text{1,2 ins}]{} Ni\text{—}CH_2\text{-}\underset{\underset{CH_3}{|}}{CH}\text{-}^{13}CH_3 \xrightarrow[\text{1,2 ins}]{}$$

$$\text{-}\text{-}CH_2\text{-}\underset{\underset{CH_3}{|}}{CH}\text{—}CH_2\text{-}\underset{\overset{CH_3}{|}}{CH}\text{—}CH_2\text{-}\underset{\overset{CH_3}{|}}{CH}\text{—}^{13}CH_3$$

$$\text{-}\text{-}CH_2\text{-}\underset{\underset{CH_3}{|}}{CH}\text{—}CH_2\text{-}\underset{\underset{CH_3}{|}}{CH}\text{—}CH_2\text{-}\underset{\overset{CH_3}{|}}{CH}\text{—}^{13}CH_3$$

Scheme 9

A primary (1,2) monomer insertion as the main propagation mode was confirmed by analyzing the end groups formed in the termination step, in which a deuterium label was introduced by quenching the polymer with deuteriated trifluoromethanol[20] (Scheme 10). The formation of a deuteriated methyl group of an isobutyl was confirmed by ^{13}C and ^{2}H NMR analysis, while in the case of vanadium catalyst the same experiment led to a deuteriated methylene of a n-propyl.[21]

$$Ni\text{—}CH_2CH(CH_3)CH_2CH(CH_3)Pol \xrightarrow{D^+} D\text{-}CH_2CH(CH_3)CH_2CH(CH_3)Pol \quad \textbf{a}$$

$$V\text{—}CH(CH_3)CH_2CH(CH_3)CH_2Pol \xrightarrow{D^+} CH_3CHDCH_2CH(CH_3)Pol \quad \textbf{b}$$

Scheme 10

A copolymer of propene with trace amounts of ^{13}C enriched ethylene was prepared also with the nickel catalyst (see the ^{13}C NMR spectrum in figure 3). In this case sequences of two, three and also four methylenes were observed in comparable amount, in agreement with the poor regioregularity of the system.

The ratio r-$S_{\alpha\gamma}$/m-$S_{\alpha\gamma}$ (6:4) is lower than the r/m ratio in the main chain (~8:2): this is not surprising considering that, since the propagation is prevailing 1,2, the $S_{\alpha\gamma}$ carbon resonance is diagnostic of the two propene units incorporated *after* the enriched ethylene unit, and that the chain end control can be affected by the presence of a less bulky chain.

In this case the region of $S_{\beta\beta}$ carbons (Figure 3) is less informative; in fact, a sharp peak at 22.50 ppm emerges from a broad band of resonances: this is the same resonance found in the case of a copolymer prepared with a highly syndiospecific C_s symmetric metallocene[6] but in no sample examined up to now there is evidence of any detectable splitting of the resonances for the *rrRrr* and *rrMrr* pseudohexads, which would allow one to establish if the steric control crosses or not the intervening ethylene unit.[6] Theoretical calculations and analysis of model compounds suggest that in this case the chemical shift difference should be significantly lower than in the case of *mmMmm* and *mmRmm* pseudohexads.

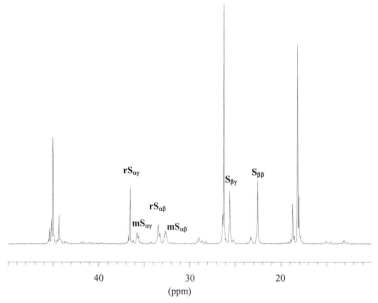

Figure 3. 100.6 MHz ^{13}C NMR spectrum ($C_2D_2Cl_4$, 100 C) of a copolymer of propene with a little amount of 1-^{13}C-ethylene prepared with the nickel catalyst-MAO. δ in ppm from hexamethyldisiloxane.

3. Bis(phenoxyimine) Group 4 Metal Catalysts

The octahedral C_2 symmetric bis(phenoxyimine) Group 4 metal complexes (Scheme 11), developed by the group of Fujita,[22] are able to produce polyethylene with activities comparable to or even exceeding those of metallocene catalysts. Production of syndiotactic polypropylene using a titanium complex of this class was first reported in the scientific literature by Coates;[23] microstructural analysis of the resulting polymer suggested that a chain-end mechanism was responsible for the observed stereocontrol.[23,24,25] The polymer stereoregularities range from moderate to high, depending on the precatalyst structure and the polymerization conditions.[26] The use of perfluorinated N-aryl groups improves the catalyst performances, concerning both activities, stereospecificity and polymer molecular weights. Very interestingly, titanium complexes bearing fluorinated N-aryl groups polymerize ethylene and propene in a living fashion, allowing the synthesis of a variety of block copolymers.[24,25,26]

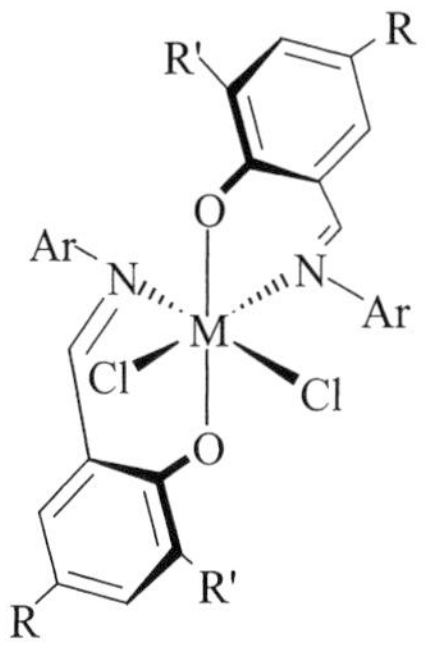

Complex 1: R = t-Bu; Ar = C_6H_5

Complex 2: R = H; Ar = C_6H_5

Complex 3: R = H; Ar = C_6F_5

Complex 4: R = t-Bu; Ar = C_6F_5

Complex 5: R = t-Bu; Ar = 3,5-F_2-C_6H_3

M = Ti; R' = t-Bu

Complex 6: M = Zr

Complex 7: M = Ti

R' = Me; R = H; Ar = C_6F_5

Scheme 11

We studied the regiochemistry and the stereochemistry of propene polyinsertion in the presence of complex 2 (Scheme 11) *via* NMR analysis of the natural abundance polymer end-groups.[27] The polymer is prevailingly syndiotactic, the content of *rr* triads being 81%, with few regioirregularly arranged monomer units (<2%). As observed in the case of the iron catalyst (see Scheme 5), allyl end groups originated from β-hydrogen elimination at the methyl of the growing chain-end after a 2,1 insertion were detected. In the saturated region of the spectrum, resonances attributable to *n*-propyl and *n*-butyl end groups were observed. The latter originates from 1,2 monomer insertion into Ti-H bonds, followed by an insertion with opposed (2,1) regiochemistry, again as observed for the iron catalyst (Scheme 6), while the former arises from at least two consecutive 1,2 monomer insertions into Ti-H bonds (Scheme 12).

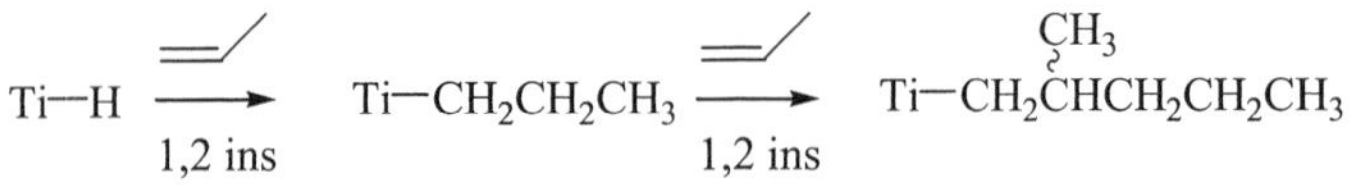

Ti—H
1,2 ins
Ti—CH2CH2CH3
1,2 ins
CH3
Ti—CH2CHCH2CH2CH3

Scheme 12

No incorporation of deuterium in the polymer was observed when the polymerization run was terminated by adding a mixture of deuteriated alcohols, therefore ruling out the production of a significant amount of *n*-propyl and *n*-butyl end groups by hydrolysis of polymer chains bound either to Ti or Al in the termination step.

The above end group analysis suggested that β-hydrogen elimination is the main termination process, generating allyl-terminated macromolecules and Ti-H bonds, into which primary insertion of propene is favoured, and is followed either by a primary (60%) or a secondary (40%) propene insertion (see Scheme 13).

Ti—Pol
2,1 ins
CH3
Ti-CHCH2CH(CH3)CH2Pol
β-H elimin
H2C=CHCH2CH(CH3)CH2Pol
+
Ti-H
1,2 ins
Ti-CH2CH2CH3
1,2 ins
CH3
Ti-CH2CHCH2CH2CH3
2,1 ins
CH3
Ti-CHCH2CH2CH2CH3

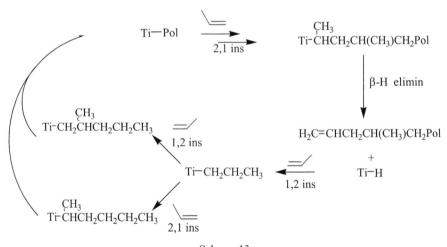

Scheme 13

Similar conclusions were reported by Coates,[28] from end group analysis of a *s*PP sample produced with complex **5** and also on the basis of the results of cyclopolymerization of 1,6-heptadiene.

The regiochemistry of polymerization has been studied also for the catalytic system based on the perfluorinated complex **3**. In this case the living nature of the polymerization made the analysis difficult, because of the absence of chain release. Low molecular weight polymer samples, suitable for ^{13}C NMR analysis of chain-end groups in natural abundance, were obtained by Fujita,[26b] quenching the reactions after short times. Only saturated end-groups were distinguished: *n*-propyl (34 %), isobutyl (33 %) and isopentyl (33 %). Isopentyl chain-end groups would originate in the initiation step from 1,2 insertion into a Ti-Me bond followed by 2,1 insertion, while *n*-propyl chain-end groups would derive from hydrolysis of a secondary growing chain. Isobutyl groups would be generated either in the initiation (*via* two consecutive propene primary insertions into a Ti-Me bond) or in the termination steps (*via* hydrolysis of a primary growing chain).

From the relative ratio of the chain-end groups formed during the initiation step (i.e. isobutyl and isopentyl) and during the termination step, Fujita estimated that 2,1 insertion is only about 70 % and proposed that the polypropylenes possess regio-block structures consisting of consecutive 2,1-insertion and 1,2-insertion segments.

Further information came from analysis of a propene-1-^{13}C-enriched ethylene copolymer prepared by catalyst **3**-MAO.[27]

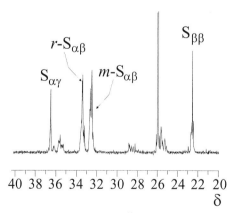

Figure 4. Aliphatic region of the 100.6 MHz ^{13}C NMR spectrum (C$_2$D$_2$Cl$_4$, 100 C) of a copolymer of propene with a little amount of 1-^{13}C-ethylene prepared with **3**-MAO. δ in ppm from hexamethyldisiloxane.

NMR analysis (see Figure 4) indicated the presence of two-methylene and three-methylene sequences. The methylene pairs outnumber the three-methylene sequences, in agreement with the previous conclusion concerning the prevailing 2,1 regiochemistry insertion of propene during the "normal" propagation, and the observation that 1,2 insertion is slightly favoured into metal-primary carbon bond.

In addition, in this case the fraction of the r-$S_{\alpha\beta}$ (Scheme 14a) enriched carbons is unexpectedly low (45%) if compared to the r diad content inside the polypropylene blocks (98%). On the basis of literature data,[29] the m-$S_{\alpha\beta}$ (Scheme 14b) resonance probably overlaps with the $S_{\alpha\beta}$ carbon flanking a regioirregular propene diad (Scheme 14c). Since two methylene sequences can derive only from a ^{13}C-enriched ethylene inserted after a secondary polypropylene block and followed by a primary propene insertion (see above, Scheme 1c), and since the secondary block is highly syndiotactic, one may conclude that either 1,2 insertion is highly isotactic or that it is prevailingly followed by a 2,1 insertion, ruling out the presence of significant amount of primary blocks. [30]

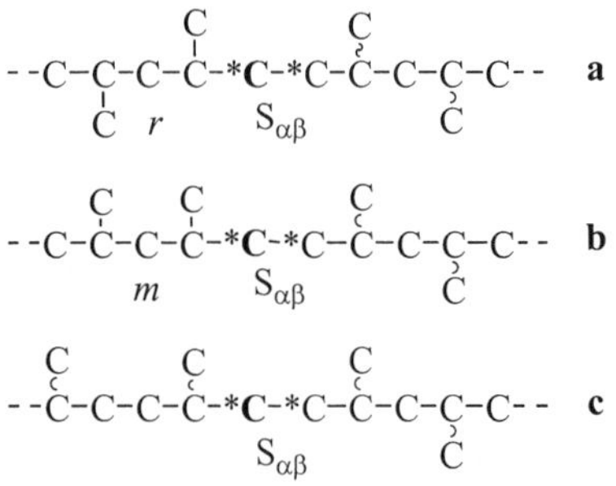

$$--C-\underset{\underset{C}{|}}{C}-C-\underset{\underset{C}{|}}{\overset{\overset{C}{|}}{C}}-*C-*C-\underset{\underset{C}{|}}{\overset{\overset{C}{|}}{C}}-C-\underset{\underset{C}{|}}{C}-C-- \quad \mathbf{a}$$
$$r \qquad S_{\alpha\beta} $$

$$--C-\underset{\underset{m}{|}}{\overset{\overset{C}{|}}{C}}-C-\underset{|}{\overset{\overset{C}{|}}{C}}-*C-*C-\underset{|}{\overset{\overset{C}{|}}{C}}-C-\underset{\underset{C}{|}}{C}-C-- \quad \mathbf{b}$$
$$m \qquad S_{\alpha\beta} $$

$$--\overset{\overset{C}{|}}{C}-C-C-\overset{\overset{C}{|}}{C}-*C-*C-\overset{\overset{C}{|}}{C}-C-\underset{\underset{C}{|}}{C}-C-- \quad \mathbf{c}$$
$$S_{\alpha\beta} $$

Scheme 14

The main resonance of the $S_{\beta\beta}$ carbons is the same observed in all the previous mentioned sPP prepared by either C_s symmetric group 4 metallocene or Ni diimine based catalysts (see above), thus not providing any additional information on the origin of the enantioselectivity.

Very recently we synthesized a zirconium(IV) complex bearing two N-(3-methylsalicylidene)-2,3,4,5,6-pentafluoroaniline and two Cl ligands (complex **6** in Scheme 11).[31] X-ray analysis (see Figure 5) indicates that **6** adopts a distorted octahedral structure with a $trans$-O, cis-N, and cis-Cl arrangement, similarly to previously reported complexes of this class.

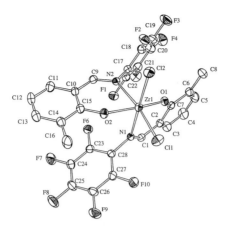

Figure 5. Ortep drawing of the molecular structure of compound **6**. Hydrogen atoms have been omitted for clarity. Thermal ellipsoids are drawn at 20% probability level.

Polymerization of propene at room temperature in the presence of **6** resulted in the production of a poorly stereoregular, prevailingly syndiotactic polypropylene (P_r = 0.67). The polymer microstructure is in agreement with a chain-end mechanism of steric control. End group analysis showed the almost exclusive presence of isobutyl end groups, indicative of a primary (1,2) propene insertion operating both in the initiation and the termination steps. The formation of isobutyl end groups during the termination step was confirmed by the presence of $CH_2DCH(CH_3)CH_2$-- end groups in the ^2H NMR spectrum of a polypropylene sample prepared with the same deuterium labelling technique used in the case of the nickel catalyst (see above, Scheme 10 a).

The reasonable polymerization mechanism involves (see Scheme 15) primary propene insertion into Zr-CH$_3$ bonds, followed by prevailingly primary insertion during the propagation, and termination *via* chain transfer of the primary growing chain to MAO (or to Al(CH$_3$)$_3$ present in the MAO solution) generating a new Zr-CH$_3$ bond for reinitiation.

Further support to the whole picture came from copolymerization of propene with a small amount of ^{13}C ethylene enriched C1 (see the spectrum in Figure 6a). In this case the three methylene sequences outnumber the methylene pairs (79 % vs 21 %), in agreement with a prevailingly primary regiochemistry: this result is unexpected in view of the opposite regiochemistry observed for related bis(phenoxyimine)titanium catalysts. In order to

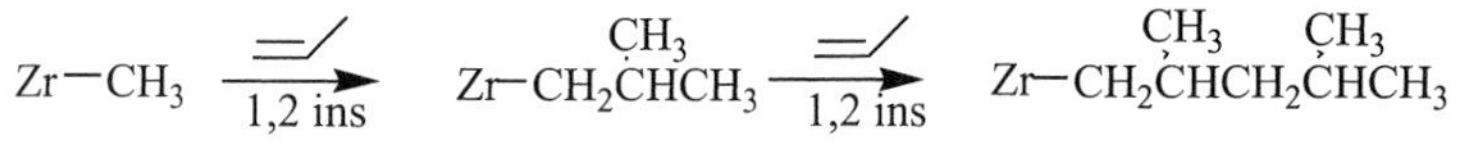

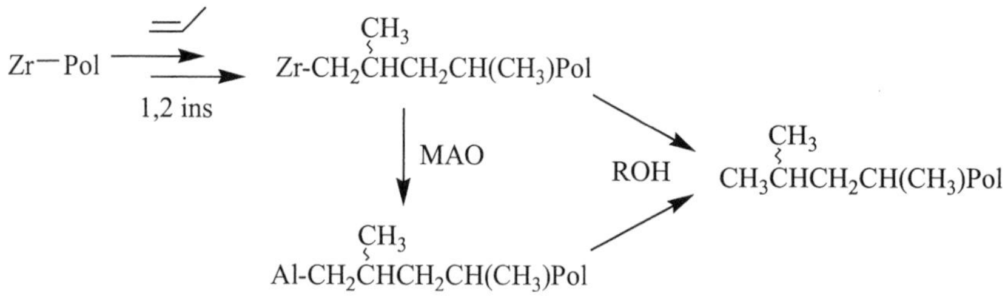

$$Zr\text{--}CH_3 \xrightarrow[\text{1,2 ins}]{=\!\!\diagup} Zr\text{--}CH_2\overset{CH_3}{\underset{\smile}{C}}HCH_3 \xrightarrow[\text{1,2 ins}]{=\!\!\diagup} Zr\text{--}CH_2\overset{CH_3}{\underset{\smile}{C}}HCH_2\overset{CH_3}{\underset{\smile}{C}}HCH_3$$

$$Zr\text{--}Pol \xrightarrow[\text{1,2 ins}]{=\!\!\diagup} Zr\text{-}CH_2\overset{CH_3}{\underset{\smile}{C}}HCH_2CH(CH_3)Pol$$

MAO

ROH $CH_3\overset{CH_3}{\underset{\smile}{C}}HCH_2CH(CH_3)Pol$

$Al\text{-}CH_2\overset{CH_3}{\underset{\smile}{C}}HCH_2CH(CH_3)Pol$

Scheme 15

discriminate the role played by the nature of the metal in determining the regiochemistry of propene insertion from any ligand effect, we tested the titanium complex (complex **7** in Scheme 11) bearing the same ligands. The polypropylene produced was very similar to that obtained with the corresponding Zr compound, according to recent literature data.[26c] End groups analysis, also confirmed by deuterium labelling data, revealed in this case the presence of n-propyl and isobutyl end groups. As discussed above, these data suggest that the polymerization is prevailingly initiated by at least two consecutive 1,2-insertions into a Ti-CH_3 bond (generating the isobutyl end groups), while the main termination step is the hydrolysis of titanium-bound secondary growing chains (generating n-propyl end groups). Conclusive results came from the structure of a copolymer of propene with a low amount of 1-^{13}C-ethylene. NMR analysis (see the spectrum in Figure 6b) showed that in this case methylene pairs outnumber the three methylene sequences (90 % vs 10 %), suggesting that 2,1 insertion of propene is the "normal" mode of propagation, and that 1,2 insertion becomes preferred after the insertion of an ethylene unit, which generates a metal-primary carbon bond.

In conclusion, zirconium and titanium complexes bearing the same ligands afford prevailingly syndiotactic polypropylenes having very similar structures, but via opposite regiochemistries.

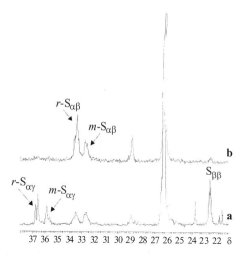

Figure 6. Aliphatic region of the 100.6 MHz ^{13}C NMR spectra (C$_2$D$_2$Cl$_4$, 100 C) of two copolymers of propene with a little amount of 1-^{13}C-ethylene prepared with a) **6**-MAO and b) **7**-MAO. δ in ppm from hexamethyldisiloxane.

Conclusions

One of the main achievements of research on metallocene catalysts involves a detailed understanding of the mechanism of stereospecific polymerization of propene and α-olefins, including a precise insight in the relationships between the precatalyst structure and the microstructure of the polymer obtained. In the last few years, several non-metallocene catalysts have been shown to afford either isotactic or syndiotactic polypropylenes, in addition to the classical syndiospecific V-based catalysts disclosed by Natta and Zambelli in the early 1960s. While the mechanistic details of the stereospecific polymerization promoted by these new catalysts are less defined, some facets have been elucidated, showing some peculiar features with respect to metallocenes. E.g., while polymerization of propene by metallocene catalysts predictably involves primary (1,2) monomer insertion as the preferential mode of propagation, practically all possible combinations of regiochemistries and stereochemistries have been observed in propene polymerization by non-metallocene catalysts. Thus, Fe(II) bis(imino)pyridine catalysts afford isotactic polypropylene via a *chain-end* mechanism of steric control and a prevailingly secondary regiochemistry of monomer insertion. On the other hand, Brookhart's Ni(II) diimine catalysts are able to produce prevailingly syndiotactic polypropylene at sub-ambient temperature through a *chain-end* control mechanism operating via primary monomer insertion. Finally, octahedral C_2-symmetric bis(phenoxyimine) titanium complexes are highly syndiospecific and living propene polymerization catalysts operating via a secondary mode of monomer insertion. Surprisingly, analogous zirconium catalysts can afford moderately syndiotactic, low molecular weight polypropylenes via primary monomer insertion.

[1] For recent reviews on metallocene catalysts see: a) H. H. Britzinger, D. Fischer, R. Mulhaupt, B. Rieger, R. M. Waymouth; *Angew. Chem. Int. Ed. Engl.*, **1995**, *34*, 1143; b) W. Kaminsky, M. Arndt; *Adv. Polym. Sci.*, **1997**, *127*, 143.
[2] a)V. C. Gibson, S. K. Spitzmesser, *Chem. Rev.* **2003**, *103*, 283; b) G. J. P. Britovsek, V. C. Gibson, D. F. Wass, *Angew. Chem. Int. Ed.* **1999**, *38*, 428.
[3] S. D. Ittel, L. K. Johnson, M Brookhart, *Chem. Rew.* **2000**, *100*, 1169.
[4] a) H. Makio, N.Kashiwa, T. Fujita, *Adv. Synth. Catal.* **2002**, *344*, 477.; b) N. Matsukawa, S. Ishii, R. Furuyama, J. Saito, M. Mitani, H. Makio, H. Tanaka, T. Fujita e-Polymers **2003**, 021
[5] The different methylenes are labelled by reporting by Greek letters the number of carbon – carbon bonds bridging the considered methylene with the methyne carbons closer on the right and the left side: C. J.Carman, C.E.Wilkes, *Rubber Chem.Technol.* **1971**, 44, 781.
[6] I. Sessa, D. R. Ferro, F. Grisi, G. Milano, D. Liguori, A. Zambelli, *Macromol Chem. Phys.* **2002**, *203*, 2604
[7] B. L. Small, M. Brookhart, A. M. A. Bennet, *J. Am. Chem. Soc.*, **1998**, *120*, 4049
[8] G. J. P. Britovsek, V. C. Gibson, B. S. Kimberley, P. J. Maddox, S. J. Mc Tavish, G. A. Solan, A. J. P. White, D. J. Williams, *Chem.Comm.*, **1998**, 849.
[9] B. L. Small, M. Brookhart, *Polym. Prep. (Am. Chem. Soc., Div. Polym. Chem.)* **1998**, *39*, 213.
[10] B. L. Small, M. Brookhart, *Macromolecules* **1999**, *32*, 2120
[11] C. Pellecchia, M. Mazzeo, D. Pappalardo, *Macromol. Rapid Commun.* **1998**, *19*, 651.
[12] J. A. Ewen, *J. Am. Chem. Soc.* **1984**, *106*, 6355
[13] W. Spaleck, F. Küber, A. Winter, J. Rohrmann, B. Bachmann, M. Antberg, V. Dolle, E. F. Paulus, *Organometallics.* **1994**, 13, 954.
[14] L. K. Johnson, C. M. Killian, M.Brookhart, *J. Am. Chem. Soc.* **1995**, *117*, 6414
[15] C. Pellecchia, A. Zambelli, *Macromol. Rapid Commun.* **1996**, *17*, 333
[16] G. Natta, A. Zambelli, I. Pasquon *J. Am. Chem. Soc.* **1962**, *84*, 1488
[17] A. Zambelli, I. Sessa, F. Grisi, R. Fusco, P. Accomazzi, *Macromol. Rapid Commun.* **2001**, *22*, 297
[18] C. Pellecchia, A. Zambelli, L. Oliva, D. Pappalardo, *Macromolecules* **1996**, *29*, 6990.
[19] P. Locatelli, M. C: Sacchi, E. Rigamonti, A. Zambelli, *Macromolecules* **1984**, *17*, 123.
[20] C. Pellecchia, A. Zambelli, M. Mazzeo, D. Pappalardo, *J. of Mol. Catal.* **1998**, *128*, 229.
[21] A. Zambelli, P. Longo, S. Terenghi, D. Recupero, G. Zannoni, *J. Mol. Catal.* **2000**, *152*, 25.
[22] a) T. Fujita, Y. Tohi, M. Mitani, S. Matsui, J. Saito, M. Nitabaru, K. Sugi, H. Makio, T. Tsutsui, *Eur. Pat. EP 0874005,* **1998** (Mitsui Chemical); *Chem. Abstr.* 129:331166; b) S. Matsui, Y. Tohi, M. Mitani, J. Saito, H. Makio, H. Tanaka, M.Nitabaru, T. Nakano, T. Fujita, *Chem. Lett.* **1999**, 1065; c) Matsui, S.; Mitani, M.; Saito, J.; Tohi, Y.; Makio, H.; Tanaka, H.; T. Fujita, *Chem. Lett.* **1999**, 1263.
[23] J. Tian, G. W. Coates, *Angew. Chem., Int. Ed.* **2000**, *39*, 3626
[24] J. Tian, P. D. Hustad, G. W. Coates, *J. Am. Chem. Soc.* **2001**, *123*, 5134
[25] J. Saito, M.Mitani, J.Mohri, S.Ishii, Y.Yoshida, T.Matsugi, S.Kojoh, N.Nashiwa, T. Fujita, *Chem. Lett.* **2001**, 576.
[26] a) J. Saito, M. Mitani, J. Mohri, Y.Yoshida, S.Matsui, S.Ishii, Kojoh, S.; Kashiwa, N.; Fujita, T. *Angew. Chem. Int. Ed.* **2001**, *40*, 2918; b) J. Saito, M. Mitani, M.Onda, J. Mohri, S. Ishii, Y. Yoshida, T. Nakano, T. Tanaka, T. Matsugi, S. Kojoh, N.Kashiwa, T.Fujita, *Macromol.Rapid Commun.* **2001**, *22*, 1072; c) M. Mitani,

R. Furuyama, J. Mohri, J. Saito, S. Ishii, H. Terao, T. Nakano, H. Tanaka, T. Fujita, *J. Am. Chem. Soc.* **2003**, *125*, 4293

[27] a) M. Lamberti, D. Pappalardo, A. Zambelli, C. Pellecchia, *Macromolecules* **2002**, *35*, 658; b) M. Lamberti, D. Pappalardo, A. Zambelli, C. Pellecchia, *Macromolecules* **2002**, *35*, 6478.

[28] P. D. Hustad, J. Tian, G. W. Coates, *J. Am. Chem. Soc.* **2002**, *124*, 3614

[29] a) A. Zambelli, G. Bajo, E. Rigamonti, *Makromol. Chem.* **1978**, *179*, 1249; b) A. Grassi, P. Ammendola, P. Longo, E. Albizzati, L. Resconi, R. Mazzocchi, *Gazz. Chim. Ital.* **1988**, 118, 539; c) P. Ammendola, S. Xiao, A. Grassi, A. Zambelli, *Gazz. Chim. Ital.* **1988**, 118, 769

[30] It is worth nothing that using 1-^{13}C ethylene, the ^{13}C-enriched methylene occupies randomly two possible locations. Therefore, the $S_{\alpha\beta}$ enriched carbons are diagnostic of the stereochemical arrangement of the two propene units incorporated immediately before and after the enriched ethylene unit. Before the incorporation of 1-^{13}C ethylene, the stereochemical arrangement of the two propene units should be statistically the same observed in the propene homopolymer (r = 99%). After the ethylene perturbation, in principle, the next two incoming propene units could generate the three different situations depicted in scheme 14 (in scheme 14 the enriched carbons adjacent to the propene units entered *after* the insertion of ethylene are in bold face). Of course, since the total fraction of r-$S_{\alpha\beta}$ carbons (accounting for the propene dyads formed either *before* or *after* the insertion of ethylene) is < 50 %, the formation of an r dyad *after* the insertion of ethylene via primary insertion (Scheme 14a) must be a poorly probable event. Thus, the presence of significant amounts of primary blocks would be not compatible with a highly syndiotactic polypropylene.

[31] M. Lamberti, R. Gliubizzi, M. Mazzeo, C. Tedesco, C. Pellecchia, *Macromolecules* in press

Macromonomers and Coordination Polymerization

Jean-François Lahitte, Frédéric Peruch,* François Isel, Pierre Joseph Lutz*

Institut Charles Sadron, CNRS, UPR22 F67083 Strasbourg Cedex, France
E-mail: lutz@ics.u-strasbg.fr; peruch@ ics.u-strasbg.fr

Summary: The present work discusses the synthesis of well-defined comb-shaped polymers or graft copolymer structures based on coordination (co)polymerization of macromonomers. Polystyrene macromonomers with various polymerizable entities were synthesized first by induced deactivation reactions. The homopolymerization of these macromonomers in the presence of selected early or late transition metal catalysts was examined. Comb-shaped polymers could be obtained over a large range of DP values. The results were compared to those obtained by anionic homopolymerization. Some results on the copolymerization of these PS macromonomers with ethylene in the presence of VERSIPOL^{TM} type catalysts were presented.

Keywords: branched polymers; coordination polymerization; copolymerization; macromonomers; polymacromonomers; polystyrene

Introduction

Macromonomers, macromolecules (usually short) with polymerizable entities at one or both chain ends, have opened new perspectives in graft or comb-shaped polymers controlled synthesis. Random radical (controlled or free) or ionic copolymerization of macromonomers with a vinyl or acrylic monomer yields graft copolymers.[1,2] Homopolymerization of macromonomers, based on a polymer-polymer reaction, has been successfully used to design a new type of comb-shaped polymer. Well defined macromolecules characterized by unusually high segment densities could be obtained either by radical or by anionic polymerization processes.[3-8] Coordination (co-)polymerization of macromonomers has attracted recently increasing interest due to its two decisive advantages: the intrinsic possibility to control the tacticity of the backbone and the ability to polymerize allylic monomers i.e. macromonomers with allylic polymerizable end groups.[9-15]

The first part of the present work deals with the synthesis of well-defined polystyrene (PS) macromonomers by induced deactivation reactions. Once characterized, these PS macromonomers will be homopolymerized in the presence of selected early or late transition

© 2004 WILEY-VCH Verlag GmbH & KGaA, Weinheim DOI: 10.1002/masy.200450923

metal catalysts and the results compared to those obtained by anionic homopolymerization. The copolymerization of these PS macromonomers with ethylene in the presence of VERSIPOL™ type catalysts will be presented and discussed in the final section.

Experimental

Synthesis of ω-allyl, ω-undecenyl, ω,ω-undecenyl or ω-vinylbenzyl Polystyrene Macromonomers. The polystyrene macromonomers with allylic, undecenyl or styrenyl end groups were prepared by induced anionic deactivation reactions as already described in several previous publications.[1,8,12]

Homopolymerization of Polystyrene Macromonomers. A Schlenk flask was charged with 1 g of PS macromonomer, dried and purged with several vacuum/argon cycles. The desired amount of methylaluminoxane (MAO) in solution in toluene was added with a syringe and the volume completed with toluene up to 10 mL. The mixture was heated to the desired temperature and the catalyst (in solution in toluene) was added. The sample was kept at that temperature for the selected polymerization time and quenched with a mixture of methanol and hydrochloric acid. The sample was recovered according to well-established procedures[12] and analyzed as indicated below.

Copolymerization of ethylene with ω-allyl, ω-undecenyl, ω,ω-undecenyl or ω-vinylbenzyl Polystyrene Macromonomers. Polymerizations were carried out in a 250 mL miniclave Büchi reactor. The reactor was charged with the PS macromonomer and purged with several vacuum/argon cycles. Then, under a low pressure of ethylene, the solvent and the catalytic system were added. Finally, the reactor was pressurized with ethylene. The resulting copolymers were precipitated three times in slightly acidified methanol, dried under vacuum and characterized by SEC, ^{1}H NMR, and IR to certify the presence of the PS sequence in the copolymer.

Characterization. Molar mass determinations were made by SEC on a Waters apparatus fitted with five PL gel columns, an autosampler Waters WISP 717, a differential refractometer Shimadzu RID 6A, a UV spectrometer Beckman 147 (λ = 254 nm), and a multi-angle laser light scattering detector Wyatt DAWN DSP (laser λ = 632.8 nm). Calibration was performed using linear PS standards. The ^{1}H NMR spectra were recorded on a Brüker AC200 apparatus.

Results and Discussion

Various methods have been developed to yield polymers quantitatively fitted with polymerizable units.[1] Among the end-functionalization reactions, those based on induced deactivation are usually efficient, provided the double bonds are not involved in side reactions. Several PS macromonomers have been prepared by quenching, under adapted experimental conditions,[12] living PS either directly with allyl bromide, 11-bromo-undecene, 4-vinylbenzyl chloride, or 4-vinylbenzyl chloride after intermediate addition of 1,1-diphenylethylene (DPE). Some data on the ω-vinylbenzyl PS macromonomers are given in Table 1. The characteristics of the ω-allyl, ω-undecenyl, and ω,ω-undecenyl macromonomers were already published.[14]

Table 1. Molecular characteristics of ω-vinylbenzyl polystyrene macromonomers prepared by induced deactivation.

Run	Method[a]	$M_{n\ th}$[b] (g.mol^{-1})	$M_{n\ exp}$[c] (g.mol^{-1})	MWD[d]	Funct.[e] (%)
1	A	1100	1120	1.05	97
2	A	1200	1200	1.04	97
3	A	5000	4800	1.05	85
4	B	1200	1150	1.05	98
5	B	1100	1120	1.05	98
6	B	2000	1980	1.06	90
7	B	5000	5000	1.04	95
8	B	10000	10300	1.05	95

Typical polymerization conditions ($M_{n,\ th.}$=1100g.mol^{-1}): toluene=100mL, styrene=15g (16.7mL, 0.144mol), BuLi [0.6N]=25 mL. Initiation at 10°C, propagation at 35°C, addition of 100mL of THF at -70°C, termination by addition of the functionalization agent (0.144 mol).
[a] A refers to a sample prepared with intermediate addition of DPE, B without DPE.
[b] Theoretical molar mass calculated from ratio monomer converted to initiator concentration.
[c] Experimental molar mass measured in THF by SEC, calibration with linear PS.
[d] Molecular Weight Distribution.
[e] Functionalization yield determined by chemical titration (Johnson and Fletcher reaction[16]).

Well-defined macromonomers with narrow molar mass distributions and free of coupling products (no shoulder detected on the SEC diagram peak) could be obtained for molar masses ranging from 1000 to 10 000 g.mol^{-1}. Experimental and theoretical molar masses are in good agreement. The functionalization yields determined by chemical titration, ^{1}H NMR analysis, and Maldi-TOF[12] are almost quantitative whatever the chain end.

Homopolymerization of ω-functionalized PS macromonomers via coordination polymerization

The homopolymerization of PS macromonomers has been investigated using hemi-metallocene catalysts: (CpTiCl₃, Cp*TiCl₃, CpZrCl₃, CpTiF₃, CGC-Ti) or the following palladium complex [(ArN=C(Me)-C(Me)=NAr)Pd(CH₂)₃(COOMe)]⁺ BAr'₄⁻ (VERSIPOL™)

The homopolymerization of PS macromonomers has been investigated using hemi-metallocene catalysts: (CpTiCl$_3$, Cp*TiCl$_3$, CpZrCl$_3$, CpTiF$_3$, CGC-Ti) or the following palladium complex [(ArN=C(Me)-C(Me)=NAr)Pd(CH$_2$)$_3$(COOMe)]$^+$ BAr'$_4^-$ (VERSIPOL^{TM})

a)   b)   c)

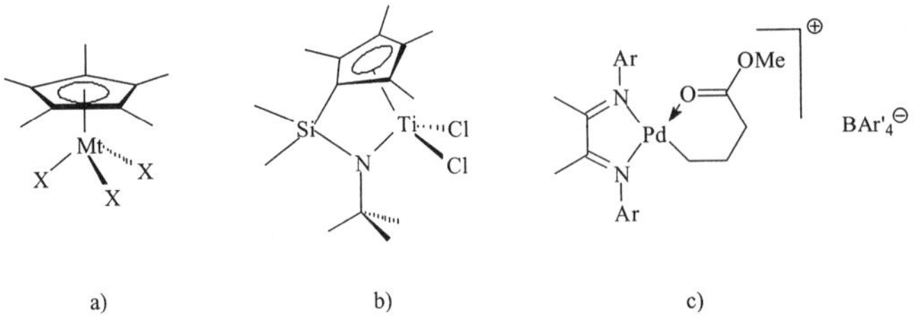

Scheme 1. Catalysts used for the polymerization of macromonomers: a) hemi-metallocene catalyst with various substituents, b) CGC catalyst (Dow), c) VERSIPOL™ catalyst.

From these attempts the following conclusions could be drawn:

- It could be confirmed that the homopolymerization of ω-vinylbenzyl PS macromonomers with CpTiCl$_3$/MAO or Cp*TiCl$_3$/MAO is possible and provides access to highly compact branched polymers as revealed by SEC with online LS characterization. In most cases, homopolymerization yields are yet far from being quantitative.

- The DP depends on the polymerization conditions. Higher molar masses have been obtained at low temperature and with low catalyst and cocatalyst concentrations. The suppression of the DPE unit in the ω-vinylbenzyl PS macromonomers has improved the homopolymerization. Best results, DP values up to 50, were obtained with CpTiF$_3$ complexes introduced by the group of Kaminsky.[17]

- CGC- (Constrained Geometry Catalyst) complexes are known to be much more efficient for the coordination polymerization of styrene. Some results obtained by homopolymerization of ω-vinylbenzyl PS macromonomers in the presence of these catalysts, under conditions almost identical to those employed with CpTiCl$_3$/MAO or Cp*TiCl$_3$/MAO, are listed in Table 2. A typical SEC diagram of the raw reaction product is given on Figure 1 together with the SEC of the polymacromonomer resulting from the polymerization of a ω-undecenyl PS macromonomer in presence of the same catalyst.

As it can be seen from these SEC diagrams, in both cases homopolymerization was possible, but the homopolymerization yields are far from being quantitative. In addition, the molecular

a)   b)

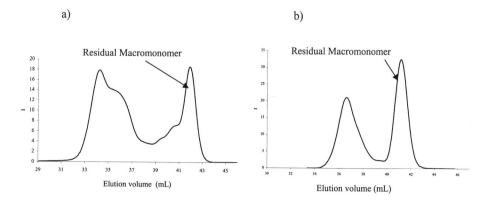

Figure 1. Typical SEC diagrams of the raw reaction product of the homopolymerization PS macromonomers in the presence of CGC catalysts: (a) ω-vinylbenzyl PS macromonomer, (b) ω-undecenyl PS macromonomer.

weight distribution is rather large for the polymacromonomers resulting from the homopolymerization of ω-vinylbenzyl PS macromonomers. A possible explanation may be the occurrence of increasing transfer reactions during the polymerization process.

When compared to earlier results observed for the anionic homopolymerization of ω-vinylbenzyl PS macromonomers,[8] the following comments can be made: for identical molar masses of the macromonomer precursor, anionic homopolymerization yields were almost quantitative and sharp molar mass distributions of the resulting polymacromonomers were noted. This is explained by the specific characteristics of the anionic polymerization (long life time of the active sites and absence of transfer reactions under appropriate conditions). On the contrary the maximum DP values are lowered by 40 for polymacromonomers originating from macromonomers of almost identical molar masses. Anionic polymerization is well known to be much more sensitive to the presence of remaining impurities.

Table 2. Polymerization of ω-vinylbenzyl PS macromonomers (with DPE) in the presence of the CGC catalyst and characteristics of the resulting polymacromonomers.

Run	Macro[a] (g)	CGC-Ti (μmol)	$M_{n,SEC}$ [b] (g.mol^{-1})	$M_{w,SEC}$ [b] (g.mol^{-1})	M_{wLS} [c] (g.mol^{-1})	MWD[d]	DP	Conv.[e] (%)
9	0.25	1	20500	36600	74000	1.8	74	70
10	0.5	1	62000	148000	420000	2.4	420	35
11	0.5	2	33000	81000	175000	2.4	175	42
12	1	2	53000	103000	290000	1.9	290	24
13	0.5	5	44000	87000	258000	2.0	258	53
14*	1	2	75000	126000	500000	1,7	100	8

Typical polymerization conditions: toluene 2ml (excepted for samples 9 and 10: 1.5 ml), [Al]/[Ti]=1000, T=50°C, polymerization time = 16 hours.
[a] Molar mass of the macromonomer: 1000 g.mol^{-1} (except sample 14* molar mass 5000 g.mol^{-1})
[b] Molar mass measured by SEC in THF, calibration with linear PS.
[c] Molar mass measured by SEC with online LS.
[d] Molecular Weight Distribution.
[e] Macromonomer conversion.

In conclusion, for the ω-vinylbenzyl PS macromonomers, the highest polymerization degrees have been obtained using the CGC-Ti complexes. For ω-vinylbenzyl PS macromonomers, free of the bulky DPE unit, homopolymerization yield and molar mass of the resulting polymacromonomer significantly increased.

Among these catalysts, the CGC-Ti catalyst was the only one able to homopolymerize ω-undecenyl PS macromonomers, giving access to a new type of graft copolymer ethylene-*g*-(ethylene-*b*-styrene). No homopolymerization of ω-allyl PS macromonomer was detected with the same CGC catalyst. As it could be anticipated, no homopolymerization at all was observed with VERSIPOL™ catalysts whatever the polymerizable entity at the PS chain end.

CpTiCl$_3$/MAO catalytic system has been reported to provide access to highly syndiotactic polystyrenes.[17] The presence of a substituent in the para position of the styrene unit does not modify the syndiotactic polymerization process. Therefore there is no reason why the backbone of polymacromonomers resulting from coordination processes should not be syndiotactic. On the contrary, these polymacromonomers with basically a syndiotactic backbone and atactic grafts are well soluble at room temperature in usual solvents (THF, toluene). They remained soluble even at rather high polymerization degrees of the main chain.

Therefore we attempted some solution NMR experiments to attest the syndiotactic nature of the backbone. In fact, the concentration of syndiotactic styrene units is too low to be really determined. As mentioned in the introduction, Senoo et al.[9] polymerized vinyl-terminated polyisoprene macromonomers in the presence of CpTiCl$_3$/MAO to high molar mass polymacromonomers. In a subsequent step, ozonolysis was achieved. The resulting polymer was no more soluble in usual solvents at room temperature and its syndiotactic character could be confirmed by NMR spectroscopy.

The solution properties of these different polymacromonomers were investigated and their compactness with respect to the linear equivalent could be confirmed. The relation between the polymerization degree of the polymacromonomers and their structures has been established. That point will be discussed in details in a forthcoming paper.

Copolymerization of PS macromonomers with ethylene

As mentioned in the introduction, radical (controlled or free) or ionic copolymerization of macromonomers with a vinyl or acrylic monomer are very efficient pathways to design graft copolymers. Only little work in this domain referred to coordination polymerization processes. On another hand higher alpha olefins can be incorporated in polyolefins via Ziegler Natta or metallocene catalysts. In addition, a part of long chains branches formed in Pd catalyzed polymerization of ethylene,[19,20] could be explained by the incorporation of higher alpha olefins i.e. of macromonomers formed during the polymerization. This led us to examine the copolymerization of ω-allyl, ω-undecenyl or ω,ω-undecenyl PS macromonomers with ethylene in the presence of the Pd dimiine catalyst introduced by Brookhart.[19]

In preliminary experiments,[11,12] copolymerization experiments of PS macromonomers with ethylene in the presence of the above mentioned Pd catalyst were performed. The reaction products were characterized by SEC with online LS. From the raw product containing unreacted macromonomer, graft copolymer species had to be isolated by repeated selective precipitation. FTIR and ^{1}H NMR confirmed unambiguously the presence of the macromonomer in the graft copolymer. This prompted us to perform a systematic investigation of that copolymerization reaction.

ω-allyl or ω–undecenyl PS macromonomers of different molar mass were copolymerized with ethylene using the VERSIPOL^{TM} catalyst under various experimental conditions with the aim of increasing the content of PS in poly(ethylene) based materials. In most cases, the average molar masses of the isolated copolymers are lower than those observed for polyethylene prepared under identical conditions. In the case of ω-allyl PS macromonomers, whatever the experimental conditions, the degree of incorporation remained low. This can be explained by the restricted accessibility of the polymerizable group. For macromonomers containing an alkyl spacer, taking the polystyrene chain away from the terminal double bond, the macromonomer content in the copolymer increases notably. In addition, higher molar mass copolymers have been obtained with average molar masses up to 158 000 g.mol^{-1}. This confirms the observations made in homopolymerization attempts on the same type of PS macromonomers. Besides the macromonomer weight percent content decreases with increasing ethylene pressure whereas the molar masses of the copolymer increases with ethylene pressure. That point will be discussed more in details in a forthcoming paper. An increase in macromonomer molar mass corresponds to a decrease of the conversion. Similar observations, were made in the case of free radical copolymerization of macromonomers with micromolecular polymerizable compounds.

To try to still increase that macromonomer content i.e the styrene content, copolymerization experiments with bifunctional PS macromonomers were performed. We examined again the influence of several factors (polymerization solvent, ethylene pressure, molar mass of the macromonomer and reaction time) on the conversion, the molar mass of the graft copolymers and the number of grafts incorporated. Some of these results are presented in Tables 3 and 4.

Table 3. Evolution of the macromonomer incorporation yield and the molar masses of the graft copolymers for the copolymerization of ethylene and ω,ω-undecenyl PS macromonomers in the presence of VERSIPOL^{TM} as a function of reaction time.

Run	Time (h)	M_{wLS}[a] g.mol^{-1}	Wt.-% of styrene	Number of grafts per chain	Yield (g)	Conv.[b] (%)
15	1	81000	21	3	0.5	11
16	2,5	210000	6.3	3	1.7	11
17	5	625000	6.8	8	4.7	27
18	7,5	673000	5.5	8	4.7	26
19	10	860000	7	12	5.5	39
20	18	gel	-	-	5.8	-

Typical polymerization conditions Ethylene pressure = 3 bars, Molar mass of the macromonomer used in copolymerization: 5000 g.mol^{-1}, amount macromonomer = 1g, cata= 10 μmol, T=25°C, toluene= 30 mL
[a] measured by SEC in THF with online LS.   [b] Macromonomer conversion.

The weight average molar mass of the copolymer, measured by light scattering, increases with increasing reaction time. After 18 hours totally insoluble materials resulted. On the contrary the samples 15-19 are well soluble in THF or toluene and do not contain significant amounts of microgel. This was confirmed by molar mass determinations on sample 19 before and after filtration: no difference in weight average molar masses was observed. The number of macromonomers units incorporated in the graft copolymer increases with reaction time. It is interesting to note that these graft copolymers still contain unsaturations. This was confirmed by copolymerization experiments between ethylene and these graft copolymers in the presence of VERSIPOL[TM] catalysts. The molar mass of the resulting product measured by SEC with online LS still increased and the presence of the styrene units could be confirmed unambiguously by SEC with UV detection.

For a given reaction time, the molar mass of the graft copolymer increases with increasing ethylene pressure (Table 4). The molar mass distribution gets larger with increasing reaction

Table 4. Evolution of the macromonomer incorporation yield and the resulting graft copolymer molar masses for the copolymerization of ethylene and ω,ω-undecenyl PS macromonomers in the presence of VERSIPOL[TM] with ethylene pressure.

Run	Time (h)	P (bar)	M_{wLS}[a] g.mol^{-1}	M_{wLS}[b] g.mol^{-1}	MWD[c]	Wt% of styrene	Number of grafts per chain
21	20	0.2	470000	510000	1.6	15.0	14
22	20	0.5	980000	1000000	3.8	10.6	21
23	20	1	1560000	1600000	4.9	8.0	26
24	20	2	1620000	4500000	4.3	8.2	26
25	20	3	gel	-	-	-	-
26	5	1	270000	300000	1.6	10.0	5
27	5	2	570000	-	2.1	7.2	9
17	5	3	620000	670000	2.0	6.8	8
28	5	4	490000	-	2.2	6.0	6
29	5	5	480000	-	1.9	5.2	5
30	3	6	430000	540000	1.9	5.0	4

Typical polymerization conditions: Catalyst = 10 µmol, Molar mass of the macromonomer used in copolymerization 5000 g.mol^{-1}, amout of macromonomer = 1g, cata= 10 µmol, T=25°C, toluene= 30 mL
[a] Molar mass measured by SEC in THF with online LS.
[b] Molar mass measured by static light scattering.
[c] Molecular weight distribution.

time due probably to increasing transfer reactions. At low ethylene pressure the macromonomer content in the copolymer is much higher.

The different samples resulting from copolymerization of ethylene with PS macromonomers were submitted to detailed investigations of their dilute solution and solid state properties. The results were compared to those obtained for polyethylenes prepared under the same conditions. The discussion of the different results is out of the scope of the present work.

Conclusion

The homopolymerization of well-defined ω-allyl, ω-undecenyl, and ω-vinylbenzyl PS macromonomers in the presence of selected early or late transition metal catalysts was examined with respect to the influence of several experimental parameters on the homopolymerization yield and the homopolymerization degree. The polymerization degrees of the resulting polymacromonomers depend strongly on the polymerization conditions. Higher molar masses have been obtained with low temperature and low catalyst and cocatalyst concentrations. The absence of the DPE unit in the ω-vinylbenzyl PS macromonomers has improved the homopolymerization results. Best results were obtained with CGC catalysts whatever the polymerizable chain end.

The Pd catalyst (VERSIPOL^{TM}) allows the incorporation of ω-allyl or ω-undecenyl PS macromonomers into PE chains giving access to a new type of copolymer structure, potential compatibilizers. The introduction of the alkyl spacer in PS macromonomers increases the macromonomer incorporation yield, which could still be improved while using bifunctional PS macromonomers.

Acknowledgments. The authors would like to thank very much Alain Rameau, Roger Meens. and Catherine Foussat for SEC measurements.

[1] K. Ito, S. Kawaguchi, *Adv. Polym. Sci.* **1999**, *142*, 129.
[2] K. Ito, *Prog. Polym. Sci.* **1998**, *23*, 581.
[3] T. Otsu, K. Yamashita, K. Tsuda, *Macromolecules* **1986**, *19*, 287.
[4] Y. Tsukahara, K. Tsutsumi, Y. Yamashita, S. Shimada, *Macromolecules* **1990**, *23*, 5201.
[5] M. Wintermantel, M. Gerle, K. Fischer, M. Schmidt, I. Wataoka, H. Urakawa, K. Kajiwara, Y. Tsukahara, *Macromolecules* **1996**, *29*, 978.
[6] F. Audouin, M. Arotçarena, D. Colombani, P. J. Lutz, *Polym. Prep.* **2002**, *41(2)*, 1889.
[7] H. Shinoda, P. J. Miller and K. Matyjaszewski, *Macromolecules* **2001**, 34, 3186.

[8]   Y. Ederlé, F. Isel, S. Grutke, P. J. Lutz, *Macromol. Symp.* **1998**, *132*, 97.

[9]   K. Senoo, K. Endo, *Macromol. Rapid. Commun.* **2000**, *21*, 1244.

[10] K. Nomura, S. Takahashi, Y. Imanishi , *Macromolecules* **2001**, *34*(14), 4712.

[11] P. J. Lutz, F. Breitling, J-F Lahitte, F. Peruch, S. Plentz Meneghetti, F. Isel, *Polym. Prep.* **2000**, *41(*2), 1889.

[12] J. F. Lahitte, F. Peruch, S. Plentz Meneghetti, F. Isel, P. J. Lutz, *Macromol. Chem. Phys.* **2002,** *203*, 2583.

[13] J. F. Lahitte, F. Pelascini, F. Peruch, S. Plentz Meneghetti, P. J. Lutz, *C. R. Chimie* **2002**, *5*, 225.

[14] J. F. Lahitte, F. Peruch,, F. Isel, P. J. Lutz, Polym. *Prep., Am. Chem. Soc. Div. Polym. Chem.* New York, (USA) **2003**, 44(2), 46.

[15] M.N. Neiser, J. Okuda, M. Schmitt, *Macromolecules* **2003**, *36,* 5437.

[16] J. B. Johnson, J. P. Fletcher, *Anal. Chem.* **1959***, 31*, 1663.

[17] W. Kaminsky, S. Lenk, V. Scholz, H. W. Roesky, A. Herzog, *Macromolecules* **1997**, *30*, 7647.

[18] N. Ishira, T. Seimiya, M. Kuramoto, M. Uoi, *Macromolecules* **1986**, *19*, 2465.

[19] L.K. Johnson, M. Killian, M. Brookhart, *J. Am. Chem. Soc.* **1995**, *117*, 6414.

[20] S. Plentz Meneghetti, J. Kress, P. J. Lutz, *Macromol. Chem. Phys.* **2000**, *201*, 1823.

Polymerization of Norbornene with Co(II) Complexes

Frédéric Pelascini,[1,2] *Frédéric Peruch,*[1] *Pierre Joseph Lutz,*[1] *Marcel Wesolek,*[2]
Jacky Kress[2]

[1]Institut Charles Sadron, 6, rue Boussingault, F-67083 Strasbourg Cedex, France
E-mail: lutz@ics.u-strasbg.fr; peruch@ics.u-strasbg.fr
[2]Laboratoire de Chimie des Métaux de Transition et de Catalyse, Institut le Bel, 4, rue Blaise Pascal, F-67070 Strasbourg Cedex, France

Summary: The norbornene polymerization was studied in the presence of 6 pyridine bis(imine) cobalt(II) complexes activated with methylaluminoxane (MAO). Norbornene was also polymerized with $CoCl_2$ associated to MAO. All these catalytic systems generate an addition polymerization of norbornene, yielding fully saturated polymers. It was shown that the polymerization yield and the molar masses are highly dependant on several reaction parameters (monomer concentration, [Al]/[Co] ratio, polymerization temperature and time) and the frame of the ligand.

Keywords: addition polymerization; catalysts; cobalt; norbornene; pyridine bis(imine)

Introduction

Bicyclo[2.2.1]hept-2-ene (norbornene) can be polymerized in three different ways (Scheme 1), each route leading to its own polymer structure. The most studied polymerization process is the Ring-Opening Metathesis Polymerization (ROMP), yielding polyalkenamers exhibiting good solubility in a wide variety of organic solvents. Cationic norbornene polymerization results in low molar mass oligomeric materials with 2,7-connectivity of the monomer. The addition polymerization of norbornene yields saturated polymers (2,3-insertion), showing high chemical resistance and very good optical properties.[1] Many catalysts based on early or late transition metals were described for this latter process.[2-5] However, few examples of cobalt based catalysts were mentioned in the literature. Cobalt acetylacetonate or cobalt neodecanoate generate ROMP process when associated to trialkylaluminum or addition process when associated to methylaluminoxane (MAO).[6] Some cobalt catalysts based on diphosphine or cyclopentadienyl ligands were developed by Heitz and coworkers.[7, 8] They lead to high molar mass addition polynorbornenes, when associated to MAO. The authors also mentioned a higher polymerization rate when the reactions were performed in chlorobenzene compared to toluene.

© 2004 WILEY-VCH Verlag GmbH & KGaA, Weinheim   DOI: 10.1002/masy.200450924

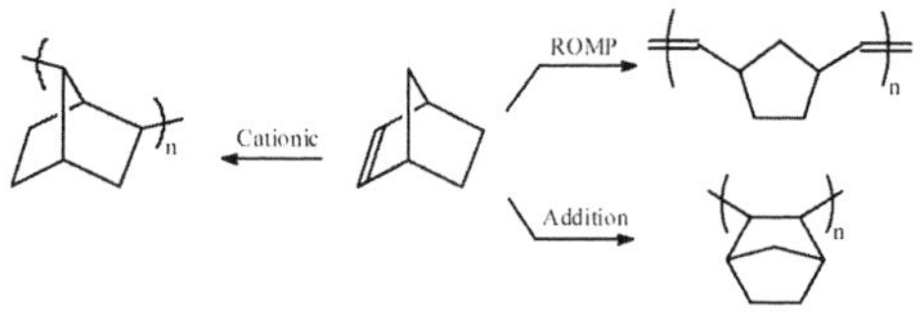

Scheme 1. Polynorbornene structure versus polymerization process.

On another hand, one of the latest breakthroughs in the olefin polymerization field was the discovery of pyridine bis(imine) iron or cobalt complexes as highly active catalysts for ethylene polymerization. Upon addition of various activators, highly linear polyethylenes of tunable molecular weight can be obtained.[9-12] Nevertheless, these catalytic systems revealed much less active towards propylene and higher α-olefins.[13-15] In a previous report with pyridine bis(imine) cobalt complexes, these catalysts proved to be not very active for norbornene polymerization under the used conditions.[16] In this paper, the ability of $CoCl_2$ and selected pyridine bis(imine) cobalt dichloride complexes (PBICo) to polymerize norbornene is studied. The influence of several reaction conditions on the polymerization is examined.

Results and Discussion

The polymerization of norbornene was investigated with 6 pyridine bis(imine) cobalt complexes and $CoCl_2$ activated with MAO. Different substituents were introduced onto the aryl group of the imine function in order to modify either the electron density on the metal or the steric hindrance around the catalytic site (Scheme 2). Their syntheses were already published.[14, 17]

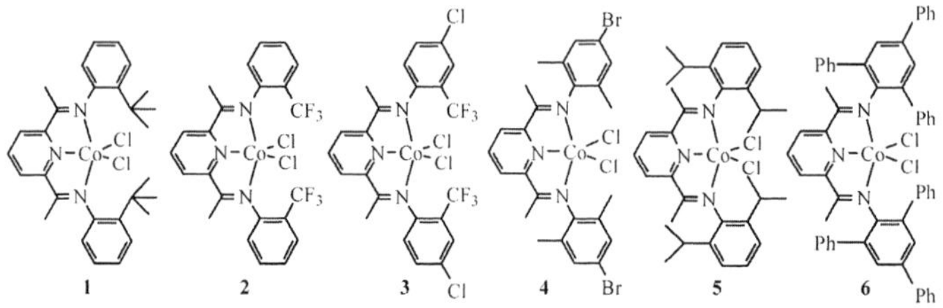

Scheme 2. Pyridine bis(imine) cobalt dichloride complexes.

All polymerizations were performed in chlorobenzene. MAO, purchased as a 10 wt.-% solution in toluene, was dried and used as a white powder (5 mol.-% of trimethylaluminum is still present). Some results are summarized in Table 1 and Figure 1. Pretty good yields were obtained within a couple of hours for all complexes. Catalyst **6** is among the most active, whereas it has revealed inactive for the ethylene polymerization (Figure 1). Moreover, to our knowledge, we were the first to describe $CoCl_2$ as an efficient catalyst for the addition polymerization of norbornene.[17] These complexes are as active as recently described palladium and nickel based complexes activated with MAO,[18, 19] but less active than some others exhibiting activities up to $5 \cdot 10^4$ g/mmol.h.[20-25] It was also checked that MAO alone was not able to polymerize norbornene under the conditions listed in Table 1. Besides, polymerization reactions were also conducted with the iron complexes and whatever the ligand or the reaction conditions, no polymer was detected. Molar masses of the polymers were determined by size exclusion chromatography (SEC) in chlorobenzene at 30°C. As indicated in Table 1, polynorbornenes exhibit high molar masses (ranging from $3.0 \cdot 10^5$ to $2.4 \cdot 10^6$ g/mol) with monomodal or bimodal distribution depending on the ligand.

Table 1. Polymerization of norbornene with several PBICo catalysts and $CoCl_2$.

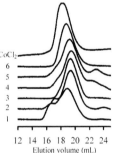

Catalyst	Loading	t_{Pol}	Yield	Activity	$M_{w\,LS}$[a]	MWD[b]
	(µmol)	(h)	(%)	(g./mmol$_{Cat}$.h)	(g/mol)	
1	9.0	4	28	9.7	$2.4\,10^6$	$3.9^{c)}$
2	9.0	4	38	13.3	$1.0\,10^6$	$2.6^{c)}$
3	9.1	4	27	9.6	$3.0\,10^5$	1.4
4	9.0	5	17	4.7	$4.9\,10^5$	2.1
5	9.8	5	25	6.7	$6.0\,10^5$	3.3
6	10.0	5	46	12.0	$4.2\,10^5$	1.4
$CoCl_2$	9.2	5	81	23.0	$1.2\,10^6$	4.0

T_{Pol}=35 °C; Solvent: Chlorobenzene (12 mL); [Al]/[Co]=400; [NBE]/[Co]=1500; [NBE]=1 mol/L
[a] Molar mass measured by SEC in chlorobenzene at 30°C with light scattering online
[b] Molecular Weight Distribution
[c] Bimodal distribution

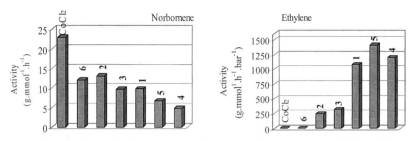

Figure 1. Comparison of ethylene and norbornene polymerization activities with several PBICo and CoCl₂.

Polynorbornenes were also characterized by ¹H and ¹³C NMR spectroscopy. As shown on Figure 2, no peaks attributable to ethylenic protons due to a metathesis process were detected. It can thus be assumed that the polynorbornenes were obtained through an addition process.

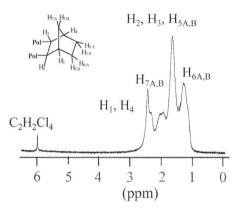

Figure 2. ¹H NMR spectrum of a polynorbornene synthesized with catalyst 1/MAO (400 MHz, C₂D₂Cl₄, 120°C).

Influence of monomer concentration

The evolution of the polymerization yield as well as the molar masses of the resulting polymers with norbornene concentration was studied. Results are summarized in Figure 3. As it can be seen, the polymerization yield increases linearly with norbornene concentration and no difference was observed for the molar masses. These observations are different from those made with nickel-based catalysts bearing phosphoraneiminato ligands.[25] Indeed, with these complexes, the molar masses were highly dependent on the monomer concentration.

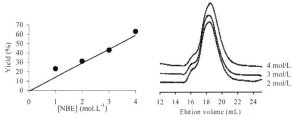

Figure 3. Evolution of norbornene polymerization yield and polynorbornene molar mass with monomer concentration (Catalyst **1** loading: 3.6 μmol; [Al]/[Co]=500; [NBE]/[Co]=1500; $T_{Pol.}$=35 °C; $t_{Pol.}$=1 h; Solvent: Chlorobenzene).

Influence of the [Al]/[Co] ratio

The influence of the [Al]/[Co] ratio on the polymerization yield was also investigated. Results are summarized in Figure 4. A different behavior is observed for the PBICo complexes and CoCl₂. Indeed, it appears clearly that CoCl₂ is fully activated for lower [Al]/[Co] ratios than the PBICo. With CoCl₂, a maximal activity is reached for low [Al]/[Co] ratios (around 100) and, after staying constant for ratios up to 1000, the activity decreases rapidly with increasing ratios. The decrease of CoCl₂ activity may be attributed to a poison effect of excess MAO. That may also be due to a poison effect of excess TMA (still present in solid MAO). With catalyst **1**, a fast increase of the activity is observed with increasing ratios up to 500 and then a slight increase is still noticeable for higher ratios. In the presence of catalyst **2**, the activity increases for [Al]/[Co] ratios up to 500 and then a plateau seems to be reached. Similar results have already been mentioned for palladium or nickel based complexes. [20, 25]

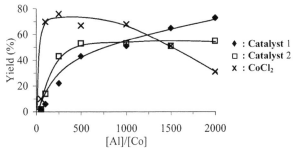

Figure 4. Evolution of norbornene polymerization yield with [Al]/[Co] ratio (Catalyst loading: 3.6 μmol; [NBE]=3 mol/L; [NBE]/[Co]=1500; $T_{Pol.}$=35 °C; $t_{Pol.}$=1 h; Solvent: Chlorobenzene).

The evolution of the molar masses with the [Al]/[Co] ratio was also investigated and results are shown in Figure 5. For catalyst **2** and CoCl$_2$, the molar mass distribution is always monomodal and the molar mass decreases with [Al]/[Co] ratio, probably due to increasing transfer reactions with higher amount of aluminum compounds. Unfortunately, the molar masses were too high to allow chain end determination by NMR spectroscopy. The molar mass distribution of polymers obtained with catalyst **1** is always bimodal and the proportion of the high molar mass fraction increases with the [Al]/[Co] ratio. Presently, we do not know exactly how these high molar mass polymer chains were produced. Nevertheless, it is probably the sign of the presence of at least two kinds of active sites in the polymerization medium.

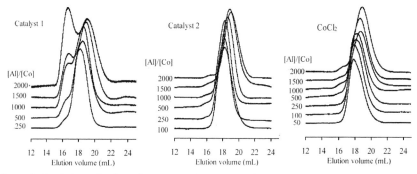

Figure 5. Evolution of polynorbornene molar masses with [Al]/[Co] ratio for catalysts **1** and **2** and CoCl$_2$.

Influence of the polymerization temperature

The influence of the polymerization temperature was also examined. Results are summarized in Figure 6. Each catalyst exhibits a peculiar behavior. With CoCl$_2$, the activity increases with temperature and reaches a plateau around 40°C. In the case of catalyst **1** the yield, quite large even for the lowest temperatures, increases with temperature to reach around 80°C the same plateau as the one observed with CoCl$_2$. With catalyst **2**, the activity increases with temperature for the lowest temperatures and reaches a plateau around 20°C. Above 60°C, the activity decreases with temperature, probably due to thermal instability. These behaviors are quite different from the behavior described for ethylene polymerization. Indeed, in this last case, the PBICo complexes always exhibit a maximal activity around 40°C and lower activities for other temperatures. [14, 26, 27]

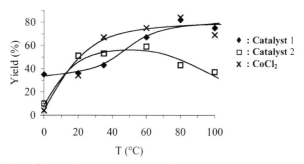

Figure 6. Evolution of the norbornene polymerization yield with polymerization temperature (Catalyst loading: 3.6 μmol; [NBE]=3 mol/L; [NBE]/[Co]=1500; [Al]/[Co]=500; $t_{Pol.}$=1 h; Solvent: Chlorobenzene).

Again, the molar masses were measured by SEC in chlorobenzene. With $CoCl_2$, the molar mass distribution is monomodal up to 80°C, with a decrease of the molar mass probably due to an increase of the transfer reactions with temperature. The shoulder appearing for temperatures higher than 80°C, is due to the polynorbornene generated by MAO alone. Indeed, at such elevated temperatures, MAO was shown to be able to polymerize NBE, with nevertheless lower yields than in the presence of a cobalt complex. Besides, the molar mass of the polynorbornenes obtained with MAO alone is exactly the same as the one of the shoulder obtained with $CoCl_2$/MAO (Figure 7). Moreover, the proportion of high molar mass polynorbornene is in good agreement with the polymerization yield observed with MAO alone. Thus, the presence of the shoulder may be attributed without any doubt to the polymerization of the monomer by MAO alone. Catalyst **2** gives monomodal molar mass distribution below 60°C and bimodal ones above 60°C. In this case, the high molar fraction is too important to be attributed to MAO. Again catalyst **1** yields bimodal molar mass distribution even at 20°C. The proportion of the high molar mass fraction increases with temperature. Concerning catalysts **1** and **2**, as the amount of high molar mass polymer chains is too important to be attributed to MAO alone, a partial decoordination of the ligand could be suggested to assume the presence of at least two kinds of active sites of different nature.

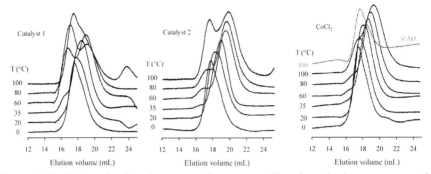

Figure 7. Evolution of polynorbornene molar masses with polymerization temperature for catalysts **1** and **2** and CoCl₂.

Kinetic study

Finally, the evolution of the polymerization yield with time was examined with 3 catalysts. Results are reported in Figure 8. The same behavior is observed for all the catalysts. Indeed, the yield increases linearly during the first hours and increases more slightly for longer reaction times. It should be noted that even after 48 h, the catalysts seem to be still active. Again, this behavior is quite different from that observed for the ethylene polymerization. Indeed, it is described in the literature that for such catalytic systems only 10% of the maximal activity remains after only 1h.[11, 26] In our case, one has to wait much longer to detect the same activity decay.

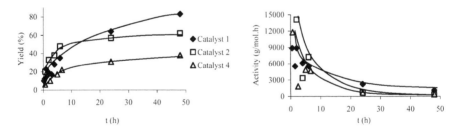

Figure 8. Evolution of the polymerization yield and activity with reaction time (Catalyst loading: 9 μmol; [NBE]=1 mol/L; [NBE]/[Co]=1500; [Al]/[Co]=400; T_{Pol}=35 °C; Solvent: Chlorobenzene).

Concerning the molar masses, again polymers obtained with catalyst **1** exhibit bimodal molar mass distribution even for the lowest reaction times, the amount of the higher molar mass fraction increasing with polymerization time. For the other 2 complexes, longer reaction times have to be waited to see the appearance of a high molar mass fraction. Again, one can suggest the simultaneous existence of several kinds of active species, due to partial decoordination of the ligand.

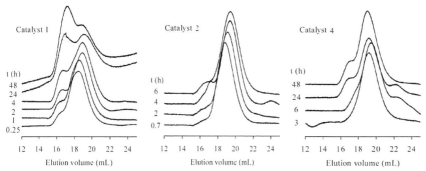

Figure 9. Evolution of polynorbornene molar masses with time for catalysts **1**, **2** and **4**.

Conclusion

The addition polymerization of norbornene was performed with 6 pyridine bis(imine) cobalt dichloride complexes, generally used for ethylene polymerization. In this study, norbornene was also polymerized for the first time with $CoCl_2$. The influence of several reactions parameters ([NBE], [Al]/[Co] ratio, polymerization temperature and time) was investigated. High molar mass polynorbornenes were obtained, exhibiting monomodal or bimodal molar mass distributions, depending on the ligand and the polymerization conditions. This behavior may be attributed to the coexistence of several active sites in the reaction medium. Moreover, the substituents on the pyridine bis(imine) ligand may influence the proportion of the different active species. Indeed, a *t*-butyl group (catalyst **1**) seems to favor the synthesis of high molar mass polynorbornene, whereas with a trifluoromethyl group (catalyst **2**) the synthesis of high molar mass polymer is disfavored. Besides, different behaviors were observed compared to ethylene polymerization. Indeed, the complexes revealed generally more robust versus temperature and less deactivation of the active species was detected with time.

274

Acknowledgements

The authors would like to thank very much Alain Rameau, Roger Meens and more specially Catherine Foussat for SEC measurements.

[1] T. F. A. Haselwander, W. Heitz, S. A. Krügel, J. H. Wendorff, *Macromol. Chem. Phys.* **1996**, *197*, 3435
[2] C. Janiak, P. G. Lassahn, *Macromol. Rapid Commun.* **2001**, *22*, 479
[3] C. Janiak, P. G. Lassahn, *J. Mol. Cat A : Chem.* **2001**, *166*, 193
[4] W. Kaminsky, A. Noll, *Polym. Bull.* **1993**, *31*, 175
[5] C. H. Bergstöm, J. Seppälä, *J. Appl. Polym. Sci.* **1997**, *63*, 1063
[6] B. L. Goodall, L. H. McIntosh III, L. H. Rhodes, *Macromol. Symp.* **1995**, *89*, 421
[7] F. P. Alt, W. Heitz, *Acta Polym.* **1998**, *49*, 477
[8] F. P. Alt, W. Heitz, *Macromol. Chem. Phys.* **1998**, *199*, 1951
[9] G. J. P. Britovsek, V. C. Gibson, B. S. Kimberley, P. J. Maddox, S. J. McTavish, G. A. Solan, A. J. P. White, D. J. Williams, *Chem. Commun.* **1998**, 849
[10] B. L. Small, M. Brookhart, A. M. A. Bennett, *J. Am. Chem. Soc.* **1998**, *120*, 4049
[11] K. R. Kumar, S. Sivaram, *Macromol. Chem. Phys.* **2000**, *201*, 1513
[12] K. Radhakrishnan, H. Cramail, A. Deffieux, P. François, A. Momtaz, *Macromol. Rapid Commun.* **2003**, *24*, 251
[13] C. Pellecchia, M. Mazzeo, D. Pappalardo, *Macromol. Rapid Commun.* **1998**, *19*, 651
[14] R. Souane, F. Isel, F. Peruch, P. J. Lutz, *C. R. Chimie* **2002**, *5*, 43
[15] J. Ramos, V. Cruz, A. Munoz-Escalona, J. Martinez-Salazar, *Polym.* **2002**, *43*, 3635
[16] M. C. Sacchi, M. Sonzogni, S. Losio, F. Forlini, P. Locatelli, I. Tritto, M. Licchelli, *Macromol. Chem. Phys.* **2001**, *202*, 2052
[17] F. Pelascini, F. Peruch, P. J. Lutz, M. Wesolek, J. Kress, *Macromol. Rapid Commun.* accepted
[18] X.-F. Li, Y.-S. Li, *J. Polym. Sci., Part A : Polym. Chem.* **2002**, *40*, 2680
[19] P. G. Lassahn, C. Janiak, J.-S. Oh, *Macromol. Rapid Commun.* **2002**, *23*, 16
[20] C. Janiak, P. G. Lassahn, *Polym. Bull.* **2002**, *47*, 539
[21] P. G. Lassahn, V. Lozan, C. Janiak, *Dalton Trans.* **2003**, 927
[22] B. Berchtold, V. Lozan, P. G. Lassahn, C. Janiak, *J. Polym. Sci., Part A : Polym. Chem.* **2002**, *40*, 3604
[23] Y.-S. Li, Y.-R. Li, X.-F. Li, *J. Organomet. Chem.* **2003**, *667*, 185
[24] D. A. Barnes, G. M. Benedikt, B. L. Goodall, S. S. Huang, H. A. Kalamarides, S. Lehnard, L. H. McIntosh III, K. T. Selvy, R. A. Shick, L. F. Rhodes, *Macromolecules* **2003**, *36*, 2623
[25] C. Mast, M. Krieger, K. Dehnicke, A. Greiner, *Macromol. Rapid Commun.* **1999**, *20*, 232
[26] G. J. P. Britovsek, M. Bruce, V. C. Gibson, B. S. Kimberley, P. J. Maddox, S. Mastroianni, S. J. McTavish, C. Redshaw, G. A. Solan, S. Strömberg, A. J. P. White, D. J. Williams, *J. Am. Chem. Soc.* **1999**, *121*, 8728
[27] J. Qiu, L. Sun, Y. Hu, Y. Li, *Chin. J. Polym. Sci.* **2000**, *18*, 509

Macromol. Symp. **2004**, *213*, 275-286

A Combined Density Functional Theory and Molecular Mechanics (QM/MM) Study of Single Site Ethylene Polymerization Catalyzed by [Cp{NC(^{t}Bu)$_2$}TiR$^+$] in the Presence of the Counterion, CH$_3$B(C$_6$F$_5$)$_3^-$

Kumar Vanka, Zhitao Xu, Tom Ziegler*

Department of Chemistry, University of Calgary, Calgary, Alberta Canada T2N 1N4, Canada

Summary: A density functional study was conducted on the approach and insertion of ethylene monomer into the Ti-C$_\alpha$ bond of the catalyst system, CpNC(^{t}Bu)$_2$RTi-μ-Me-B(C$_6$F$_5$)$_3$ (R= methyl, propyl). A validated QM/MM model was used to represent the counterion. Solvent effects were incorporated with single point solvent calculations done with cyclohexane (ε = 2.023) as the solvent. For R=Me (the initiation step), approach and insertion of the ethylene was found to be endothermic, with the barrier for insertion being 12.7 kcal/mol for the most favourable case. For R=Pr (the propagation step), the insertion barrier was found to decrease slightly (11.5 kcal/mol for the most favorable case), corroborating experimental evidence of decrease in insertion barrier with increase in chain length. Termination by chain transfer to monomer was also considered, and found to be unfavourable, in comparison to insertion, by 8.6 kcal/mol for the propagation step. Solvent effects were found to be significant for the propagation step, changing the rate determining step from insertion to uptake for the most favorable case of insertion.

Keywords: catalysts; computer modeling; initiation; propagation; termination

Introduction

Single-site homogeneous catalysts have received increasing attention as alternatives to traditional Ziegler-Natta type heterogeneous catalysts. This is primarily due to their ability to achieve higher stereoselectivity, narrower molecular weight distribution, and higher activity. Among the more highly active homogenous catalysts are metallocenes and related organometallic compounds containing a Group IV transition metal. The general structure of these complexes contains a Group IV transition metal center (M) coordinated to two ligands (L) and two alkyl (R) groups: L$_2$MR$_2$.

These complexes by themselves are not very effective as polymerization catalysts, but require activation by a co-catalyst or a counterion. The co-catalysts are generally Lewis acids, such as tris(pentafluorophenyl)borane or methylalumoxane (MAO), whose function is to abstract one of

          DOI: 10.1002/masy.200450925

the alkyl groups as R⁻ to produce the charge separated contact ion-pair, $L_2MR-\mu R-A$. This contact ion-pair can then dissociate to form the form the cationic complex, L_2MR^+, which is generally accepted [1-11] to be the activated metal catalyst, and the negatively charged counterion, AR^-.

Several theoretical studies, [12-19] investigating the insertion process, have been conducted, using the bare cation, L_2MR^+, as the model for the single site catalyst. However, recent findings indicate [7-10,20-21] that the anionic counterion plays an important role in the polymerization process. Calculations showed [20-21] that the energy required to separate the ions in the contact ion-pair was very high, which implied that the counterion would remain in the vicinity of the cation during the insertion process. Figure 1 below shows the mechanism for monomer (ethylene) insertion into the M-R bond of the catalyst, with the inclusion of the counterion.

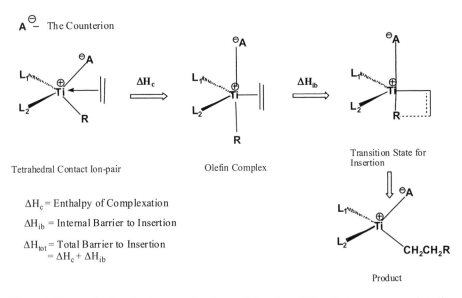

Figure 1. The mechanism for the complexation and insertion of the ethylene monomer into the metal-carbon bond of the cationic catalyst, L_2TiR^+, with the inclusion of the counterion.

Figure 1, however, shows only one of the ways in which the ethylene monomer can approach the contact ion-pair, $L_2TiR-\mu R-A$. Depending on the position of the ethylene monomer relative to the

μ – alkyl bridge of the ion-pair, there are two modes of approach of the ethylene. It can attack the ion-pair cis to the μ-alkyl group (as in Figure 1), an approach which will be denoted as the "cis" approach. The alternate mode of attack of the monomer, trans to the μ-alkyl group, will be denoted as the "trans" attack.

Unfortunately, due to the size of the counterions, theoretical studies of the mechanism outlined in Figure 1, where the counterion is incorporated, are difficult and involve considerable computational effort. Hence few examples of such studies are found in the literature.[20-31] Nifant'ev et al.[22] have studied ethylene insertion for the $Cp_2ZrEt^+A^-$ (A^- = $CH_3B(C_6F_5)_3^-$, $B(C_6F_5)_4^-$) systems. Lanza [23] investigated olefin insertion into the Ti-CH_3 and the Ti-C_3H_7 bond of the $H_2Si(C_5H_4)(^tBuN)TiCH_3$-$CH_3B(C_6F_5)_3$ ion-pair; Fusco [24] and Bernardi [25] studied the same process for $Cp_2(Ti/Zr)CH_3$-$Cl_2Al[O(Al(CH_3)_3AlHCH_3]_2$ and $Cl_2TiCH_3(\mu$-$Cl)_2AlH_2$ respectively. Chan et al.[26] have investigated the formation of ethylene separated ion-pair complexes for ion-pairs formed between different catalysts and the counterion $B(C_6F_5)_3CH_3^-$, as well as ethylene insertion [20] into the Zr-CH_3 bond in the Cp_2ZrCH_3-μ-CH_3-$B(C_6F_5)_3$ system.

In an earlier theoretical study, [31] we investigated the insertion of the ethylene monomer into the methyl chain of different catalyst systems of the type $L_1L_2TiCH_3^+$, in the presence of the counterion. The counterion used for the purpose of the investigation was $B(C_6F_5)_3CH_3^-$. For the sake of computational expediency, a QM/MM model was used for the counterion. This allowed us to study a large number of different systems. In the QM/MM model, the perfluorophenyl (C_6F_5) groups in $B(C_6F_5)_3CH_3^-$ were replaced with molecular mechanics (MM) atoms, and Cl atoms are used to cap the quantum mechanical (QM) system. Thus, $B(C_6F_5)_3CH_3^-$ was replaced with $BCl_3CH_3^-$ in the QM system. This model has been validated.[32-35]

In this present study, we investigate the approach of ethylene and its insertion into the Ti-methyl (initiation step) and Ti-propyl (propagation step) bonds for the $[CpNC(^tBu)_2TiR^+][B(C_6F_5)_3CH_3^-]$ (R = CH_3 and C_3H_7) ion-pair systems. The different approaches of the ethylene monomer, cis and trans to the μ-methyl group of the ion-pair, will be considered for each case. The purpose of the investigation will be to isolate the important factors that contribute to the barrier of insertion, and to compare and contrast the initiation and propagation steps for the same ion-pair system. The effect of the solvent on the insertion process will also be considered, with single point solvation calculations to determine the energy of solvation for the ion-pair, the olefin complexes and the transition states. Cyclohexane ($\varepsilon = 2.023$) will be used as the solvent.

Computational Details

The density functional theory calculations were carried out using the Amsterdam Density Functional (ADF) program version 2000.01, developed by Baerends et al. [36-39] and vectorized by Ravenek. [40] The electronic configurations of the atoms were described by a triple-ζ basis set on titanium (n = 3) for ns, np, nd and (n+1)s, augmented with a single (n+1)p polarization function. Double-ζ slater type orbitals (STO) basis were used for carbon (2s,2p), hydrogen (1s) and nitrogen (2s,2p), augmented with a single 3d polarisation function except for hydrogen where a 2p polarisation function was used. The gas phase energy difference was calculated by augmenting the local density approximation energy with Perdew and Wang's non-local correlation and exchange corrections (PWB91). [41] The solvation energies were obtained from a single point full QM calculation using the Conductor-like Screening Model (COSMO) [42,43] and optimized geometries from QM/MM calculations. A dielectric constant of 2.023 was used to represent cyclohexane as the solvent. The MM atoms were described using the SYBYL/TRIPOS 5.2 force field constants.[44] The code for QM/MM in ADF has been implemented by Woo et al. [45]

First Insertion Study

The insertion of the ethylene monomer was first considered for the system $CpNC(^tBu)_2TiMe^+$ in the absence of the counterion. QM/MM atoms were used to model the tertiary butyl groups, with hydrogens used as capping atoms. Hence the QM portion of the cation was $CpNCH_2TiMe^+$. The values of the ethylene complexation energy, ΔH_c, and the internal insertion barrier, ΔH_{ib}, were found to be –20.0 kcal/mol and 3.1 kcal/mol respectively. The total barrier for insertion, found by summing up the two values, was calculated to be –16.9 kcal/mol, indicating that the process is highly exothermic in nature.

Therefore, using just the naked cation as a model for the catalyst, one would be led to conclude that the cationic $CpNC(^tBu)_2TiMe^+$ system is an excellent polymerization catalyst. This view is modified substantially when the counterion, $B(C_6F_5)_3CH_3^-$, is included in the calculations. Figure

2 below shows the cis approach of the ethylene monomer towards the ion-pair, [CpNC(^{t}Bu)$_2$MeTi-μ-MeB(C$_6$F$_5$)$_3$. The reaction coordinate used for this approach is $\Delta R = R$(Ti-μ-CH$_3$) - R(Ti-X): the difference between the distance of the midpoint of the ethylene monomer and the carbon of the methyl bridge from the metal centre. The energy values denoted in the figures are solvent corrected, with the corresponding gas phase values denoted in parenthesis.

As shown in Figure 2, the ethylene forms a π complex at a distance of about 2.57 Å, at an energy cost of 7.7 kcal/mol in comparison to the separated species. The ethylene moiety in the π complex lies perpendicular to the plane formed by the metal centre, the methyl chain and the μ carbon of the counterion. The reason for the endothermic nature of this complexation is the displacement of the counterion from the metal centre, in order to accommodate the ethylene molecule.

The ethylene molecule, then proceeds to rotate itself into the insertion plane. There is a barrier to this rotation of about 2.8 kcal/mol, after which it forms the "in plane" olefin complex, which lies 10.3 kcal/mol above the separated reactant species. After formation of the olefin complex the ethylene then approaches the methyl chain, to enable insertion into the Ti-C bond. The system is slightly destabilized as a result, leading eventually to the transition state for insertion, lying 12.7 kcal/mol above the reactants. The counterion is only slightly displaced from the cation at this stage, by only about 2.5 Å.

After formation of the transition state, the ethylene inserts and proceeds to form the product with the propyl chain, which lies at –24.0 kcal/mol below the separated species. Hence the total barrier to insertion for the cis approach is 12.7 kcal/mol. This is significantly different from the findings in the naked cation case, where the insertion transition state was found to be lower in energy than the reactants.

The trans approach of the ethylene was also studied, and also found to lead to the formation of a π complex, followed by insertion into the metal carbon bond. Like in the cis case, the displacement of the counterion during insertion was quite small: 2.54 Å, which suggests that for the first insertion, the counterion is only slightly displaced from the metal centre. The total barrier to insertion for this case was 15.5 kcal/mol. This implies that, for the initiation step of the monomer, the ethylene would insert via the cis mode of approach in preference to the trans.

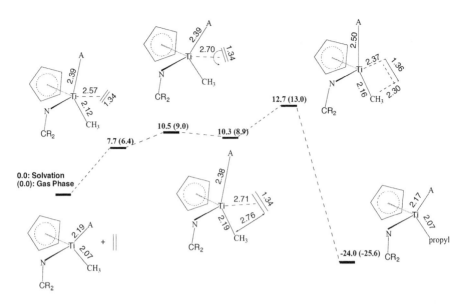

Figure 2. The energy profile for the cis approach of the ethylene monomer towards the contact ion-pair $[Cp\{NC(^tBu)_2\}TiMe^+][CH_3B(C_6F_5)_3^-]$.

Second Insertion Study

The results of the first insertion study indicated that the counterion had to be included to get a clear understanding of the insertion process. Therefore, for the second insertion investigation (the propagation step), we have ignored the naked cation case and focussed our attention on the approach and insertion of the ethylene monomer toward the contact ion-pair, $[Cp\{NC(^tBu)_2\}TiPr^+][CH_3B(C_6F_5)_3^-]$: the product from the initiation step, as discussed in the previous section.

The propyl chain in the contact ion-pair, $[Cp\{NC(^tBu)_2\}TiPr^+][CH_3B(C_6F_5)_3^-]$, can orient itself in different conformations. It is necessary to study these different conformations in order to determine the most stable resting state. This was done by a linear transit wherein the dihedral angle, θ, formed between the C_β-C_α-Ti and the C_α-Ti-μC planes was altered (see Figure 3 below) from -180 to 180 degrees, in order to rotate the chain so as to determine the most stable resting states. The accompanying energy profile in Figure 3 shows that the most stable resting states of the ion-pair correspond to the staggered conformations of the chain, with θ at -55° (**A**), 55° (**C**) and 178° (**B**) respectively. In contrast, the eclipsed conformation, with θ equaling 0° (**D**) was found to be an energy maxima, lying 7.6 kcal/mol above the most stable conformation, **A**.

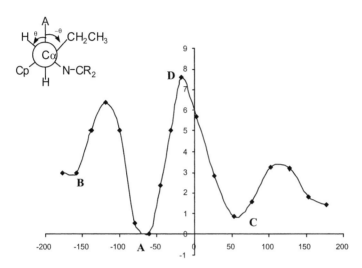

Figure 3. The relative stabilities of the different chain conformations, formed by rotation of the propyl chain in the contact ion-pair $[Cp\{NC(^tBu)_2\}TiPr^+][CH_3B(C_6F_5)_3^-]$.

For the insertion studies, the four different conformations, **A**, **B**, **C**, **D** shown in Figure 3 were considered. Beginning with the ion-pair in these four states, the approach of the ethylene monomer cis and trans to the μ-methyl bridge was studied. This gave rise, therefore, to eight possibilities, which will be discussed. The relative energies of the olefin complexes and insertion barriers were calculated with respect to the totally separated species: ethylene monomer and **A**, the most stable conformation of the ion-pair in its resting state.

Figure 4 below shows the cis approach of the ethylene monomer towards the contact ion-pair in the conformation **C**. This was found to be the most favourable insertion pathway for the propagation step. A π complex is formed similiar to the case for the first insertion. Insertion follows as the ethylene moiety approaches the α carbon.

282

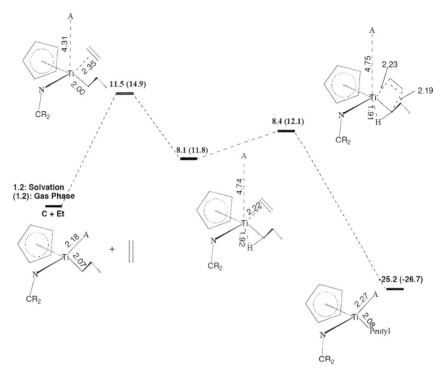

Figure 4. The energy profile for the cis approach of the ethylene monomer towards the contact ion-pair [Cp{NC(^{t}Bu)$_2$}TiPr$^+$][CH$_3$B(C$_6$F$_5$)$_3$$^-$], with the propyl chain lying out of the insertion plane, for the case **C**.

It is interesting to note, from Figure 4 above, that the counterion is much displaced from the metal centre during the insertion (from 2.18 Å to 4.75 Å), leading to significant charge separation. Hence solvent effects become significant, and play a role in influencing the rate determining step. In the gas phase, it is the insertion step that is rate determining, as seen in Figure 4. However, after solvent effects have been taken into account, the uptake of the monomer prior to formation of the π complex becomes the rate determining step. The barrier to insertion is, therefore, the uptake barrier: 11.5 kcal/mol. This is 1.2 kcal/mol lower than the barrier obtained during the initiation step (12.7 kcal/mol), which implies that the insertion of the monomer becomes easier with increase in the chain length, as has been observed experimentally by Liu et al.[46]

Two of the eight cases of ethylene approach do not lead to insertion but to termination. These are (i) the cis approach towards conformation **D** and (ii) trans approach towards **B**. In these two cases, the approach led to the chain transfer from the alkyl chain to the monomer. As an illustration, the trans approach towards **B** is shown in Figure 5 below. After the formation of the π complex, the insertion of the ethylene monomer would involve the rotation of the β carbon of the propyl chain out of the insertion plane, to facilitate the approach of the ethylene towards the α carbon. The barrier to this rotation was found to be high (~22 kcal/mol). Instead, the ethylene preferred to extract the β hydrogen from the chain, leading to termination (see Figure 5). The barrier to the chain termination was found to be 20.1 kcal/mol, which is higher than the insertion barrier by 8.6 kcal/mol.

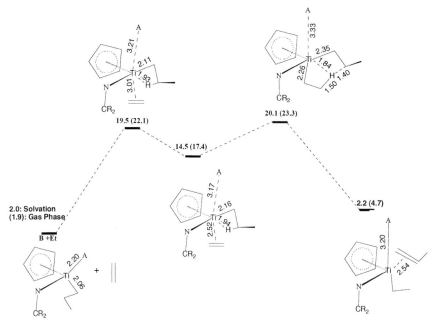

Figure 5. The energy profile for the trans approach of the ethylene monomer towards the contact ionpair $[Cp\{NC(^tBu)_2\}TiPr^+][CH_3B(C_6F_5)_3^-]$, for the case **B**.

Collected in Figure 6 are the results for the eight cases of monomer approach, cis and trans to the conformers **A**, **B**, **C** and **D**, leading either to insertion or termination. The gas phase values are also included in the parenthesis for comparison. Cis and trans approaches for the out-of-plane conformations, **A** and **C,** both lead to insertion, and have lower barriers than for the in-plane cases, **B** and **D** (see Figure 6). This suggests that the alkyl chain in the ion-pair would prefer to sit out of the insertion plane during monomer approach. The most likely mode of insertion would be through the cis attack for the **C** case (discussed in Figure 4 above). It is also clear from the values obtained that the ion-pair would strongly prefer insertion to termination, since the two approaches leading to termination have barriers much higher than the insertion barrier for the most favourable insertion case - cis approach, **C**.

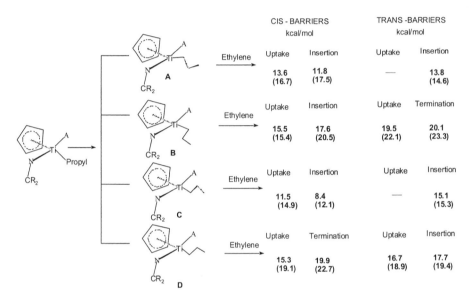

Figure 6. Overall picture of the Second Insertion Step – the barriers obtained from the cis and trans approaches of the ethylene monomer towards the contact ion-pair $[Cp\{NC(^tBu)_2\}TiPr^+][CH_3B(C_6F_5)_3^-]$, with the propyl chain in different conformations.

Conclusions

A density functional study was conducted on the approach and insertion of the ethylene monomer into the $Ti-C_\alpha$ bond for the ion-pair $CpNC(^tBu)_2RTi-\mu-Me-B(C_6F_5)_3$(R = methyl, propyl). The chief conclusions from the study are summarized below.

(a) The counterion plays a very significant role in the insertion process for both the first (R = methyl) and second (R = propyl) insertions of the ethylene monomer. The incorporation of the counterion makes the insertion process highly endothermic, in contrast to the corresponding naked cationic case, where the insertion was found to be exothermic. Moreover, for the second insertion, the presence of the counterion influences the alkyl chain conformation of the ion-pair, making the out-of-plane chain conformations more favourable for insertion than the in-plane ones.

(b) The barrier to second insertion for the most favorable approach (11.5 kcal/mol) is found to be 1.2 kcal/mol lower than the barrier for the first insertion (12.7 kcal/mol). This reduction in the height of the barrier, upon going from first to second insertion, corroborates experimental findings. [46]

(c) Solvent effects become important with the increase in the chain length. The displacement of the counterion increases during most approaches of the ethylene monomer during the second insertion, thereby increasing the charge separation between cation and anion. Therefore, incorporation of the solvent effects reduces the barriers to uptake and insertion to a significant extent. In fact, for the cis approach of ethylene towards **A**, inclusion of solvent effects reverses the rate determining step, making the uptake barrier higher than the insertion.

Acknowledgement

This investigation was supported by the Natural Science and Engineering Research council of Canada (NSERC) and by Nova Research and Technology Corporation (NRTC). Tom Ziegler would like to thank the Canadian Government for a Canada Research Chair.

[1] M. J. Bochmann. *Chem. Soc. Dalton Trans.* **1996**, 225.
[2] S. Pasynkiewicz. *Polyhedron* **1990**, *9*, 429.
[3] M. R Mason, J. M. Smith, S. G.Bott, A. R. Barron, *J. Am. Chem. Soc. 115*, 4971.
[4] J. L. Atwood, D. C. Hrncir, R. D. Priester, R. D. Rogers, *Organometallics* **1983**, *2*, 985.
[5] C. J. Harlan, S. G. Bott, A. R. Barron, *J. Am. Chem. Soc.* **1995**, *117*, 6465.
[6] A. R. Barron. 218'th ACS national meeting, New Orleans, August 22-26, 1999.
[7] X. Yang, C. L. Stern, T. J. Marks, *J. Am. Chem. Soc.* **1994**, *116*, 10015.
[8] P. A. Deck, T. J. Marks, *J. Am. Chem. Soc.* **1995**, *117*, 6128.
[9] L. Jia, C. L. Stern, T. J. Marks, *Organometallics* **1997**, *16*, 842.
[10] L. Li, T. J.Marks, *Organometallics* **1998**, *17*, 3996.
[11] P.Deck, C. L.Beswick, T. J. Marks, *J. Am. Chem. Soc.* **1998**, *120*, 1772.
[12] T. K.Woo, P. M. Margl, T. Ziegler, P. E. Blöchl, *Organometallics* **1997**, *16*, 3454.
[13] T. K. Woo, P. M.Margl, J. C. W. Lohrenz, P. E. Blöchl, T. Ziegler, *J. Am. Chem. Soc.* **1996**, *118*, 13021.
[14] P. M. Margl, J. C. W. Lohrenz, P. E. Blöchl, T. Ziegler, *J. Am. Chem. Soc.* **1996**, *118*, 44.
[15] L. Fan, D. Harrison, T. K.Woo, T. Ziegler, *Organometallics* **1995**, *14*, 2018.
[16] R. J. Meier, G. H. J. V. Doremaele, S.Tarlori, F. Buda, *J. Am. Chem. Soc.* **1994**, *116*, 7274.

286

[17] T. Yoshida, N. Koga, K. Morokuma, *Organometallics* **1995**, *14*, 746.
[18] H.Weiss, M. Ehrig, R. Ahlrichs, *J. Am. Chem. Soc.* **1994**, *116*, 4919.
[19] E. P.Bierwagen, J. E.Bercaw, W. A. Goddard III, *J. Am. Chem. Soc.* **1994**, *116*, 1481.
[20] M. S. W. Chan, T. Ziegler, Organometallics **2000**, *19*, 5182.
[21] K.Vanka, M. S. W. Chan, C. C. Pye, T. Ziegler, *Organometallics* **2000**, *19*, 1841.
[22] I. E. Nifant'ev, L. Y.Ustynyuk, D. N.Laikov, *Organometallics* **2001**, *20*, 5375.
[23] G. Lanza, I. L. Fragala, T. J. Marks, *Organometallics* **2002**, *21*, 5594.
[24] R.Fusco, L.Longo, F. Masi, F. Garbassi, *Macromol. Rapid Commun.* **1998**, *19*, 257.
[25] F. Bernardi, A.Bottoni, G. P. Miscione, *Organometallics* **1998**, *17*, 16.
[26] M. S. W. Chan, K.Vanka, C. C. Pye, T. Ziegler, *Organometallics* **1999**, *18*, 4624.
[27] K.Vanka, T. Ziegler, *Organometallics*, **2001**, *20*, 905.
[28] E. Zurek, T. K. Woo, T. Firman, T. Ziegler, *Inorganic Chemistry*, **2001**, *40*, 361.
[29] E. Zurek, T. Ziegler, *Inorganic Chemistry* **2001**, *40*, 3279.
[30] E. Zurek, T. Ziegler, *Organometallics* accepted for publication.
[31] K. Vanka, Z. Xu, T. Ziegler, *Israeli Journal of Chemistry* submitted for publication.
[32] Z. Xu, K. Vanka, T. Firman., A. Michalak, E. Zurek, C. Zhu, T. Ziegler, *Organometallics* **2002**, *21*, 2444.
[33] Z. Xu, K. Vanka, T. Ziegler, to be submitted.
[34] T. Ziegler, K.Vanka, Z. Xu, to be submitted.
[35] E. Zurek, T. Ziegler, to be submitted.
[36] E. J. Baerends, D. E. Ellis, P. Ros, *Chem. Phys.* **1973**, *2*, 41.
[37] E.J. Baerends, P.Ros, *Chem. Phys.* **1973**, *2*, 52.
[38] G. te Velde, E. J. Baerends, *J. Comp.Phys.* **1992**, *92*, 84.
[39] C. G. Fonseca, O. Visser, J. G. Snijders, G. te Velde, E. J. Baerends, in "Methods and Techniques in Computational Chemistry", *METECC-95;* ClementiE., Corongiu, G., Eds.; STEF; Cagliari,; **1995**, p.305.
[40] W. Ravenek, in "Algorithms and Applications on Vector and Parallel Computers"; H. J. J. te Riele, T. J. Dekker, and H. A. vand de Horst, Eds.; Elservier, Amsterdam, The Netherlands **1987**.
[41] J.P. Perdew, *Phys. Rev. B.* **1992**, *46*, 6671.
[42] A. Klamt, G. Schuurmann, *J. Chem. Soc. Perkin Trans.* **1993**, *2*, 799.
[43] C. C. Pye, T. Ziegler, *Theor. Chem. Acc.* **1999**, *101*, 396.
[44] M. Clark, R.D. Cramer III, N.van Opdenbosch, *J. Comp. Chem.* **1989**, *10*, 982.
[45] T. K. Woo, L. Cavallo, T. Ziegler, *Theor. Chim. Acta.* **1998**, *100*, 307.
[46] Z. Liu, E. Somsook, C. B.White, K.A. Rosaaen, C. R. Landis, *J. Am. Chem. Soc.* **2001**, *123*, 1119.

Macromol. Symp. **2004**, *213*, 287-301   287

Experimental and Calculated Vibrational Spectra and Structure of Ziegler-Natta Catalyst Precursor: 50/1 Comilled MgCl$_2$-TiCl$_4$

Luigi Brambilla,[1] *Giuseppe Zerbi,*[*1] *Stefano Nascetti,*[2] *Fabrizio Piemontesi,*[2] *Giampiero Morini*[2]

[1]Dipartimento di Materiali, Chimica e Ingegneria Chimica G. Natta del Politecnico di Milano, Piazza Leonardo da Vinci, 32 20133 Milano, Italy
[2]Basell Polyolefins, Centro Ricerche G. Natta, 44100 Ferrara, Italy

Summary: The vibrational Infrared and Raman Spectra of a MgCl$_2$-TiCl$_4$ Ziegler-Natta catalyst precursor with a 50/1 MgCl$_2$/TiCl$_4$ ratio have been recorded. The Raman spectrum of this catalyst precursor, in the range 50-500 cm^{-1}, shows clear scattering lines which can be assigned to the complex MgCl$_2$-TiCl$_4$, well separated from those of the initial species. Analogous, but less clear signals can be found in the infrared spectrum. Vibrational symmetry analysis and quantum chemical calculations of suitable models of MgCl$_2$-TiCl$_4$ complex have been made for the interpretation of the experimentally recorded spectra. The observed spectroscopic signals can be explained in terms of the existence of only one type of MgCl$_2$-TiCl$_4$ complex where the TiCl$_4$ molecules are complexed on the MgCl$_2$ along the (110) lateral cuts in a local C$_{2v}$ symmetry with the Ti atoms in an octahedral coordination.

Keywords: infrared spectroscopy; quantum chemical calculations; Raman spectroscopy; TiCl$_4$; Ziegler-Natta catalyst

Introduction

TiCl$_4$ supported on activated MgCl$_2$ plays a fundamental role in the catalysis for the production of polyolefins[1]. On the basis of experimental evidence, suggesting that preferential lateral cuts corresponding to the (100) and (110) planes are formed during MgCl$_2$ activation[2], a variety of catalytic species have been proposed to occur when TiCl$_4$ is chemisorbed on coordinatively unsaturated Mg^{2+} ions located on the above side faces[3,4]. However, experimental evidence on the nature of the catalytic complexes are still lacking.

For the solution of this problem, several vibrational spectroscopic studies on Ziegler-Natta catalyst precursor have been reported[5-10]. Among these, most of the Infrared studies[5-8] have

© 2004 WILEY-VCH Verlag GmbH & KGaA, Weinheim   DOI: 10.1002/masy.200450926

been devoted to elucidate the nature of the complexes between the catalysts and the electron donors used in order to improve the stereospecificity in 1-olefin polymerization. Only Ystenes[9,10] reports that in the Infrared transmission spectra of $MgCl_2$-$TiCl_4$ precatalyst distinct bands near 450 cm^{-1} appear which can be assigned to the stretching of the terminal Ti-Cl bonds. For the first time, the present study attempts to clarify the molecular structure of the $TiCl_4$ species complexed on $MgCl_2$ combining experimental Infrared and Raman vibrational spectroscopy data with quantum chemical calculations. In particular we focus our attention at one $MgCl_2$-$TiCl_4$ catalyst precursor, obtained by co-milling the two components in a 50/1 molar ratio.

Experimental

Pure $MgCl_2$ and $TiCl_4$ were purchased from Aldrich and used without further purification. $MgCl_2$, $TiCl_4$ and the catalyst precursor sample have been always handled and prepared in an inert atmosphere. The $MgCl_2$-$TiCl_4$ precatalyst with a 50/1 magnesium to titanium ratio was prepared by co-milling 12 g of $MgCl_2$ and 0.5 g of $TiCl_4$ in a 330 ml centrifugal mill containing 4 porcelain balls. The co-milling time was 4 hours. The Ti content in this catalyst precursor is comparable with that of some industrial catalysts for polyolefin polymerization[1].

Samples for Raman spectroscopy have been sealed in quartz tubes with a diameter of 5 mm; to record the Far Infrared spectra, the materials have been suspended in nujol and sandwiched between non additivated polyethylene sheets. Raman spectra were recorded with a "Modular XY" Dilor spectrometer in a backscattering geometry. The green laser line at 514,5 nm from an Ar+ laser of Spectra Physics, model 2030, was used. The laser power on the sample has been always less that 5 mw. The far infrared spectra have been recorded with a Nicolet Magna 760 with a solid substrate FAR IR (® Nicolet) beam splitter and equipped with a polyethylene window DTGS detector.

Calculations

Quantum chemical calculations have been carried out to predict the equilibrium geometry of $TiCl_4$-$MgCl_2$ complex models and to calculate their vibrational spectra. Normal modes frequencies as well as infrared and Raman intensities have been included in the calculations.

The spectroscopic observables have been calculated always by *ab initio* RHF/6-31g using Gamess[11] and Gaussian 98[12] programs. A full energy minimization calculation has been adopted only for the free $TiCl_4$ molecule in the tetrahedral geometry. In the other cases a constrained geometry energy minimization has been used to simulate the $MgCl_2$ crystal or to predict the vibrational properties of the $TiCl_4$ molecules when the symmetry changes from T_d to C_{2v}. Since the values of the calculated normal mode frequencies and absolute vibrational infrared and Raman intensities have been shown to be basis set dependent, we use as a reliable structural tool only the qualitative trend and the qualitative intensity ratios.

Results and Discussion

Useful information derive from Raman spectra of the $MgCl_2$-$TiCl_4$ complexes because for this class of materials their Raman scattering lines are easy to be located and are well separated from those of isolated $TiCl_4$ or $MgCl_2$ species. On the contrary, the interpretation of the Far-Ir spectra is not so easy since the signals characteristic of the complex are weak if compared to the very strong vibrational transitions of $MgCl_2$ which dominate a large range of the Far Infrared frequencies.

In table 1 we report the experimental frequencies of the Raman scattering lines and Infrared absorption bands for the compounds studied.

Liquid TiCl₄

The experimental Raman spectrum of $TiCl_4$ at room temperature, reported in Figure 1a, shows the characteristic Raman lines at 500, 389, 139 and 122 cm^{-1}.

Table 1. Infrared and Raman experimental vibrational frequencies for $TiCl_4$, $MgCl_2$ and $MgCl_2$-$TiCl_4$ (50/1).

Compound		Observed vibrational frequencies in cm^{-1}
TiCl₄	Raman	122 m, 139 m, 389 s, 500 vw
	Far-Infrared	500 s
MgCl₂	Raman	157 v , 243 vs
	Far-Infrared	273 vs, 363 w, 405 w
MgCl₂-TiCl₄ 50/1	Raman	84 vw, 125 w, 157 ** w, 175 vw, 243** s, 293 w, 389 * vw, 449 m, 464 m
	Far-Infrared	262 ** vs, 363 ** m, 405 ** m, 475 vw
Relative intensity: vs=very strong, s=strong, m=medium, w=weak, vw=very weak . Line assigned to: * free $TiCl_4$ or vibration of the complex ; ** bulk $MgCl_2$		

Figure 1b gives the calculated Raman spectrum for the free and isolated molecule of $TiCl_4$ in a tetrahedral geometry. For the $TiCl_4$ molecule belonging to the T_d symmetry point group the irreducible representation for its 9 vibrational normal modes and their spectroscopic activity is the following:

$\Gamma_{vib} = A_1$ (Raman pol.) + E (Raman dep.) + $2F_2$ (IR, Raman dep.)

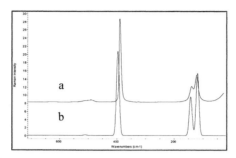

Figure 1. 1a) Experimental Raman spectrum of liquid $TiCl_4$ (λexc = 514.5 nm); 1b) ab initio calculated Raman spectrum of $TiCl_4$ in the tetrahedral structure.

For sake of simplicity we label the calculated vibrational frequencies as:

- v_1, v_2 and v_3 the triply degenerate F_2 stretching mode of $TiCl_4$ at 511 cm^{-1} ;

- v_4 the A_1 totally symmetric $TiCl_4$ stretching mode at 396 cm^{-1} ;

- v_5, v_6 and v_7 the second triply degenerate F_2 mode at 143 cm^{-1};

- v_8 and v_9 the doubly degenerate E mode at 119 cm^{-1}.

The observed Raman signals are in full agreement with the predictions from group theory and the *ab initio* calculated spectra are in an exceptionally good agreement with the experiments.

The vibrational assignment then turns out to be straightforward and in agreement with those proposed previously in the literature[13,14].

Milled $MgCl_2$

Figure 2 reports the experimental infrared and Raman spectra of $MgCl_2$ milled for 8 hours, in the limited spectral range studied, with the characteristic sharp Raman lines at 157 and 243 cm^{-1} and the intense and broad infrared absorption bands at 273, 363 and 405 cm^{-1}.

Because all these vibrations are observed in the Raman and Far-Infrared spectra of the catalyst precursors we think that the bulk of the $MgCl_2$ is only slightly perturbed by the complexation process. Raman spectra are slightly influenced by the milling process while the Far-Infrared spectra show a general broadening of the vibrational bands with the increasing milling time logically connected with the electronic confinement due to the small size of the particles.

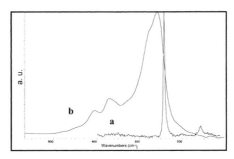

Figure 2.   Experimental 2a) Raman spectrum (λexc = 514.5 nm) and 2b) Far-Infrared spectrum of 8 hours milled $MgCl_2$.

$MgCl_2$-$TiCl_4$ (50/1) catalyst precursor

The vibrational spectra of the 50/1 precursor have been considered; data on different mixtures will be reported elsewhere. In the Raman spectrum shown in Figure 3, in addition to the two lines at 243 and 157 cm^{-1}, ascribed to $MgCl_2$ and the very weak band at 389 cm^{-1} ascribed to free $TiCl_4$ or probably due to a vibration of the complex (see table I), the following lines are clearly observed: a broad Raman scattering centred near 464 cm^{-1} from which a stronger and sharp component appears at 449 cm^{-1}, a weak line at 293 cm^{-1}, very weak and hardly detectable lines near 84, 125 and probably at 175 cm^{-1}.

The Far IR spectrum (Figure 4) shows only a very weak and broad band centered near 475 cm^{-1} (expanded in the insert) floating on the wing of the strong absorption of $MgCl_2$. The observation of other IR transitions is made practically impossible by the strong and broad absorption of $MgCl_2$ which dominates the whole spectral range.

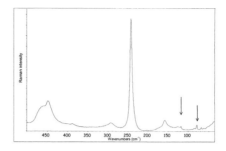

Figure 3. Raman spectrum (λexc = 514.5 nm) of $MgCl_2$-$TiCl_4$ (50/1) complex. The sharp and narrow peaks at low wavenumbers, indicated with arrows in Figure 4, are plasma laser lines.

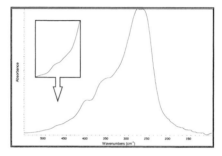

Figure 4. Far-Infrared spectrum of $MgCl_2$-$TiCl_4$ (50/1) complex.

The origin of the new vibrational transitions observed in the Raman and infrared spectra can be ascribed to the vibrational motions mostly localized on the $TiCl_4$ molecule which must have taken up another geometry as a consequence of the formation of the complex on the $MgCl_2$ surface.

It is generally accepted that lateral cuts corresponding to (110) and (100) $MgCl_2$ crystal planes are preferentially formed during precatalyst preparation[2] and different sites on activated $MgCl_2$ surface have been proposed for the $TiCl_4$ complexation[15-17]. When $TiCl_4$ molecules react with undercoordinated Mg^{2+} ions belonging to these two lateral cuts (tetracoordinated along the (110) and pentacoordinated along the (100)) can form complexes of different geometry.

For this reason many theoretical models[1] for the $TiCl_4$-$MgCl_2$ complexes have been proposed, both on (110) and (100) lateral cuts with the $TiCl_4$ molecules either in monomeric or dimeric species[18]

In our theoretical analysis we have studied the spectroscopic properties of $TiCl_4$ molecules involved in the simplest and energetically most favoured complexes of the monomeric species proposed in literature[19]; in particular we have considered:

i) a geometry with $TiCl_4$ complexed along the $MgCl_2$ (110) lateral cut with the Ti atom in an octahedral coordination directly interacting with the $MgCl_2$, Figure 5a;

ii) a geometry with $TiCl_4$ complexed along the $MgCl_2$ (110) lateral cut with the Ti atom in a slightly distorted tetrahedral coordination interacting with $MgCl_2$ by two Cl atoms, Figure 5b;

iii) a geometry with $TiCl_4$ complexed along the $MgCl_2$ (100) lateral cut in a distorted trigonal bipyramid geometry, Figure 5c.

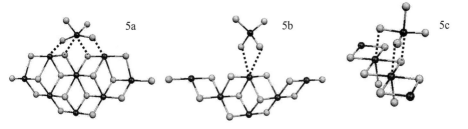

Figure 5. Models studied for the complexation of monomeric $TiCl_4$ on $MgCl_2$ crystal.

For case i) various authors[3,4] have proposed that upon complexation on the (110) lateral cut of $MgCl_2$, $TiCl_4$ tends to reach an octahedral coordination (Figure 5a). In such complex $TiCl_4$ takes up a local C_{2v} symmetry with the following spectroscopic activity:

$\Gamma_{vib} = 4A_1$ (IR, Raman) $+ A_2$ (Raman) $+ 2B_1$ (IR, Raman) $+ 2B_2$ (IR, Raman)

We have followed the evolution of the vibrational spectra from the free molecule (T_d symmetry) to the molecule in the complex (C_{2v} symmetry) by carrying out several *ab initio* calculations of the vibrational spectra where one of the Cl-Ti-Cl valence angles is increased step-wise at a fixed value starting from the tetrahedral angle and reaching 180°. For each value of the angle energy minimization has been carried out and then in this constrained optimized geometry the vibrational spectra, infrared and Raman, have been calculated. The evolution of the calculated vibrational frequencies (vs. Cl-Ti-Cl valence bond angle) and their symmetry species are reported in Figure 6.

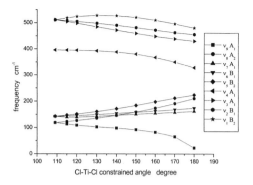

Figure 6. *Ab initio* calculated dispersion of the vibrational frequencies vs. Cl-Ti-Cl valence bond angle for TiCl$_4$ molecule from T$_d$ to C$_{2v}$ symmetry by stepwise opening of one Cl-Ti-Cl valence bond angle.

The symmetry properties of the normal modes of the starting tetrahedral TiCl$_4$ molecule are obvious when considering the structure of the irreducible representation and the degeneracy of the modes. Because of the lowering of the symmetry the F$_2$ mode near 511 cm^{-1} splits into a triplet v_1, v_2 and v_3. The A$_1$ mode v_4 obviously does not show splitting. The second F$_2$ mode splits in a triplet v_5, v_6 and v_7, while the E mode splits in a doublet v_8 and v_9. It is obvious that the calculated changes of the normal vibrations v_i, which are the solutions of the vibrational secular equation, describe the changes of the curvature of the vibrational potential

$$\left(\frac{\partial^2 V}{\partial Q_i^2} \right)_{eq.} \tag{1}$$

in the normal coordinate space at each equilibrium configuration, i.e.

$$2V = \sum_i \left(4\pi^2 c^2 v_i^2 \right) Q_i^2 \tag{2}$$

This is precisely the additional information we derive from the vibrational spectra on the evolution of the electronic structure of TiCl$_4$ when approaching MgCl$_2$.

If it is assumed that the vibrational motions in the complex are still localized, such that the symmetry properties of C$_{2v}$ point groups are satisfactorily maintained, it is worth looking at the repulsion of the vibrational levels which belong to the same symmetry species, as reported in Figure 6.

The calculated Raman spectra reported in Figure 7a show the increase in intensity of the v_3 (A_1 component) of the F_2 vibration at 511 cm^{-1} with the parallel decrease in intensity of the v_4 (A_1) vibration when the Cl-Ti-Cl bond opened. The calculated infrared spectra reported in Figure 7b show that when the Cl-Ti-Cl valence angle is opened the vibration at 511 cm^{-1} (IR strong) splits into a triplet with a redistribution of the intensities among the three components.

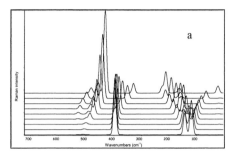
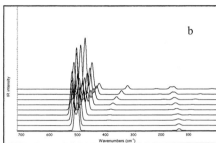

Figure 7. Evolution of *ab initio* calculated spectra for TiCl$_4$ molecule from T$_d$ to C$_{2v}$ symmetry by stepwise opening of one Cl-Ti-Cl valence bond angle: a) raman spectra and b) infrared spectra.

To take in account the influence of the MgCl$_2$ crystal on the vibrational properties of TiCl$_4$ we have carried out a calculation on the whole model 5a. The TiCl$_4$ molecule in a tetrahedral geometry was placed over the MgCl$_2$ crystal where the Cl and Mg atom have been keep fixed except the six atoms near the TiCl$_4$ molecule. After the constrained energy minimization the initial tetrahedral geometry of the TiCl$_4$ complexed with MgCl$_2$ was changed in a new C$_{2v}$ symmetry, with the Ti atom in an octahedral coordination. The calculated Cl-Ti-Cl valence bond angle of the atoms interacting with the crystal was 159.73 degree.

The calculated Raman spectrum, in Figure 8a, shows a number of lines larger than that observed in Figure 7. Some of these bands, near 280, 310 and 3870 cm^{-1} (indicated with * in Figure 8a), do not seem to have any correspondence in the experimental spectra; this is obviously related to the fact that the model of MgCl$_2$ crystal considered in the calculations has finite dimensions. However a comparison of the calculated and experimental spectra in Figure 8b reveals the that vibrations mostly localized on the TiCl$_4$ molecule or due to collective motions of the MgCl$_2$ (indicated by arrows) are satisfactorily predicted.

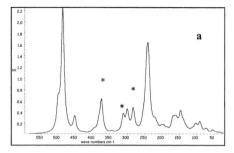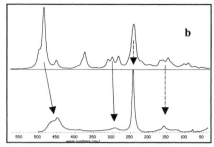

Figure 8. 8a) calculated Raman spectrum of model 5a; 8b) comparisons of 8a with the experimental Raman spectrum of $MgCl_2$-$TiCl_4$ 50/1. Solid arrows indicate vibrational transitions localized of $TiCl_4$, dashed on $MgCl_2$.

The role of the $MgCl_2$ lattice is an important issue in our work and has been carefully considered in this work. Several calculations have been carried out on a variety of models ranging from the free $TiCl_4$ which changes its geometry to various $TiCl_4$-$MgCl_2$ complexes where the number of $MgCl_2$ molecules in the chunks of lattice considered has been changed. Obviously an infinite perfect crystal of $MgCl_2$ could not be considered in the calculations; moreover because of milling the nanoparticles of $MgCl_2$ could show logical confinement effects, causing perturbations of its electron and phonons.

Calculations have shown that the forces between $TiCl_4$ and $MgCl_2$ lattice are not so strong to significantly perturb the dynamics of either the $TiCl_4$ molecule or the $MgCl_2$ cluster. The dynamical coupling of $TiCl_4$ in the complex is restricted on the surface of the $MgCl_2$, thus removing any worries about the size of the support made up by $MgCl_2$.

On the basis of the above calculations, it becomes possible to reach an interpretation of the experimental infrared and Raman spectra of the $MgCl_2$-$TiCl_4$ complex. Concerning the "isolated" $TiCl_4$ in a C_{2v} symmetry (Figure 7), for the triplet in the 400-500 cm^{-1} range, only two of the three components v_1 and v_2 are observed near 460 cm^{-1} in the infrared spectrum while the third component v_3 at 449 cm^{-1} is observed in the Raman spectrum. This assignment is also supported by the fact that v_3 is an A_1 mode (generally strong in the Raman) and must also be coupled with the other strong A_1 mode (v_4) at 387 cm^{-1}, as clearly shown by the repulsion of the two levels when the geometry is changed (Figure 6). Moreover the experimental Raman spectrum shows the right intensities redistribution between the A_1 vibrations (v_3 and v_4) predicted by the quantum chemical calculations.

Turn to the Raman spectrum of the complex 50/1 (Figure 8a, 8b) it shows practically the same features of TiCl4 in C2v symmetry as discussed before (Figure 7).

The experimental and theoretical results turn out undoubtedly to be in very good agreement and the reported evidence suggests that, when $TiCl_4$ is complexed on $MgCl_2$ at the 50/1 $MgCl_2/TiCl_4$ molar ratio, its geometry evolves from the initial tetrahedral structure to a new C_{2v} symmetry with the Ti approaching an octahedral coordination.

For case ii) the vibrational properties of a $TiCl_4$ molecule in a slightly distorted tetrahedral geometry have been considered, based on a $TiCl_4$ complexed on $MgCl_2$ along the (110) lateral cut with the Ti atom tetracoordinated[19] (Figure 5b). In such complex $TiCl_4$ takes up a local C_{2v} symmetry as in the previous case. Stepwise increases of the bond distance Ti-Cl of a pair of Ti-Cl bonds have been introduced in the calculations by constraining the Ti-Cl bond length from a $d_{Ti-Cl} = 2.202$ Å to 2.43 Å. At every step the infrared and Raman spectra have been re-calculated after geometry re-optimization. Within this range of values for d_{Ti-Cl}, the Cl-Ti-Cl valence angles of $TiCl_4$ turn out to remain nearly tetrahedral.

The evolution of the calculated vibrational frequencies (vs. Cl-Ti-Cl valence bond length) and their symmetry species are reported in Figure 9. The calculated Raman and infrared spectra reported in Figures 10a and 10b show that:

- within the F_2 triplet near 551 cm^{-1} v_1 shifts upward, v_2 (A_1) stays almost unchanged while v_3 shows a strong dispersion toward lower frequencies without an appreciable decrease of the infrared intensity;

- v_2 (A_1) increases its Raman intensities only for large values of the Ti-Cl bond length;

- v_4 (A_1) shifts substantially towards lower frequencies without any appreciable decrease of its Raman intensity and remains the strongest Raman line for any value of the Ti-Cl bond lengths considered;

- v_5 to v_9 keep their frequencies practically unchanged.

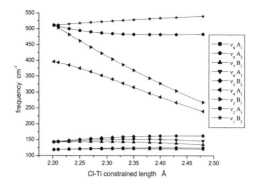

Figure 9. *Ab initio* calculated dispersion of the vibrational frequencies vs. Ti-Cl bonds lengths for TiCl$_4$ molecule by stepwise increasing of two Ti-Cl bonds from 2.202 Å (calculated length for the tetrahedral symmetry) to 2.48 Å.

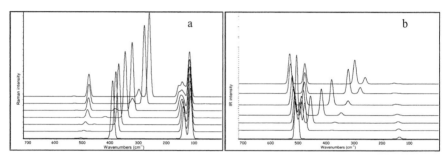

Figure 10. Evolution of: 10a) calculated Raman spectra and 10b) calculated infrared spectra of a TiCl$_4$ molecule from T$_d$ to C$_{2v}$ symmetry by stepwise increasing of two Ti-Cl bonds.

The calculated Raman spectrum of model 5b is reported in Figure 11a. Figure 11b reports the comparison between calculated and experimental spectra.

From the calculations it turns out that the complexed TiCl$_4$ molecule keeps substantially the tetrahedral geometry: the calculated Cl-Ti-Cl angle of the two Cl atoms interacting with the MgCl$_2$ has the values of 102.78 degree with a Ti-Cl bond length of 2.221 Å. This is reflected in the Raman spectrum where the vibrations localized on TiCl$_4$ molecule occur very close to those of the free TiCl$_4$.

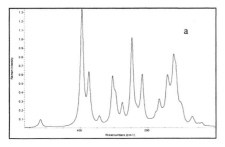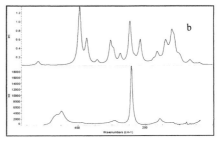

Figure 11. 11a) calculated Raman spectrum of model 5b; 11b) comparisons of 11a) with the experimental Raman spectrum of MgCl$_2$-TiCl$_4$ 50/1.

Evidence is gathered which shows that the experimental spectra of the MgCl$_2$-TiCl$_4$ (50/1) precursor do not show an acceptable agreement with the spectra of the model considered.

For the case iii) we have calculated only the infrared and Raman spectra of the molecule complexed on the (100) lateral cut, model 5c, because there is no unique way (contrary to the previous cases) to reach the desired geometry starting from the T$_d$ structure of the free TiCl$_4$.

In such complex TiCl$_4$ takes up a local C$_s$ symmetry with the following spectroscopic activity:

Γ_{vib} = 6A'(IR, Raman) + 3A"(IR, Raman)

The calculated infrared and Raman spectra are shown in Figure 12.

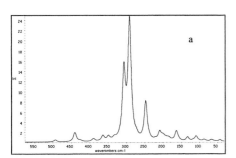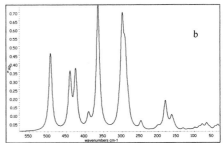

Figure 12. 12a) calculated Raman spectrum and 12b) calculated infrared spectrum of model 5c.

Band frequencies and intensities do not seem in a good agreement with the experimental results. As a result, only the structure of the TiCl$_4$ complex described in i) and represented in Figure 5a seems to be supported by the experimental vibrational spectra of the 50/1 complex.

Conclusions

In this paper we focused our attention on the structure of the complex $MgCl_2$-$TiCl_4$ in the 50/1catalyst precursor. Our conclusion is based both on the experimental results and on *ab initio* calculations.

The vibrational transitions observed in the Raman and infrared spectra can be ascribed to the vibrational motions mostly localized on the $TiCl_4$ molecule; it turns out that the vibrational motions of $TiCl_4$ are not strongly coupled with the modes of the $MgCl_2$ crystal.

At this low $TiCl_4$ concentrations (50/1) only one type of complex seems to occur on $MgCl_2$. The $TiCl_4$ molecule is forced to lower its symmetry from a full symmetrical T_d free species to almost an octahedral coordination of the Ti atom in the complex.

For the first time a clear indication of the octahedral geometry of $TiCl_4$ molecules absorbed on $MgCl_2$ is obtained from vibrational spectroscopy suggesting that the new species are bonded at the (110) lateral cut of $MgCl_2$ also in agreement with recent calculations [19].

The formation of octahedral $TiCl_4$ species on $MgCl_2$ on supported Ziegler Natta catalysts is usually accepted as model of aspecific catalytic sites in propylene polymerization (Corradini model, see ref. 3 and 20). Corradini indicated the monomeric titanium species on (110) lateral surface as possible aspecific sites producing the amorphous polymer.[20]

It is well known that $MgCl_2$-$TiCl_4$ catalysts are poorly isospecific in the polymerization of propylene as only about 40-50 % of isotactic polymer is obtained in the absence of Lewis bases acting either as internal or as external donor. Signals of different catalytic sites (possibly isospecific) may be expected, but the presence of a very low amount of titanium in our model system may not give enough Raman intensity to make them detectable.

The influence of the amount of titanium bonded on $MgCl_2$ on both infrared and Raman spectra and on the performance of the corresponding Ziegler-Natta catalysts in propylene polymerization is under evaluation and it will be presented in a future work.

Acknowledgement

Antonio Cristofori and Antonella Coraini are gratefully acknowledged for sample preparation, Leonardo Contado for the recording of the far infrared spectra and Matteo Tommasini for the calculated Raman spectra.

In the early stages of this work we have enjoyed the competent and thoughtful discussions with the late Professor Ugo Giannini we all miss.

[1] E. Albizzati, U. Giannini, G. Collina, L. Noristi, L. Resconi, in: "*Polypropylene Handbook*", Moore E.P. Jr., Eds., Carl Hanser Verlag, Munich **1996**, p. 11.
[2] U. Giannini, G. Giunchi, E. Albizzati, P.C. Barbè, in: "*Recent Advances in Mechanistic and Synthetic Aspects of Polymerization*", NATO ASI Sect. 215, Fontanille, M., Guyot, A., Eds., Reidel **1987**, p. 473.
[3] P. Corradini, V. Busico, G. Guerra, in: "*Comprehesive Polymer Science*", G.C. Eastmann, A. Ledwith, S. Russo, P. Sigwalt, Eds., Pergamon, **1989**, Vol. *4*, p. 29.
[4] L. Barino, R. Scordamaglia, *Macromol. Chem. Theory Simul.* **1998**, *7*, 407.
[5] E. Ritter, Ø. Nirisen, S. Kvisle, M. Ystenes, H.A. Øye, in "*Transition Metal Catalyzed Polymerizations*", R.P. Quirk Eds., Cambridge University Press, Cambridge **1988**, p. 292.
[6] M. Terano, T. Kataoka, T. J. Keii, *Polym.* Sci., Part A: Polym. Chem. **1990**, *28*, 2035.
[7] R. Spitz, J.L. Lacombe, A. J. Guyot, *Polym. Sci., Part A: Polym. Chem.* **1984**, *22*, 264.
[8] J.C.W. Chien, J.C. Wu, C.I.J. Kuo, *Polym. Sci., Part A: Polym. Chem.* **1983**, *21*, 725.
[9] Ø. Bache, M.Ystenes, *Appl. Spectrosc.* **1994**, *48*, 985.
[10] Ø. Bache, M. Ystenes, Intern. *Symposium on Metallorganic Catalysts for Synthesis and Polymerization*, September 13, Hamburg **1998**.
[11] M.W. Schmidt, K.K. Baldridge, J.A. Boatz, S.T. Elbert, M.S. Gordon, J.H. Jensen, S. Koseki, N. Matsunaga, K.A. Nguyen, S. Su, T.L. Windus, M. Dupuis, J.A. Jr Mongomery, *J. Comput. Chem.* **1993**, *14*, 1347.
[12] M.J. Frisch, G.W. Trucks, H.B. Schlegel, G.E. Scuseria, M.A. Robb, J.R. Cheeseman, V.G. Zakrzewski, J.A. Montgomery, R.E. Stratmann, J.C. Burant, S. Dapprich, J.M. Millam, A.D. Daniels, K.N. Kudin, M.C. Strain, O. Farkas, J. Tomasi, V. Barone, M. Cossi, R. Cammi, B. Mennucci, C. Pomelli, C. Adamo, S. Clifford, J. Ochterski, G.A. Petersson, P.Y. Ayala, Q. Cui, K. Morokuma, D.K. Malick, A.D. Rabuck, K. Raghavachari, J.B. Foresman, J. Cioslowski, J.V. Ortiz, B.B. Stefanov, G. Liu, A. Liashenko, P. Piskorz, I. Komaromi, R. Gomperts, R.L. Martin, D.J. Fox, T. Keith, M.A. Al-Laham, C.Y. Peng, A. Nanayakkara, C. Gonzalez, M. Challacombe, P.M.W. Gill, B.G. Johnson, W. Chen, M.W. Wong, J.L. Andres, M. Head-Gordon, E.S. Replogle, J.A. Pople, *Gaussian 98, Revision A.7*, Gaussian Inc., Pittsburgh, PA, **1998**.
[13] K. Nakamoto, in "*Infrared and Raman spectra of inorganic and coordination compounds*", Wiley, New York **1997**.
[14] R. J. H. Clark, *J. Chem. Soc (A)*, **1971**, 2999.
[15] E. Puakka, T. T. Pakkanen, T. A. Pakkanen, *Surf. Sci.* **1995**, *334*, 289.
[16] U. Giannini, *Makromol. Chem. Suppol.* **1981**, *4*, 216.
[17] P. Corradini, V. Barone, R. Fusco, G. Guerra, *Gazz. Chim. Ital.* **1983**, *113*, 601.
[18] A. G. Potapov, V. V. Kriventsov, D. I. Kochubey, G. D. Bukatov, V. A. Zakharov, *Macromol. Chem. Phys.* **1997**, *198*, 3477.
[19] G. Monaco, M. Toto, G. Guerra, P. Corradini, L. Cavallo, *Macromolecules* **2000**, *33*, 8953.
[20] P. Corradini, V. Barone, R. Fusco, Gazzetta Chimica Italiana **1983**, *113*, 601.

Macromol. Symp. **2004**, *213*, 303-313   303

Defects Distribution of Metallocene and MgCl$_2$-Supported Ziegler-Natta Isotactic Poly(propylenes) as Revealed by Fractionation and Crystallization Behaviors

Rufina G. Alamo

Florida Agricultural and Mechanical University and Florida State University College of Engineering. Department of Chemical Engineering. Tallahassee, FL 32310, USA

Summary: The inter and intramolecular distribution of defects of poly(propylenes) of the Ziegler-Natta (ZN) and metallocene (M) types, with matched molar masses and overall defect concentrations, are inferred from the crystallization and polymorphic behavior of their narrow molecular mass fractions. The fractions obtained from the M-iPP display a range in molecular masses but the same concentration of defects and provide direct evidence of the uniform intermolecular defect distribution and the "single site" nature of the catalyst. The stereodefects of the ZN-iPP fractions are more concentrated in the low molecular mass fractions, corroborating a broad interchain distribution of the nonisotactic content. In addition, the invariance of the linear growth rates among the ZN fractions and very low contents of the gamma polymorph, developed even by the most defected ZN fraction, are consistent with a stereo blocky intramolecular distribution of defects in the ZN-iPP molecules. In contrast to the linear growth rates, which are sensitive to the defect microstructure, the overall crystallization rates correlate with nucleation density and not necessarily with the iPP chain microstructure.

Keywords: crystallization; defect distribution; growth rates; metallocene; poly(propylene); Ziegler-Natta

Introduction

The type and activity of reactive centers that control the mechanisms of stereo control, and hence the absolute distribution of chain defects, in the polymerization of propene with Ziegler-Natta (ZN) catalysts has been an issue of intense study.[1-4] These highly reactive species are practically inaccessible to direct observation[5] and, almost invariably, their nature has been inferred by a ^{13}C NMR analysis of the polymer stereosequence distribution. The generality of the defect distribution has been hampered because, historically, ZN catalysts have been modified to increase the isotacticity level. Coupled to these modifications is a possible change in the mechanism that dictates stereospecificity and a reduced sensitivity in

                    DOI: 10.1002/masy.200450927

the spectroscopic detection of stereosequences. Consequently, the influence of a catalyst in the defect microstructure is often reevaluated as higher field NMR spectrometers, with higher resolution and sensitivity, become available.[6-13]

The statistics of stereo control in the polymerization are often inferred by fitting the experimental ^{13}C NMR stereosequences (triads, pentads, heptads, nonads) with combinations of simple limiting statistical models, each model using random statistics.[14-15] Two states models proposed independently by Chujo[16] and Doi[17], are most frequently used to compute the components of symmetric (syndiotactic and atactic) and asymmetric (isotactic) sequences that are associated with sites that exhibit "chain-end" control or enantiomorphic site control respectively. More recently, this simple two state combination model has been found to be inappropriate to describe high resolution NMR data of ZN fractions.[18, 19] Moreover, a better agreement between the experimental and calculated stereosequence distribution was obtained in terms of three-state models that allow for reversible switches between two enantiomorphic control sites.[18-19] Such switches were associated with C1 symmetric-like active species that display 2 sites, one of them highly enantioselective.[18, 20] Fitting the experimental stereosequences to such a three state model, it was concluded that high and low tactic fractions of late generation ZN-iPPs display the same three types of stereosequences or building blocks. The difference between the fractions lies in the content of each stereosequence but not the type.

The influence of the type of catalyst and polymerization process on microstructures of industrial ZN poly(propylene)s has been classically studied by fractionation and ^{13}C NMR analysis of the fractions.[21-24] TREF[22, 23] and solvent gradient extraction[11] led to fractions of increasing isotacticity and increasing molar mass, a result that clearly indicated a non uniform inter chain composition of defects, otherwise then expected from the inferred multiplicity of sites. It was also observed that most of the fractions contained syndiotactic *rrrr* pentads and, as in other works,[10-13, 21, 25] it was proposed that some of the propagation errors were "blocky" in nature. Thus, the presence of stereoblocks in the ZN-iPP has been inferred or suspected by most previous investigators and emphasized in Busico et al.'s latest work.[18]

On the basis of TREF elution profiles and degree of isotacticity of the individual fractions, Morini et al.[22] subdivided ZN-based iPPs into four major inter-molecular components: highly isotactic, mainly isotactic, stereoblocks and atactic. The fractional contents of each component in the poly(propylenes) were found to be a strong function of the type of donor. However, the properties of the individual fractions were not studied and, as a consequence, nothing could be inferred about the defect distribution within each fraction.

The question that is most difficult to address by analysis of the NMR spectra or from the fractionation results is the nature of the intramolecular defect distribution, i.e. if the defects within a given chain deviate strongly from the random distribution. To address this issue some investigators have fitted the experimental stereosequences with models that use first order Markovian statistics for at least the symmetric chain component,[13] while in other examples the data were analyzed using a Coleman-Fox "two-state" statistical model.[10] A drawback in any of these fittings is that the experimental NMR data of the stereosequences comes from very short sequences (up to a nonad distribution). Thus stereodefects that terminate continuous isotactic sequences of 10 units or 1,000 units are indistinguishable. Any inference of a non-random intramolecular distribution of stereodefects must be obtained from the properties of the poly(propylene) fractions. We found these indirect analyses missing in the majority of earlier works dealing with elucidating distributions of stereo defects in ZN poly(propylenes). Only in one recent work was the distribution of defects in ZN-iPP considered to be blocky after comparing the content of gamma phase developed in ZN-iPP fractions and metallocene-iPPs with matched average defects compositions.[26]

The metallocene catalyst, on the other hand, leads to poly(propylenes) with uniform interchain defect composition and random intramolecular distribution.[27-29] These poly(propylenes), or their fractions, serve as models with which the properties of ZN fractions can be compared. Hence, in addition to the inferred intermolecular distribution by TREF, GPC and NMR data of ZN-iPP fractions, their adherence to or deviation from the behavior of a matched metallocene-iPP can probe the intramolecular defect distribution of the ZN-iPP molecules. In this work we use a property that is highly sensitive to defect distribution, the crystallization rates, to infer the intramolecular distribution of defects in the ZN-iPP. Moreover, the behavior of fractions from the metallocene-iPP should provide direct evidence of the uniform interchain distribution of defects and the single site nature of the catalyst.

From a crystallization perspective, the defects of the poly(propylene) chain confer a copolymeric character to the molecule. Theoretical accounts of the crystallization and melting of these "copolymers" have been undertaken from phase equilibrium consideration involving more than one species.[30] When defects are rejected from the crystal, the equilibrium melting/crystallization temperature does not depend directly on the composition of the copolymer but rather upon the sequence distribution of crystallizable sequences. The sequence distribution probability (p) of a block copolymer is one. This requires that there is no depression in the melting or crystallization of block copolymers with increasing lengths or numbers of blocks. The unchanged undercooling leads to constant crystallization rates. Conversely, the parameter p of a random copolymer is less than one and, on equilibrium considerations, melting or crystallization are largely depressed with increasing concentration of defects in the chain. The change in undercooling affects the crystallization kinetics of these copolymers accordingly. These theoretical accounts give the basis for the above mentioned comparisons. Experimental work was aimed at distinguishing intramolecular stereoblock-like versus a random-like distribution of defects in the ZN fractions from analysis of their crystallization rates compared with those from matched fractions of a metallocene-iPP. In addition and as separate work, ^{13}C NMR stereosequences of fractions were theoretically analyzed using two state and three state statistical models for conformity with the observed crystallization rates.[19]

Experimental

A ZN-iPP obtained with a fourth generation MgCl$_2$ – supported industrial catalyst using an alkylphathalate-type internal donor and alkoxysilane-type external donor, and a metallocene catalyzed iPP were fractionated via supercritical fluid extraction in n-propane by increasing pressure, at a constant temperature of 150°C.[31, 32] Both poly(propylenes) have matched mass average molecular mass and the same 0.51 mol% overall concentration of all type of defects, the latter measured from the ^{13}C NMR spectra. The content of stereo defects was taken as half the fraction of *mmmr* pentads and the *erythro* 2,1 regio defects calculated from the ^{13}C NMR resonances characteristic of this group. Characterization data of the unfractionated parents and their fractions are listed in Table 1.

The linear growth rates (G) were followed in an Olympus BH2 polarized optical microscope fitted with a Linkam hot stage, TP-93. The calculated uncertainty in the value of G was very low ($\pm$ 0.01 x 10^{-6} cm/s). Overall crystallization rates were taken as the inverse of the time required for 50% of the transformation to take place. The degree of transformation, at fixed isothermal temperature, was followed by the variation of the heat flow vs. time in a Perkin Elmer DSC-7. WAXD patterns were recorded at room temperature in a Phillips X' Pert PW 3040 MRD diffractometer.

Table 1. Characterization of parent polymers and fractions from Metallocene and ZN-iPP.

Sample	Mw (g/mol)	Mw/Mn	stereo defects (mol%)	regio defects (mol%)	Total defects (mol%)	Tm (°C)[a]	ΔH (J/g[a]
M203K0.51 parent	203,900	2.00	0.11	0.40	0.51	155.0	76.0
Mf86K0.56	86,000	1.43	0.16	0.40	0.56	156.2	85.7
Mf121K0.50	121,000	1.32	0.11	0.39	0.50	156.3	75.2
Mf143K0.54	143,000	1.24	0.07	0.47	0.54	156.1	86.0
Mf200K0.46	200,000	1.23	0.08	0.38	0.46	155.8	80.4
Mf235K0.45	235,000	1.24	0.07	0.38	0.45	154.6	93.1
Mf358K0.41	358,000	1.34	0.07	0.34	0.41	154.0	73.5
Mf383K0.41	383,000	1.47	na[b]	na	na	na	
Z263K0.51 parent	262,600	3.19	0.51	---	0.51	161.4	79.4
Zf97K1.03	97,000	1.31	1.03	---	1.03	159.0	89.6
Zf163Kxxx	157,000	1.27	Na		na	na	
Zf163K0.60	163,000	1.97	0.60	---	0.60	161.3	80.5
Zf204K0.41	204,000	2.15	0.41	---	0.41	160.9	81.6
Zf328K0.36	328,000	1.77	0.36	---	0.36	162.3	81.5

(a) Rapidly quenched samples to 25 °C. Tm: Peak melting temperature, ΔH: Heat of fusion
(b) Data not available.

Results and Discussion

Both fractionations of the parents iPPs display a systematic increase of the molar mass of the fractions with increasing pressure, as seen in Table 1, indicating that the fractionation took place preferentially by molecular mass. The concentration of defects of the metallocene fractions is basically constant within the series, providing direct evidence of the narrow inter-chain composition distribution of the M-iPP. However, as the molar mass of the ZN-iPP fractions increase, the defect concentration decreases from 1.03 to 0.36 mol% respectively, evidencing the expected inter-chain heterogeneity of defect distribution in the parent ZN-iPP. Similar trends have been observed in ZN-iPP using other fractionation procedures.[21-24] The ZN-iPP contains a fraction of highly isotactic long molecules which are not present in the metallocene-iPP. The results from each fractionation explain the observed differences in isothermal linear growth rates of the unfractionated matched po(lypropylenes) (Figure 1A). At any crystallization temperature, the spherulites of the unfractionated ZN-iPP grow at about twice the rate of the metallocene-iPP. The ZN-iPP crystallites melt at ~8°C higher than metallocene-iPP crystals (Figure 1B). Thus, it appears that the longer isotactic sequences found in the ZN-iPP are selected earlier during crystallization and lead to a fraction of thicker crystallites that melt at higher temperatures than the crystals formed from the more uniformly distributed M-iPP. Moreover, it was found that the nucleation density of M-iPP is significantly higher than that of ZN-iPP, perhaps due to residues in the melt from the metallocene catalyst. Since overall crystallization rates are a direct function of the primary nucleation density, the overall crystallization rates of M-iPP were found faster than those of the ZN-iPP. As a consequence of nucleation differences, on cooling from the melt at a constant rate of 10 °C/min, the heat of crystallization evolves faster for M-iPP as seen in Figure 1C. Any measured overall crystallization rate would correlate with nucleation density and not necessarily with the microstructure of the iPP molecules.

The fractionation results alone do not allow conclusions about the nature of the intramolecular defect distribution. This issue can only be inferred after analysis of the properties of the individual fractions.

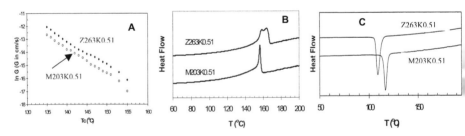

Figure 1. Properties of unfractionated iPPs. A. Isothermal linear growth rates, B. DSC melting C. DSC crystallization at 10 °C/min.

Crystallization of Metallocene iPPs

Figure 2 shows the spherulitic linear growth rates of the metallocene fractions, measured isothermally in a temperature range between 134°C and 155°C. The rates follow the strong negative temperature coefficient dictated by nucleation and growth theories.[33, 34] The observed small but systematic decrease of the rate with molecular mass, at a fixed crystallization temperature, follows the pattern of other homopolymers and copolymers.[34-36] The change in the rate is pronounced only when crystallization can be followed at low undercoolings;[33,34] large variations with molar mass are not expected at the relatively high undercoolings (>30°C) at which crystallization of iPP is observed. The variation of the rate with molar mass is attributed to the increased number of entanglements per chain in the melt, which affect segmental transport in the crystallization process.

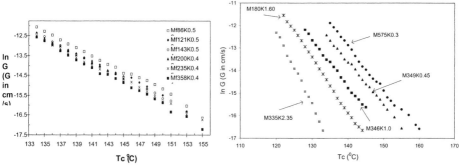

Figure 2. Linear growth rates of molecular mass fractions of M-iPP as function of Tc.

Figure 3. Linear growth rates of M-iPPs with different concentration of defects.

Homogeneous metallocene poly(propylenes), with stereo (~mrrm~) and regio (2, 1 type) defects were studied to guide the variation of growth rates with concentration of defects.[37] The linear growth rates for a series of M-iPPs with total defect content increasing from 0.30 to 2.35 mol% are given in Figure 3. The growth rate decreases about two orders of magnitude when the total defect content is increased in the range studied. This strong effect on the rate reflects the overall inhibition of the crystallization process imposed by the stereo (~mmrrm~) and 2,1 regio defects despite their partial inclusion in the lattice.[38] An increase in non-crystallizable structural irregularities leads to a reduction of the amount of crystallizable sequences and, thus, to a decrease in the crystallization rate. [35, 37] This behavior is analogous to the overall crystallization rates of a series of random ethylene copolymers.[35] For these copolymers the rate was also found to decrease about 2 orders of magnitude with increasing concentration of branch points from 1.21 to 2.3 mol %.

Crystallization of ZN-iPP fractions

Isothermal linear growth rates of the ZN-iPP fractions are given in Figure 4. Compared to those of the metallocene fractions there is no systematic variation of G with either molar mass or concentration of defects, despite a variation in defects from 0.36 to 1.03 mol% among the ZN fractions. Should the defects in each ZN fraction be randomly distributed, one would have expected a significantly lower G for the fraction with 1.03 mol% defects, as indicated by the behavior of metallocene-iPPs in Figure 3. Furthermore, there are only minor differences between the growth rates of any ZN-iPP fraction at a fixed crystallization temperature. This experimental fact can only be reconciled with a defect distribution that deviates from the random defect distribution of the metallocene fractions and that is propagated among all the ZN-iPP fractions. The invariance of the linear growth rates among the ZN fractions is consistent with a stereo blocky intramolecular distribution of defects in all ZN-iPP molecules. One could argue that the expected decrease of G with increasing defect content in the ZN fractions may be compensated by an opposite effect of molar mass on G. However, appropriate corrections to the data of figure 4 for constant molecular mass did not cause any significant variation in the data or trends of the rates shown in figure 4.[39]

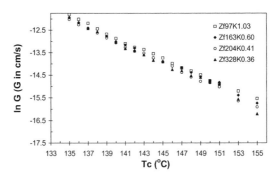

Figure 4. Linear growth rates of molecular mass fractions of ZN-iPP as function of Tc.

Additional evidence, which supports the stereoblock nature of the ZN-iPP chains, is found in the significantly low contents of gamma polymorph that ZN-iPP fractions develop compared to metallocene-iPPs with matched concentrations of defects. A metallocene-iPP with a concentration of 1.0 mol% of defects homogeneously distributed, crystallized at 125°C, develops about 50% crystallites in the γ polymorph and the other half in the α form.[39, 40] The more defected ZN fractions (Z97K1.03 and Zf163K0.6) develop only about 10% of the γ polymorph at the same crystallization temperature. These results prove that the required presence of short crystallizable sequences for the formation of the γ polymorph is fulfilled in the metallocene-iPP with a random distribution of defects but not in the ZN fractions. Significantly lower contents of the gamma phase are expected in molecules where long isotactic sequences join blocks of poorly stereo regular sequences, as schematically shown in Figure 5. In this microstructural model for ZN-iPP, the intermolecular defect composition is non-uniform with the shorter chains having a higher concentration of non isotactic units so as to comply with the fractionation behavior. However, the intramolecular distribution of defects is blocky. All chains display similar types of long crystallizable sequences, which are highlighted in the schematic figure. Accordingly, molecular fractions from this model are expected to show negligible differences in crystallization rates. The experimental growth rates and polymorphic behavior of ZN-iPP fractions conform with this structural model. Stereoblock distributions in various types of iPPs have been also inferred in other works by analysis of their polymorphic behavior.[26, 41]

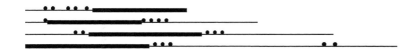

Figure 5. Model for stereosequence distribution in parent ZN-iPP.

Conclusions

Classical fractionation and ^{13}C NMR characterization of individual fractions of matched ZN-iPP and M-iPP have been carried out to determine the intermolecular distribution of defects in the iPP chains. These methods have been combined with analysis of the linear growth rates of the fractions, a useful tool to infer the intramolecular defect distribution in the ZN-iPP chains. The fractions from the M-iPP display a range in molecular mass but the same defect concentration. These fractions provide evidence of the uniform inter-chain defect distribution in this poly(propylene). The linear growth rates of the M-iPP fractions only reflect differences due to the molecular mass effect on the crystallization rate and, thus, are consistent with a random intramolecular distribution of defects.

The molar fractions obtained from the ZN-iPP confirmed that the content of defects is broadly distributed on an intermolecular basis. In addition, the basically identical growth rates found in all ZN fractions and the formation of very low contents of the gamma polymorph, even in the most defected fraction, is consistent with a stereoblocky intramolecular distribution of defects. A schematic model of this distribution has been proposed.

Acknowledgements

This work was supported by the National Science Foundation (DMR-0094485). Contributions to this work by J.C. Randall, J.A. Blanco, S. Putcha, C. Chi, P.K. Agarwal, C.J. Ruff, E. Ritchson and D. Li are acknowledged.

[1] E. Albizzati, U. Giannini, G. Collina, L. Noristi, L. Resconi in *Polypropylene Handbook. Part I. Catalysts and Polymerizations* E.P. Moore Jr., Hanser Pub. **1996**.
[2] V. Busico, R. Cipullo, *Prog. Polym. Sci.* **2001**, *26*, 443.
[3] J.C. Chadwick, G. Morini, G. Balbontin, I. Camurati, J.J.R. Heere, I. Mingozzi, F. Testoni, *Macrom. Chem. Phys.* **2001**, *202*, 1995.
[4] M.C. Sacchi, F. Forlini, I. Tritto, P. Locatelli, G. Morini, L. Noristi, E. Albizzati, *Macromolecules* **1996**, *29*, 3341.
[5] V. Busico, R. Cipullo, F. Cutillo, M. Vacatello, *Macromolecules* **2002**, *35*, 349.
[6] A. Zambelli, P. Locatelli, G. Bajo, F.A. Bovey, *Macromolecules* **1975**, *8*, 687.
[7] A. Zambelli, P. Locatelli, A. Provasoli, D.R. Ferro, *Macromolecules* **1980**, *13*, 267.
[8] H. Kawamura, T. Hayashi, Y. Inoue, R. Chujo, *Macromolecules* **1989**, *22*, 2181.
[9] V. Busico, P. Corradini, L. De Martino, F. Graziano, A.M. Iadicicco, *Makromal. Chem.* **1991**, 192, 49.
[10] V. Busico, P. Corradini, R. De Biasio, L. Landriani, A.L. Segre, *Macromolecules* **1994**, *27*, 4521.
[11] V. Busico, R. Cipullo, P. Corradini, L. Landriani, M. Vacatello, *Macromolecules* **1995**, *28*, 1887.
[12] V. Busico, R. Cipullo, G. Talarico, A.L. Segre, J.C. Chadwick, *Macromolecules* **1997**, *30*, 4786.
[13] J.C. Randall, *Macromolecules* **1997**, *30*, 803.
[14] R.A. Shelden, T. Fueno, T. Tsunetsugu, J. Furukawa, *J. Polym. Sci., Part A*, **1965**, *3*, 23.
[15] F.A. Bovey, G.V.D. Tiers, *J. Polym. Sci.* **1960**, *44*, 173.
[16] R. Chûjô, *Kagaku* **1981**, *36*, 420.
[17] Y. Doi, *Makromol. Chem., Rapid Commun.* **1982**, *3*, 635.
[18] V. Busico, R. Cipullo, G. Monaco, G. Talarico, M. Vacatello, J.C. Chadwick, A.L. Segre, O. Sudmeijer, *Macromolecules* **1999**, *32*, 4173.
[19] J.C. Randall, R.G. Alamo, P.K. Agarwal, C.J. Ruff, *Macromolecules* **2003**, *36*, 1572.
[20] V. Busico, R. Cipullo, C. Polzone, G. Talarico, J.C. Chadwick, *Macromolecules* **2003**, *62*, 2616.
[21] R. Paukkeri, T. Vaananen, A. Lehtinen, *Polymer* **1993**, *34*, 2488.
[22] G. Morini, E. Albizzati, G. Balbontin, I. Mingizzi, M.C. Sacchi, F. Forlini, I. Tritto, *Macromolecules* **1996**, *29*, 5770.
[23] P. Viville, D. Daoust, A.M. Jonas, B. Nysten, R. Legras, M. Dupire, J. Michel, G. Debras, *Polymer*, **2001**, *42*, 1953.
[24] J. Xu, L. Feng, S. Yang, X. Kong, *Eur. Polym. J.* **1998**, *34*, 431.
[25] J.C. Chadwick, A. Miedema, B.J. Ruisch, O. Sudmeijer, *Makromol. Chem.* **1992**, *193*, 1463.
[26] F.P.T.J. van der Burgt, S. Rastogi, J.C. Chadwick, B. Rieger, *J. Macromol. Sci. Phys.* **2002**, *B41*, 1091.

[27] M. Farina, G. Di Silvestro, A. Terragni, *Macromol. Chem. Phys.* **1995**, *196*, 353.
[28] H.H. Brintzinger, D. Fischer, R. Mulhaupt, B. Rieger, R.M. Waymouth, *Angew Chem. Int. Ed. Engl.* **1995**, *34*, 1143.
[29] L. Resconi, L. Cavallo, A. Fait, F. Piemontesi, *Chem. Rev.* **2000**, *100*, 1253.
[30] P.J. Flory, *Trans. Faraday Soc.* **1955**, *51*, 848.
[31] J.J. Watkins, V.J. Krukonis, P.D. Condo, Jr., D. Pradhan, P. Ehrlich, *J. Supercritical Fluids* **1991**, *4*, 24.
[32] L.J.D. Britto, J.B. P. Soares, A. Peulidis, V. Krukonis, *J. Polym. Sci., Polym. Phys. Ed.* **1999**, *37*, 553.
[33] J.D. Hoffman, L.J. Frolen, G.S. Ross, J.I. Lauritzen, *J. Res. Natl. Bur. Stand.* **1975**, *79A*, 671.
[34] E. Ergoz, J.G. Fatou, L. Mandelkern, *Macromolecules* **1972**, *5*, 147.
[35] R.G. Alamo, L. Mandelkern, *Macromolecules* **1991**, *24*, 6480.
[36] L.C. Lopez, G.L. Wilkes, *Polymer* **1988**, *29*, 106.
[37] R.G. Alamo, C. Chi, Crystallization Behavior and Properties of Polyolefins. In *Molecular Interactions and Time-Space Organization in Macromolecular Systems*; Y. Morishima, T. Norisuye, K. Tashiro, Eds.; Springer: Berlin **1999**; p. 29.
[38] D.L. VanderHart, R.G. Alamo, MR. Nyden, M.H. Kim, L. Mandelkern, *Macromolecules* **2000**, *22*, 6078.
[39] R.G. Alamo, J.A. Blanco, P.K. Agarwal, J.C. Randall, *Macromolecules* **2003**, *36*, 1559.
[40] R.G. Alamo, M.H. Kim, M.J. Galante, J.R. Isasi, L. Mandelkern, *Macromolecules* **1999**, *32*, 4050.
[41] C. De Rosa, F. Auriemma, T. Circelli, R.M. Waymouth, *Macromolecules* **2002**, *35*, 3622.

Macromol. Symp. **2004**, *213*, 315-325   315

Structural Characterization and Mechanical Behavior of Metallocenic Copolymers of Ethylene and 5,7-Dimethylocta-1,6-diene

María L. Cerrada,[1] *Rosario Benavente,*[1] *José M. Pereña,*[1] *Ernesto Pérez,*[1] *Jorge Moniz-Santos,*[2] *M. Rosário Ribeiro*[2]

[1]Instituto de Ciencia y Tecnología de Polímeros (CSIC), C/ Juan de la Cierva 3, 28006 Madrid, Spain
[2]Grupo de Estudos de Catálise Heterogénea, Instituto Superior Técnico, Av. Rovisco País, 1049-001 Lisboa, Portugal

Summary: The relationships between structure and mechanical properties have been established in several copolymers of ethylene and 5,7-dimethylocta-1,6-diene synthesized by a metallocene catalyst. A dependence with composition and polymerization temperature has been found for different structural and mechanical parameters. The branches cannot be incorporated into the orthorhombic crystal lattice and, consequently, structural parameters such as crystallinity and crystal size are considerably affected as 5,7-dimethylocta-1,6-diene content increases in the copolymer. These structural changes influence significantly the rigidity of the copolymers and a decrease of this parameter, determined from either storage modulus or microhardness, with increasing 5,7-dimethylocta-1,6-diene molar fraction is found. The location of the different viscoelastic relaxations is also strongly dependent on composition.

Keywords: crystallinity; ethylene-5,7-dimethylocta-1,6-diene copolymers; metallocene catalysts; microhardness; relaxation

Introduction

Many efforts have been made to use metallocene catalysts for the copolymerization of olefins with functional monomers since their advent. However, a strong deactivation of the catalyst by the functional monomer is observed. Different procedures have been developed in order to defeat this problem. They include the use of vinyl monomers bearing functional groups that are distant to the double bond,[1] or sterically hindered,[1-3] or protected by a pre-complexation reaction.[4-6]

Another approach to the production of functionalized polyolefins less harmful to catalyst activity involves the copolymerization of olefins with pro-functional monomers followed by

    DOI: 10.1002/masy.200450928

the chemical modification of these pre-formed polymers. Boranes and dienes[7-13] are common examples. Concerning the second mentioned choice, unsaturated polyolefins are obtained by copolymerization of α–olefins with non-conjugated dienes. They present pendent olefin unsaturations that can be easily reacted with different functional groups. Reported methods include bromination, epoxidation, radical addition and hydrosilylation.[11-13] These polymers are, therefore, useful intermediates in the production of side functional polyolefins, which can then act as interfacial modifiers to enhance adhesion and compatibility of polyolefins with other polymers, fillers or even fibers.

We have, recently, obtained polyethylenes with high incorporations of 5,7-dimethylocta-1,6-diene, labelled as DMO. The structure of this diene helps preventing side-reactions like cyclization and crosslinking.[14] Therefore, the objective of this work is to characterize the structure in the solid state of these precursors of functionalized polyolefins, that are a series of copolymers of ethylene and DMO in a wide range of compositions. In addition to the previously mentioned structural characterization of these copolymers, named as CEDMO, the analysis of their mechanical behavior by means of either dynamic mechanical or microhardness measurements is presented.

Experimental

Polymerization reactions were carried out as follows: toluene, ethylene and DMO comonomer were charged into the reactor. A constrained geometry metallocene catalyst [14] (CGC) (η_5-C_5Me_4)SiMe$_2$(NtertBu)TiCl$_2$ was pre-contacted with methylaluminoxane (MAO) for 15 min prior introduction of this pre-activated catalyst into the reactor. After 1 h, polymerization was stopped and the polymers were carefully washed and dried preceding to characterization. Two different polymerization temperatures were utilized: 25°C and 60°C, to learn the effect of this variable on the final properties. Several CEDMO copolymers with different comonomer content, ranging from 1.4 to 10 mol % in DMO, were obtained, as seen in Table 1. The DMO molar fraction was estimated from 1H NMR spectra performed in 1,1,2,2-tetrachloroethane at 107°C with a Bruker MSL 300P spectrometer.

The different CEDMO copolymers and the corresponding homopolymers were prepared as films by compression molding in a Collin press at 140°C at a pressure of 1.5 MPa for 5 min and a further quenching within Teflon plates cooled with water.

Wide and small-angle X-ray diffraction patterns, WAXS and SAXS respectively, were recorded in the transmission mode at room temperature employing synchrotron radiation (λ= 0.150 nm) in the beamline A2 at HASYLAB (Hamburg, Germany). Two linear position-sensitive detectors were used. The SAXS one, at a distance of 235 cm from the sample, was calibrated with the different orders of the long spacing of rat-tail cornea (L = 65 nm). It was found to cover a spacings range from 5 to 55 nm. The WAXS detector, covering a 2θ range from about 10 to 30 degrees, was calibrated with the different diffractions of crystalline PET.

Table 1. DMO composition and sample characteristics. Experimental conditions with CGC catalyst for polymerization at 25 °C: [Ti] = $7\text{-}10\cdot10^{-5}$ mol/L, Al/Ti = 1000, P_{et} = 3.4 bar and for polymerization at 60 °C with CGC catalyst: [Ti] = $3\text{-}6\cdot10^{-5}$ mol/L, Al/Ti = 1000, P_{et} = 1.0 bar

Sample	DMO content (mol-%)	M_w/M_n	M_w (10^3 g $\cdot$ mol^{-1})
Polymerization Temperature: 25°C			
25HDPE	0.0	3.1	149
25CEDMO1	1.4	2.0	79
25CEDMO2	4.0	2.3	75
25CEDMO3	5.8	6.3	74
Polymerization Temperature: 60°C			
60HDPE	0	2.3	57
60CEDMO1	2	2.8	56
60CEDMO2	4.4	7.5	25
60CEDMO3	7.8	n. d.	n. d.
60CEDMO4	10	n. d.	n. d.

Calorimetric analyses were carried out in a Perkin-Elmer DSC7 calorimeter, connected to a cooling system and calibrated with different standards. The sample weights ranged from 5 to 7.5 mg. A temperature range from -70°C to 150°C has been studied with a heating rate of 10°C min^{-1}. For crystallinity determinations, a value of 290 J g^{-1} has been taken as the enthalpy of fusion of a perfectly crystalline material.[15]

Viscoelastic properties were measured with a Polymer Laboratories MK II dynamic mechanical thermal analyzer working in the tensile mode. The real (E′) and imaginary (E″)

components of the complex modulus and the loss tangent (tan δ) of each sample were determined at 1, 3, 10 and 30 Hz, over a temperature range from -150 to 150°C, at a heating rate of 1.5°C min^{-1}.

A Vickers indentor attached to a Leitz microhardness tester was used to carry out microindentation measurements undertaken at 23°C. A contact load of 0.98 N and a contact time of 25 s were employed. Microhardness, MH, values (in MPa) were calculated according to the relationship:

$$MH = 2 \sin 68° \, P / d^2 \qquad [1]$$

where P (in N) is the contact load and d (in mm) is the diagonal length of the projected indentation area.

Results and Discussion

Crystalline structure and thermal properties

Polyethylene crystallizes in an orthorhombic lattice under usual conditions. This crystalline cell is characterized by two main diffraction peaks: the (110) and the (200). It is known that in general, the alkyl branches cannot enter into the polyethylene crystal lattice, although the methyl branches are included in the crystalline cell at a substantial degree.[16,17] The DMO comonomer units copolymerized here with ethylene are bulkier than common alkyl ones and, therefore, the capability of being incorporated into the crystalline lattice does not practically exist and the hindrance imposed for the PE crystallization is expected to be high. Consequently, the presence of the DMO comonomer leads to a significant decrease of the (110) and (200) diffraction intensities and, accordingly, the crystallinity in the copolymers diminishes as DMO content increases, as depicted in Figure 1. In addition, a shift of the positions of the maxima to lower angles and a broadening of crystalline diffractions are observed. This broadening indicates a decrease in the crystallite size with the increase of DMO content. It is important to note that all of the samples analyzed display a clear long spacing, L, as listed in Table 2, which is attributed to the presence of lamellar crystals though some of the specimens are practically amorphous, as observed from WAXS patterns. In these cases, for high DMO contents, bundles of crystallites are expected. Moreover, long spacing strongly depends upon both comonomer composition and polymerization temperature. On the one hand, a decrease of long spacing with the increase of DMO content is observed and, on the other hand, larger long spacings are found in the copolymers synthesized at 25°C due to their higher molecular weight.

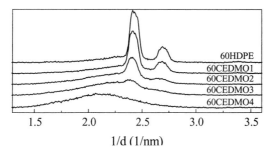

Figure 1. WAXS patterns of the samples polymerized at 60°C

The crystallite thickness, l_c, has been determined from the long spacing and the WAXS crystallinity by assuming a simple two-phase model. This method will be only applicable for those cases where uniform lamellae stacks are present. Anyway, decrease of the crystallite thickness values is observed as DMO content is increased in the copolymers (Table 2).

Table 2. Melting temperatures (T_m), enthalpies of melting (ΔH_m), crystallinities estimated from either D.S.C. or WAXS data (f_c^{DSC} and f_c^{WAXS}), long spacings (L), crystal thickness (l_c), relaxation temperatures (E'' basis, at 3 Hz), storage modulus (E' at 25°C and at 3Hz) and microhardness of the different specimens

Sample	T_m (°C)	ΔH_m (J g^{-1})	f_c^{DSC}	f_c^{WAXS}	L (nm)	l_c (nm)	T_γ (°C)	T_β (°C)	T_α (°C)	E'$_{25}$ (MPa)	MH (MPa)
Polymerization Temperature: 25°C											
25HDPE	132	194	0.67	0.64	27.8	17.8	-118	-	54	1432	37
25CEDMO1	124	151	0.52	0.55	15.2	8.4	-119	-22	42	605	17
25CEDMO2	91	81	0.28	0.26	12.2	3.2	-134	-28	12	90	8
25CEDMO3	62	41	0.14	0.05	11.4	0.6	-126	-6	-	21	4
Polymerization Temperature: 60°C											
60HDPE	132	209	0.72	0.72	20.1	14.5	-118	-	48	1670	38
60CEDMO1	126	168	0.58	0.60	17.7	10.6	-118	-15	45	845	23
60CEDMO2	107	84	0.29	0.35	11.6	4.1	-133	-16	18	100	8
60CEDMO3	69	41	0.14	0.10	10.3	1.0	-134	-26	-5	12	3
60CEDMO4	61	12	0.04	-	10.2	-	-129	-12	-	3.7	1

The d.s.c. traces for the different samples support the X ray diffraction results just mentioned, as represented in Figure 2. A depression of the melting temperature (T_m) is observed, as expected, as DMO content increases owing to the smaller crystallite sizes (assuming that Tm only depends on the composition of the melt). In addition, a broadening of the melting region is also found in the copolymers with higher DMO compositions. On the other hand, the melting enthalpy is also significantly reduced with increasing DMO content in the copolymer.

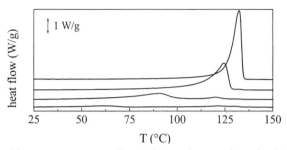

Figure 2. D.s.c. melting curves corresponding to the specimens polymerized at 25°C. From top to bottom: 25HDPE, 25CEDMO1, 25CEDMO2, 25CEDMO3

Moreover, at a given comonomer content, the distortion on the crystallites is larger in the current copolymers than in cases where the branches are 1-hexene, 1-octene[18] or even longer alkyl lateral chains,[19] as seen in Figure 3.

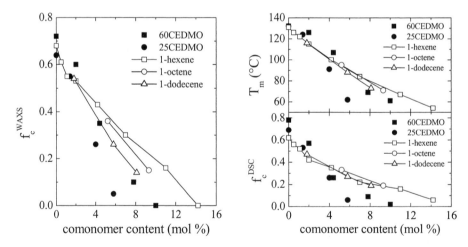

Figure 3. Dependence of crystallinity estimated by WAXS experiments (left picture), melting temperature and crystallinity estimated by DSC measurements (right picture: upper and lower plot, respectively) with comonomer content of the different CEDMO specimens under study and other ethylene copolymer samples

This feature may be attributed to the dimethyl substitutions within the branches and, accordingly, to a higher global volume of the comonomer unit. Therefore, similar values of crystallinity, melting temperature and long spacing are observed compared to those obtained for 1-hexene, 1-octene or 1-octadecene copolymers but in the current ethylene-DMO specimens those values are shown at lower comonomer incorporation.

Viscoelastic behavior and microhardness measurements

The viscoelastic behavior of polyethylene is strongly influenced by variables characteristic of the crystalline regions, such as crystallinity and lamellar thickness.[20] Therefore, a decrease of the storage modulus is seen as DMO content increases and, accordingly, crystallinity and crystal size diminish, this feature being related to the reduction of rigidity (Figure 4).

Such a diminution in stiffness with DMO composition has also been found in the microhardness values, as shown in Figure 4. On the other hand, it seems that the MH values are slightly higher in the copolymers obtained at 60°C at a similar comonomer content because of their lower molecular weights.

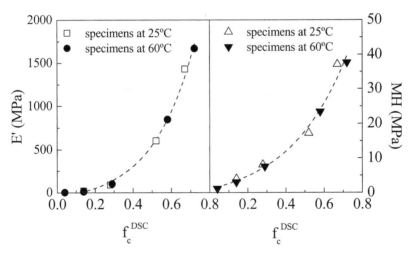

Figure 4. Dependence of E' (left plot) and MH (right plot) on crystallinity for the different specimens

MH measures mostly the resistance of the material to plastic deformation. A direct correlation is commonly found between the elastic modulus (here represented by storage modulus, E') and MH and the following empirical equation has been proposed:[21]

$$MH = a \cdot E^b \qquad [2]$$

where a and b are constants. As seen in Figure 4, both magnitudes, storage modulus and MH, display an analogous variation with crystallinity for the two copolymerization temperatures corroborating the mentioned direct relationship between these two mechanical parameters for these CEDMO copolymers.

Concerning the viscoelastic response several relaxation processes are observed depending upon composition: γ, β and α in order of increasing temperatures.

The γ relaxation in polyethylene was firstly attributed to crankshaft movements of polymethylenic chains. Figure 5 displays a shift to lower temperatures of the γ peak as the DMO content rises in the copolymers since this process is associated to the amorphous phase and there is a higher mobility in the amorphous regions, as DMO incorporation increases.

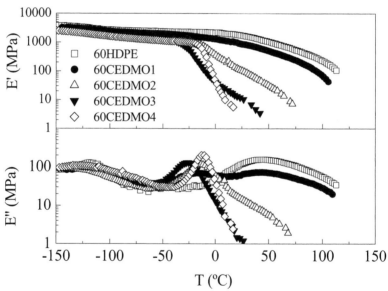

Figure 5. Temperature dependence of the two components to the complex modulus for the specimens polymerized at 60°C

The β relaxation process in the CEDMO copolymers is also moved to lower temperatures and its intensity becomes higher as comonomer molar fraction increases. However, a shift to higher temperatures is observed in the copolymers with the highest comonomer content either that prepared at 25°C or at 60°C. This feature in those two copolymers with the highest DMO molar fractions seems to point that other additional effect is taking place. A certain degree of crosslinking might be possible for those highest DMO contents.

The α relaxation in polyethylene has been associated to vibrational and reorientational motions within the crystallites. The location of the α relaxation depends strongly on the comonomer content. Thus, this process is only clearly observed in the homopolymers and in copolymers with low DMO incorporations. For higher contents, the lower crystallinity joined to the smaller and more imperfect crystallites lead to a decrease of the intensity and a shift to lower temperatures of this α relaxation and an overlap with the β process is observed. A good linear relationship is found between the location of this relaxation, T_{α}, and the crystallinity (Figure 6). However, its correlation with the most probable crystal thickness, l_c, is not linear (Figure 6). A very similar non-linear variation has been reported before[22,23] for other ethylene copolymers.

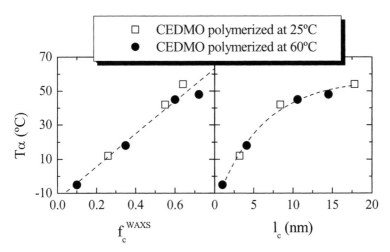

Figure 6. Variation of location of α relaxation, T_{α}, with crystallinity (left plot) and crystal thickness (right plot) for the different specimens

Conclusions

The most important factor affecting the structure and properties of these CEDMO copolymers is, evidently, the comonomer content and its influence on crystallinity and crystallite thickness. The distortion on the crystallites is larger in the current copolymers than in cases where the branches are common alkyl lateral chains. Higher crystallinities and smaller crystal sizes have been found in copolymers synthesized at 60°C at a similar comonomer content, probably due to their lower molecular weights. A relationship has been found between storage modulus and microhardness, showing a similar dependence with structural parameters. A shift to lower temperatures is generally found for the γ and β relaxations as DMO content increases. Finally, the α relaxation is also moved to lower temperatures as crystallinity and crystal thickness decrease and a linear dependence is found between its location and the former parameter.

Acknowledgements

The financial support of the Comunidad Autónoma de Madrid, Ministerio de Ciencia y Tecnología, Comisiones Mixtas CSIC/ICCTI and CSCI/CNR, and Fundação para a Ciência e Tecnologia (Projects 07G/0038/2000, MAT2001-2321, 2002PT0001-Proc. 423, 2003IT0006 and POCTI/CTM/41408/2001, respectively) is gratefully acknowledged. The IHP Programme "Access to Research Infrastructures" of the European Commission has supported the synchrotron work in the polymer line of Hasylab at DESY, Hamburg (Contract HPRI-CT-1999-00040). We thank the collaboration of the Hasylab personnel, and specially Dr. S. Funari, responsible of the polymer beamline.

[1]   M. R. Kesti, G. W. Coates, R. M. Waymouth, *J. Am. Chem. Soc.* **1992**, *114*, 9679.
[2]   R. Goretzki, G. Fink, *Macromol.Rapid Commun.* **1998**, *19*, 511.
[3]   C. Wilen, H. Luttikhedde, T. Hjertberg, J. H. Nasman, *Macromolecules* **1996**, *29*, 8569.
[4]   P. Aaltonen, G. Fink, B. Lofgren, J. Seppala, *Macromolecules* **1996**, *29*, 5255.
[5]   R. Goretzki, G. Fink, *Macromol. Chem. Phys.* **1999**, *200*, 881.
[6]   J. M. Santos, M. R. Ribeiro, M. F. Portela, S.G. Pereira, T. G. Nunes, A. Deffieux, *Macromol. Chem. Phys.* **2001**, *202*, 2195.
[7]   W. Kaminsky, D. Arrowsmith, H.R. Winkelbach, *Polym. Bull.* **1996**, *36*, 577.
[8]   M. Dolatkhani, H. Cramail, A. Deffieux, *Macromol. Chem. Phys.* **1996**, *197*, 2481.
[9]   P. Pietikäinen, P. Stark, J.V. Seppälä, *J. Polym. Sci., Part A: Polym. Chem.* **1999**, *37*, 2379.
[10] P. Pietikäinen, J.V. Seppälä, L. Ahjopalo, L-O Pietilä, *Eur. Polym. J.* **2000**, *36*, 183.
[11] M. Hackmann, T. Repo, G. Jany, B. Rieger, *Macromol. Chem. Phys.* **1998**, *199*, 1511.

[12] J. Suzuki, Y. Kino, T. Uozumi, T. Sano, T. Teranishi, J. Jin, K. Soga, T. Shiono, *J. Appl. Pol. Sci.* **1999**, *72*, 103.

[13] T. Uozumi, G.L. Tian, C.H. Ahn, J.Z. Jin, S. Tsubaki, T. Sano, K. Soga, *J. Polym. Sci. Part A: Polym. Chem.* **2000**, *38*, 1844.

[14] J. M. Santos, M. R. Ribeiro, M. F. Portela, H. Cramail, A. Deffieux., A. Antiñolo, A. Otero, S. Prashar, *Macromol. Chem. Phys.* **2002**, *203*, 139.

[15] B. Wunderlich, *"Macromolecular Physics"*, Academic Press, New York **1980**, vol. *3*, p. 42.

[16] E. Pérez, D. L. VanderHart, *J. Polym. Sci., Part B: Polym. Phys.* **1987**, *25*, 1637.

[17] R. Alamo, R. Domszy, L Manderkern, *J. Phys. Chem.* **1984**, *88*, 6587.

[18] M. L. Cerrada, R. Benavente, E. Pérez, *J. Mater. Res.* **2001**, *16*, 1103.

[19] R. Benavente, E. Pérez, R. Quijada, *J. Polym. Sci., Part B: Polym. Phys.* **2001**, *39*, 277.

[20] I.M. Ward, *"Mechanical Properties of Solids Polymers"*, 2nd edition, J. Wiley and Sons. Chichester, **1985**.

[21] F. J. Baltá-Calleja, *Adv. Polym. Sci.* **1985**, *66*, 117.

[22] R. Popli, M. Glotin, l. Mandelkern, R. S. Benson, *J. Polym. Sci., Part B: Polym. Phys.* **1984**, *22*, 407.

[23] J. M. Martínez-Burgos, R. Benavente, E. Pérez, M. L. Cerrada. *J. Polym. Sci., Part B: Polym. Phys.* **2003**, *41*, 1244.

Macromol. Symp. **2004**, *213*, 327-333

On-Line Monitoring and Fingerprint Technology: New Tools for the Development of New Catalysts and Polyolefin Materials

*Arno Tuchbreiter, Jürgen Marquardt, Bernd Kappler, Josef Honerkamp,
Rolf Mülhaupt**

Freiburger Materialforschungszentrum (FMF) der Albert-Ludwigs Universität
Freiburg, Stefan-Meier-Str. 21, D-79104 Freiburg, Germany
E-Mail: rolf.muelhaupt@makro.uni-freiburg.de

Summary: The High-Output Polymer Screening (HOPS) combines process-relevant automated reactor systems and rapid polymer characterization with on-line polymerization monitoring and automated data acquisition ("electronic notebook") in order to make effective use of advanced data mining tools. This has led to the development of fingerprint technology based upon correlations between spectroscopic data and polymerization process conditions, catalyst compositions, as well as polymer end-use properties. Infrared spectroscopic fingerprints proved to be very useful for accelerating polymer analyses including characterization of polymer molecular architectures as well as non-destructive testing of the mechanical, thermal and other end-use polymer properties. Such spectroscopic fingerprints represent important components of effective on-line quality control systems. With ATR-FT-IR probes on-line monitoring of catalytic olefin copolymerization was performed in solution to measure in real time copolymerization kinetics, catalyst productivities, catalyst deactivation as well as copolymerization parameters and copolymer sequence distributions. Monomer consumption and comonomer incorporation were monitored simultaneously. Advanced fingerprint technology can reduce significantly the need for time- and money consuming polymer testing and can also stimulate the search for new catalyst systems and polymeric materials.

Keywords: combinatorial chemistry; data mining; high throughput screening; IR spectroscopy; multivariate calibration; polypropylene

Introduction

An important objective in catalyst as well as polymer development is to reduce the costs and the time required for identifying new products and bringing them from the lab to the market place. High throughput methods were introduced in drug development to prepare and screen a

                DOI: 10.1002/masy.200450929

large number of small molecules with respect to their biological activities.[1] High throughput methods and combinatorial chemistry was also introduced in materials development.[2] While automated reactors and data mining tools are established in polymer reaction engineering and process research at pilot plant facilities, many catalyst and polymer research labs are still employing manually operated reactors with inadequate process control and tedious, time-consuming data acquisition, especially with respect to the characterization of polymer properties. Today inexpensive computing power is at hand to facilitate work in research labs by means of automated reactors equipped with computer guided process control and automated acquisition of polymerization data as well as product properties.

High-Output Polymer Screening (HOPS)[3]

Most high throughput systems increase screening speed at the expense of process-relevant information. The High-Output Polymer Screening (HOPS)[3], developed at the Freiburg Materials Research Center jointly with BASF AG and Basell, employs process-relevant reactors equipped with precise process control and data acquisition of both polymerization data as well as polymer properties in a data base system ("electronic notebook"). The basic concept of HOPS is displayed in Figure 1. Virtually any lab reactor can be upgraded to perform HOPS. Special miniaturized reactors and sophisticated reactor arrays are not required. This is important to assure process relevant results. Since polymerization reactions are very well controlled, it is possible to generate a large number of well-defined and well-characterized polymer samples within very short time. This enhances the yield of useful information per experiment, which can be employed in data analysis. Also this information can be used to improve the templates for molecular modeling of catalysts and to facilitate simulations of polymerization processes. An important objective of this research was to search correlations of spectroscopic data with polymer architectures and polymer properties.

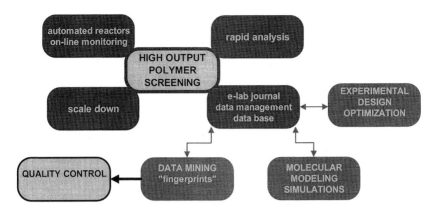

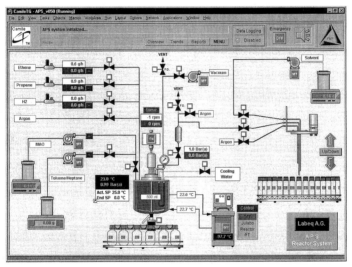

Figure 1. High-Output Polymer Screening (HOPS)

Fingerprint Technology and Rapid Polymer Characterization

Automated reactors significantly improve the effectiveness of lab reactor operations. However, the increased number of samples caused severe bottleneck situation in polymer characterization. Especially NMR spectroscopic investigations require measurements requiring several hours per sample. Infrared (IR) spectroscopy is being used for polyolefin characterization since the pioneering days of Natta's group.[4,5] At that time the preparation of IR samples was rather tedious because it required either time-consuming compression molding of KBr pellets or film formation, e.,g. by casting solutions onto mercury surfaces. This traditional bottleneck was eliminated by employing the attenuated total reflection IR spectroscopy, also known as "Golden Gate".[6] As shown in Figure 2, the polymer sample is added to the top of a dense crystal platform and the IR beam is directed to the bottom of the crystal at an angle greater than the systems critical angle, which depends on the refractive properties of both the crystal and the polymer. This causes a total reflection of the wave, minus what is absorbed by the sample. It was demonstrated for ethene/1-hexene copolymers that 40 samples can be processed per hour. Using multivariate calibration ("fingerprint") of model copolymers, analyzed by means of NMR spectroscopy, the accuracy of this methods was improved considerably. Also MIR and NIR spectroscopy were used successfully to establish fingerprints for copolymer microstructure analysis.[7]

Another time- and money-consuming bottleneck relates to the measurement of end-use properties such as mechanical and thermal properties, density, crystallinity, hardness, etc. . Frequently, also wet chemical analysis must be performed to measure properties such as gel content and soluble fractions. Here multivariate calibration combined with a well-characterized large number of polymer samples, obtained by means of HOPS, offers interesting opportunities. The approach is rather simple. Instead of expensive high-resolution NMR, inexpensive spectroscopy such as IR and industrial magnetic resonance are being used to correlate the spectroscopic data with polymer properties in order to identify spectroscopic "fingerprints". Clearly the versatility of this fingerprint technology depends on the individual polymer systems and the quality and easy availability of a fairly large number of "model" samples. With HOPS such model samples are readily available. Fingerprints were established for the non-destructive rapid determination of tensile strength, Shore hardness, flexural modulus and density of polypropylenes.[3] In a similar approach, Pandy, Kumar and Garg evaluated melt flow index, Izod impact strength, and flexural modulus by means of multivariate statistical analysis.[8] Karlsson and Camacho measured number average weight and crystallinity of recycled HDPE by means of non-destructive testing and fingerprints originating from NIR and MIR spectroscopy combined with multivariate partial least square calibration.[9] With HOPS technology it is very easy to adopt fingerprint technology in research labs as well as pilot plants because of the availability of large sample numbers required to sharpen the data mining tools. As seen in Figure 4, also fingerprint correlations between catalyst compositions and polymer properties are feasible.

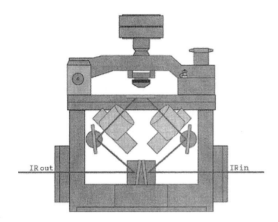

Figure 2. ATR-FTIR analysis of polymer powders and pellets

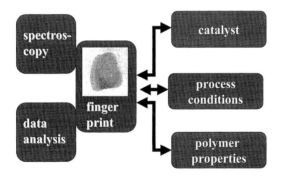

Figure 3. Fingerprint technology: "fishing for correlations"

On-Line Monitoring of Solution Copolymerizations[10]

For solution polymerizations and copolymerizations it is possible to insert ATR-FT-IR probes, e.g., ReactIR from Mettler) into the reactor. The reactor system is displayed in Figure 4. Although IR – in contrast to Raman - does not detect ethylene, it is possible to monitor the consumption of 1-hexene and simultaneously the incorporation of ethene and 1-hexene into the copolymer. Fingerprint technology and data processing were employed successfully in order to monitor polymerization kinetics as well as copolymer composition and copolymer sequence analysis. The results are in good agreement with time-consuming NMR measurements. When the differential equations of copolymerization reactions were modified to take into account the time-dependent variation of catalyst activity during polymerization it is also possible to identify living polymerization processes and catalyst deactivations. This HOPS system proved to be useful for the development of solution processes and the formation of elastomers.

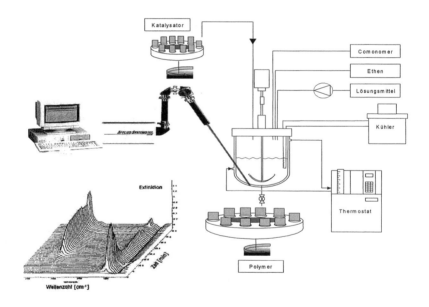

Figure 4. On-line solution copolymerization monitoring by means of ATR-FTIR-probes

Outlook

HOPS and fingerprint technology make research labs much more effective with respect to rapid feedback of experimental results and lower costs and less environmental pollution due to the non-destructive evaluation of polymeric materials. The large amount of polymers, which are readily available using process- relevant and well-controlled polymerization conditions, facilitate implementation of fingerprint technology. On one hand, fingerprint technology speeds up polymer characterization. On the other hand, the fingerprint technology offers new opportunities of pinpointing the influence of catalyst variations and changes of process conditions on the end-use properties of polymers. This information can be used to achieve better quality control and also to stimulate catalyst and polymer developments.

Acknowledgement

The authors gratefully acknowledge the financial support from the German Federal Ministry of Education and Research (BMBF) as part of project no. 03N8012 and also from BASF AG and Basell GmbH. Special thanks is also granted to Bent Tschernia from Labeq AG in Zurich/Switzerland for his support with respect to engineering of the HOPS reactor systems.

[1] G. Jung, *"Combinatorial Chemistry"*, 1st edition, Wiley-VCH, Weinheim **1999.**

[2] J.N. Cawse, *Acc. Chem. Res.* **2001,** *34*, 213.

[3] A. Tuchbreiter, J. Marquardt, B. Kappler, J. Honerkamp. M.O. Kristen, R. Mülhaupt, *Macromol. Rapid. Commun.* **2003,** *24*, 47-62.

[4] G. Natta, P. Pino, E. Mantica, F. Danusso, G. Mazzanti, M. Peraldo, *Chimica e l'Industria (Milan, Italy)* **1956,** *38*, 124-7.

[5] E. I. Pokrovskii, M. V. Vol'kenshtein, *Doklady Akademii Nauk SSSR* **1957,** *115*, 552-3.

[6] A. Tuchbreiter, J. Marquardt, J. Zimmermann, P. Walter, R. Mülhaupt, *J. Comb. Chem.* **2001,** *3*, 598.

[7] A. Tuchbreiter, B. Kappler, R. Stockmann, R. Mülhaupt, J. Honerkamp, *Macromol. Mater. Eng.* **2003,** *288*, 29-34.

[8] G.C. Pandey, A. Kumar, R.K. Garg, *Eur. Polym. J.* **2002,** *38*, 745.

[9] W. Camacho, S. Karlsson, *J. Appl. Polym. Sci.* **2002,** *85*, 321

[10] B. Kappler, A. Tuchbreiter, D. Faller, P. Liebetraut, W. Horbelt, J. Timmer, J. Honerkamp, R. Mülhaupt, *Polymer*, **2003,** *44*, 6179.

Polymacromonomers with Polyolefin Branches Synthesized by Free-Radical Homopolymerization of Polyolefin Macromonomer with a Methacryloyl End Group

Hideyuki Kaneko, Shin-ichi Kojoh, Nobuo Kawahara, Shingo Matsuo, Tomoaki Matsugi, Norio Kashiwa*

R&D Center, Mitsui Chemicals, Inc., 580-32 Nagaura, Sodegaura, Chiba 299-0265, Japan

Summary: Polymacromonomers with polyolefin branches were successfully synthesized by free-radical homopolymerization of polyolefin macromonomer with a methacryloyl end group. Propylene-ethylene random copolymer (PER) with a vinylidene end group was prepared by polymerization using a metallocene catalyst. Then, the unsaturated end group was converted to a hydroxy end group via hydroalumination and oxidation. The PER with the hydroxy end group was easily reacted with methacryloyl chloride to produce methacryloyl-terminated PER (PER macromonomer; PERM). The free-radical polymerization of thus-obtained PERM was done using 2,2'-azobis(isobutyronitorile) (AIBN) as a free-radical initiator. From NMR analyses, the obtained polymers were identified as poly(PERM). Based on gel permeation chromatography (GPC), the estimated degree of polymerization (D_p) of these polymers were about 30. Thus, new class of polymacromonomers with polyolefin branches was synthesized.

Keywords: functionalization of polymers; graft copolymers; macromonomers; polyolefins; radical polymerization

Introduction

Macromonomers, which consist of a macromolecular segment and a polymerizable chain end segment, are useful as a starting material for producing new polymer architectures. Homopolymerization and copolymerization of macromonomer yield various polymers that have unique topologies and properties. For example, polymacromonomers, which are produced by homopolymerization of macromonomer, can have unique molecular morphologies ranging from star-shaped spheres to rodlike cylinders by controlling the degree of polymerization (D_p) of the backbone and the length of the branch chains.[1] However, previous studies on macromonomers have been limited to polymers obtained by radical, cationic, and anionic polymerizations, such as polymethacrylate, polystyrene, and poly(ethylene oxide).

Synthesis and copolymerization of macromonomer possessing a polyolefin segment obtained by coordination polymerization have also been reported. For example, Mülhaupt *et al.* reported the synthesis of methacryloyl-terminated polypropylene via vinylidene-terminated polypropylene obtained by metallocene-catalyzed polymerization and the free-radical copolymerization of this polyolefin macromonomer with methyl methacrylate.[2] Matyjaszewski *et al.* reported the synthesis of methacryloyl-terminated polyethylene obtained by Pd-mediated living polymerization and the copolymerization of this polyethylene macromonomer with *n*-butyl acrylate by atom transfer radical polymerization.[3] However, there have been no reports on homopolymerization of polyolefin macromonomer to give a graft copolymer consisting of polar polymer backbone and polyolefin branches.

In this study, we focused on the macromonomers based on lower molecular weight PER as a polyolefin segment. These PER macromonomers (PERMs) were expected to have the advantage of facile homopolymerization because those have low viscosity and are easily soluble by many hydrocarbon solvents. By using such new class of polyolefin macromonomers obtained by the functionalization of vinylidene-terminated PER, we successfully synthesized polymacromonomers possessing polyolefin branches by the free-radical homopolymerization. Furthermore, the nature of the obtained polymacromonomer that consisted of a polymethacrylate backbone and PER branch chains was investigated. This is a first report of a polyolefin-based polymacromonomer with unique polymer architecture.

Experimental Section

Materials. Dicyclopentadienylzirconium dichloride (Cp$_2$ZrCl$_2$), Bis(1,3-dimethylcyclopentadienyl)zirconium dichloride ((1,3-Me$_2$Cp)$_2$ZrCl$_2$), methacryloyl chloride, triethylamine, and AIBN were commercially obtained and used without further purification. Ethylene was purchased from Sumitomo Seika Co., Ltd., and propylene was obtained from Mitsui Chemicals, Inc. Methylalminoxane (MAO) was purchased from Albemarle as a 1.2M toluene solution, and the trimethylaluminum, which is considered an impurity was evaporated in vacuo prior to use. Diisobutylaluminum hydride (DIBAL-H) was purchased from Tosoh-Finechem Co. Ltd. All other chemicals were obtained commercially and used as received.

Preparation of PER. PER with vinylidene end group has been prepared by copolymerization of propylene with ethylene using Cp_2ZrCl_2/MAO (Al/Zr=1000) catalyst system at 50°C for 5 h.

Hydroxylation of PER. PER (50 g, 0.071 mol as vinylidene end group) and toluene (250 ml) were placed in 500-ml glass reactor and stirred. Then DIBAL-H (50 ml, 0.28 mol) was added, and the system was then heated at 110 °C for 6 h. Dried air at 110 °C was then continuously fed (100 l/h) into the system. After 3 h, the reaction mixture was washed with *aq*.HCl and distilled water. After evaporation of solvent, hydroxylated PER was obtained as a yellow viscous liquid.

Synthesis of Methacryloyl-terminated PER (PERM). After hydroxylated PER (20 g, 12.7 mmol as hydroxy group) and toluene (30 ml) was placed in a 100-ml Schlenk tube, triethylamine (2.7 ml) and methacryloyl chloride (2.5 ml) were added to the tube, and then the mixture was stirred at room temperature for 3.5 h. The reaction mixture was washed with *aq*.HCl and distilled water. After the solvent was evaporated, the product was purified by liquid chromatography.

Homopolymerization of PERM. After PERM (3.0 g, 2.98 mmol as methacryloyl group) was placed in a 100-ml Schlenk tube, toluene (20 ml) and AIBN (75 mg) were added to the tube, and the mixture was heated at 70 °C for 30 h. The reaction mixture was washed with *aq*.HCl and distilled water. After the solvent was evaporated, a viscous liquid was obtained.

Polymer Fractionation. First, the polymer obtained by homopolymerization of PERM was dissolved in *n*-hexane and poured into a glass column packed with silica gel. Then, the first fraction was eluted by *n*-hexane and the second fraction by *n*-hexane/CH_2Cl_2 (2:1) mixed solvent. After evaporation of the solvent, each fraction was analyzed by GPC, ^1H NMR, and ^{13}C NMR.

Analytical Procedures. ^1H NMR and ^{13}C NMR spectra were recorded by JEOL GSX-270 or GSX-400 spectrometers with tetramethylsilane as the internal standard using chloroform-*d* or 1,2-dichlorobenzene-d_4 as a solvent. The gel permeation chromatograms (GPC) of the polymers were recorded by using a Waters Alliance GPC2000 equipped with four TSKgel columns (two sets of TSKgelGMH$_6$-HT and two sets of TSKgelGMH$_6$-HTL) at 140 °C and 1,2-dichlorobenzene as the solvent. The molecular weights of the polymers were determined by polystyrene standard.

Results and Discussion

Synthesis of PERM.

The synthetic route of PERM developed in this study is shown in Scheme 1. Vinylidene-terminated PERs were obtained by propylene copolymerization with ethylene by using typical metallocene catalysts. Unsaturated chain ends of the polymer were converted into hydroxy

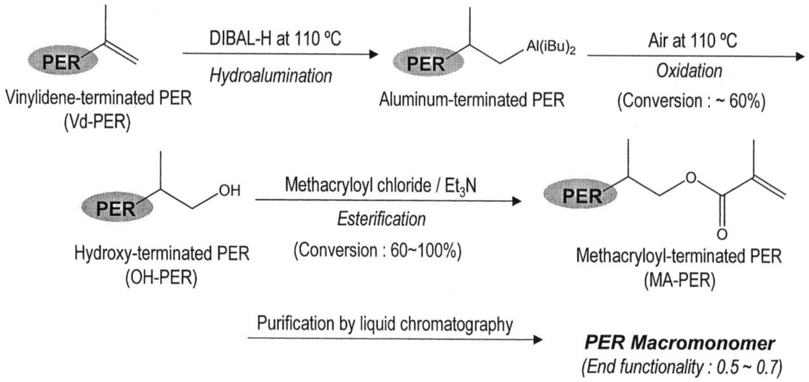

Scheme 1. Synthetic route of PERM

end groups by hydroalumination or hydroboration and subsequent oxidation, as previously reported.[4] Hydroalumination of vinylidene end groups by DIBAL-H was then carried out at 110 °C in toluene, as recommended in the literature.[4] Then, dried air was fed into the resulting mixture at 110 °C to convert the alkylaluminum end groups to hydroxy end groups. The obtained hydroxy end groups easily reacted with methacryloyl chloride in the presence of a Lewis base at room temperature to produce methacryloyl end groups. The PER with a methacryloyl end group was useful as PERM owing to the polymerizable methacryloyl end group. By selecting the polymerization conditions and catalyst, we prepared two types of functionalized PERs with different molecular weights (PER1 and PER2).

Figures 1(i), (ii), and (iii) show the ^{1}H NMR spectra of PER1 with a vinylidene end group (Vd-PER1), with a hydroxy end group (OH-PER1), and with a methacryloyl end group (MA-PER1), respectively. In Figure 1(i), the signals assigned to the vinylidene group protons (c; 4.6-4.8 ppm), were detected in addition to the signals of PER main chain protons at 0.7 to 1.8 ppm. These additional signals are generated from the chain transfer reaction induced by monomers when the propagating chain end was a propylene unit.[5] The content of vinylidene, ethylene, and propylene units was 4.0, 38, and 58 mol%, respectively, calculated from the relative intensities of the protons of each unit in the ^{1}H NMR spectrum. In Figure 1(ii), the new signals assigned to the hydroxymethylene group protons (*e*; 3.3-3.6 ppm) were detected in addition to the signals of unreacted vinylidene protons. The calculated content of the hydroxymethylene and vinylidene groups in OH-PER1 was 2.4 mol% and 0.09 mol%, respectively, indicating that about 60% of the vinylidene end groups in Vd-PER1 were hydroxylated and that the other vinylidene end groups were converted to the saturated end groups.

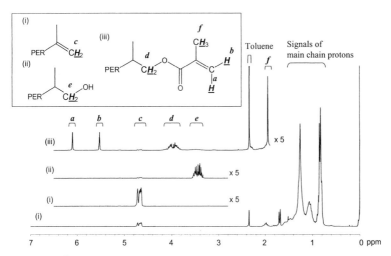

Figure 1. ^{1}H NMR spectra of (i)Vd-PER1, (ii)OH-PER1 and (iii) MA-PER1 (270 MHz, in CDCl$_3$)

This hydroxylation efficiency of 60% would be reasonable, because we previously reported that the hydroxylation efficiency of alkylaluminum-terminated PP was 52%.[6] In Figure 1(iii), the new signals assigned to the methacryloyl end group protons (*a*, *b*, *d* and *f*; 6.1, 5.5, 3.8-4.2 and 1.95 ppm, respectively) were detected. From the relative intensities of the protons of each group, the calculated content of the methacryloyl, hydroxymethylene, and vinylidene groups was 2.4, 0.37, and 0.08 mol%, respectively, indicating that almost all hydroxy end groups in OH-PER1 were converted to methacryloyl end groups.

To remove the undesirable PER that has a saturated vinylidene or hydroxy end group, the obtained MA-PER1 was purified by liquid chromatography. The content of the methacryloyl, ethylene, and propylene units of the purified MA-PER1 was 3.9, 35, and 61 mol%, respectively, calculated from ^{1}H NMR analysis, and the number average molecular weight (M_n) was 670 estimated from the GPC measurement.

Therefore, the estimated end functionality (f_{MA}) of the obtained MA-PER1 was 0.68. The absence of vinylidene and hydroxymethylene end groups in the ^{1}H NMR spectrum indicates that residual chain ends were saturated end groups. Using the same purification method as for MA- PER1, MA-PER2 with higher M_n than MA-PER1 was obtained.

Table 1. Functionalization Results for PER Chain Ends

		Vd-PER [a]	OH-PER	MA-PER	
				before purification	after purification
PER1	M_n [b]	520	520	690	670
	Content of				
	vinylidene end group (mol%)[c]	4.0	0.09	0.08	n.d. [d]
	hydroxy end group (mol%)[c]	-	2.4	0.37	n.d. [d]
	methacryloyl end group (mol%)[c]	-	-	2.4	3.9
	(End functionality; f_{MA})				(0.68)
PER2	M_n [b]	1,550	1,650	1,960	1,330
	Content of				
	vinylidene end group (mol%)[c]	1.4	0.04	n.d. [d]	n.d. [d]
	hydroxy end group (mol%)[c]	-	0.85	0.01	0.10
	methacryloyl end group (mol%)[c]	-	-	0.52	1.6
	(End functionality; f_{MA})				(0.53)

[a] Polymerization conditions: (PER1) 0.01 mmol of Cp_2ZrCl_2, 10 mmol of MAO, ethylene/propylene flowrate of 10/90 (l/h), in 800 ml of toluene, 50 °C, 5 h, (PER2) 0.02 mmol of $(1,3-Me_2Cp)_2ZrCl_2$, 20 mmol of MAO, ethylene/propylene flowrate of 20/80 (l/h), in 800 ml of toluene, 50 °C, 2 h.
[b] Determined by GPC.
[c] Calculated from ^{1}H NMR.
[d] Not detected.

Table 1 summarizes these functionalization results. For the synthesis of MA-PER2, both the conversion of the vinylidene end groups to hydroxy end groups and that of the hydroxy end groups to methacryloyl end groups were about 60%. Such low conversion of the hydroxy end groups to methacryloyl end groups in PER2 is probably due to the higher viscosity of PER2 than that of PER1. After purification, the estimated f_{MA} of the obtained MA-PER2 was 0.53 from ^{1}H NMR analysis and GPC measurement. The obtained MA-PER1 (f_{MA} = 0.68) and MA-PER2 (f_{MA} = 0.53) were then used as PERMs for free-radical polymerization without further purification.

Homopolymerization of PERM.

Homopolymerization of the PER1 macromonomer (PERM1) and PER2 macromonomer (PERM2) was carried out at 70°C and 60°C, respectively, in toluene in the presence of AIBN as a radical initiator. Each resulting product was pale yellow viscous oil. Figure 2 shows the GPC traces for PERMs and the corresponding homopolymerized products. Each GPC trace of the homopolymerized product showed a new peak at a higher M_n region (over 10^4 mol/g) in addition to a peak of PERM at a lower M_n region (under 10^4 mol/g). These peaks indicate the formation of polymacromonomer. The estimated M_n of poly(PERM1) and poly(PERM2) was 18,760 and 45,920, respectively, from the higher M_n part of GPC traces. From a calculation based on M_n values for the two PERMs (670 for PERM1, and 1,330 for PERM2), the estimated D_p of these poly(PERM1) and poly(PERM2) was about 28 and 35, respectively.

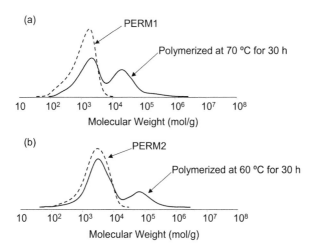

Figure 2. GPC traces for homopolymerization of (a)PERM1 and (b)PERM2

Although these two D_p values cannot be directly compared due to differences in polymerization conditions, these two macromonomers indicate nearly equal efficiency of polymerization, despite the different M_n.

Fractionation and Characterization of Poly(PERM1).

To isolate the higher M_n region, the mixture was fractionated by liquid chromatography using a silica gel column into two fractions, n-hexane and n-hexane / dichloromethane (2:1) eluates. The GPC traces of these two fractions were shown in Figure 3. From a calculation based on M_n value (670) for the PERM1, the first fraction was estimated to be the poly(PERM1) (D_p = 30) and the second fraction the unpolymerized PERM1 containing the oligomer of PERM1 (D_p= 2~3). Based on the ^1H NMR spectrum, the second fraction contained unreacted methacryloyl end groups.

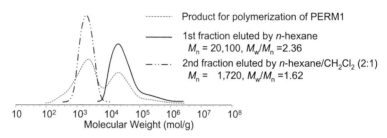

Figure 3. GPC traces of the product for polymerization of PERM1
and fractions obtained by liquid chromatography

Figures 4 and 5 show the ^1H and ^{13}C NMR spectra of the isolated poly(PERM1), respectively. In the ^1H NMR spectrum, the signals of methacryloyl group protons of PERM1 (1.95, 5.5, and 6.1 ppm) disappeared and the signals of methylene protons adjoining ester group shifted from 3.8-4.2 ppm to 3.4-4.1 ppm (a) in comparison with the ^1H NMR spectrum of PERM1 in Figure 1(iii) (MA-PER1). In addition, the signals of methylene protons in polymethacrylate backbone appeared at 1.7-2.0 ppm (b). In the ^{13}C NMR spectrum, in addition to the signals of PER chain carbons, signals appeared at 15-20, 45, 52-55, 70, and 176-178 ppm. Based on the ^{13}C NMR measured in DEPT mode, these signals are respectively assigned to methyl carbon (e), quaternary carbon (d) and methylene carbon (c) in the polymethacrylate main chain, methylene carbon (b) adjoining ester group, and carbonyl carbon (a), respectively.

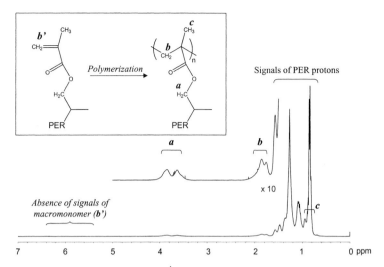

Figure 4. ^{1}H NMR spectrum of poly(PERM1)
(400 MHz, in 1,2-dichlorobenzene-d_4 at 118 °C)

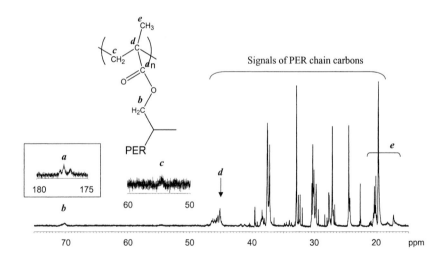

Figure 5. ^{13}C NMR spectrum of poly(PERM1)
(100 MHz, in 1,2-dichlorobenzene-d_4 at 118 °C)

Thus, NMR analyses revealed the formation of poly(PERM1).

Based on the D_p obtained by GPC measurement, this poly(PERM1) consists of a polymethacrylate backbone (M_n = 2,550) and 30 PER branch chains (M_n = 585). This is the first report of such unique polymer architecture, which is expected to be the hybrid materials consisting of polyolefins and polar polymers. Despite having the polymethacrylate backbone, this poly(PERM1) is a viscous liquid polymer and soluble even in non-polar hydrocarbon solvents such as n-hexane, due to the high branch density and the high content of PER segment (87 wt%).

In addition, the result of liquid chromatography indicates that the polarity of this poly(PERM1) was lower than that of unpolymerized macromonomers in n-hexane, despite having the same composition (the molar ratio of methacryloyl segment to PER segment was considered to be equal). One interpretation of this lower polarity for poly(PERM1) than that for PERM1 is as follows. This poly(PERM1) consists of two different segments, polar polymethacrylate and non-polar PER. Therefore, in non-polar solvent such as n-hexane, this poly(PERM1) would form a core-shell type structure, in which the polar core of the polymethacrylate backbone is inside the non-polar shell formed by PER side branches. Consequently, the apparent polarity of the poly(PERM1) in n-hexane is decreased.

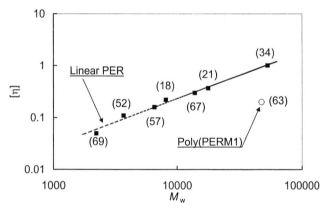

Figure 6. Double-logarithmic polts of [η] vs M_w for linear PERs and poly(PERM1).
(Values in parentheses are the propylene content (mol%) in PER.)

The relationship between M_w and the intrinsic viscosity ([η]; measured in decalin at 135°C) of PERs obtained by using metallocene catalysts is shown in Figure 6. In the case of linear PER, a linear relationship was observed independent of monomer composition. On the other hand, the obtained poly(PERM1) located obviously below this line. This result shows that poly(PERM1) has a lower viscosity than the linear PER with the same molecular weight owing to its unique polymer topology and composition.

Conclusions

Polymacromonomers with polyolefin branches were successfully synthesized by homopolymerization of methacryloyl-terminated PER macromonomer, which were synthesized by the conversion of the vinylidene-terminated PER obtained by metallocene-catalyzed polymerization. The obtained polymacromonomer could be isolated by liquid chromatography with a silica gel column. Based on ^{1}H and ^{13}C NMR analyses and GPC measurement, the obtained polymer consisted of a polymethacrylate backbone and 30 PER branch chains. This novel polymacromonomer exhibited the nature of PER rather than the nature of polymethacrylate because of its unique polymer architecture such as the high concentration of polyolefin branches. It is expected that various unique polymers having both polar polymer segment and polyolefin segment could be synthesized by applying this synthetic route.

[1] Y. Tsukahara, K. Mizuno, A. Segawa, Y. Yamashita, *Macromolecules* **1989**, *22*, 1546. K. Terao, T. Hokajo, Y. Nakamura, T. Norisuye, *Macromolecules* **1999**, *32*, 3690. M. Wintermantel, M. Gerie, K. Fisher, M. Schmidt, I. Wataoka, K. Urakawa, Y. Kajiwara, Y. Tsukahara, *Macromolecules* **1996**, *29*, 978. M. Gerle, K. Fisher, S. Roos, A. H. E. Müller, M. Schmidt, S. S. Sheiko, S. Prokhorova, M. Möller, *Macromolecules* **1999**, *32*, 2629. S. S. Sheiko, M. Gerle, K. Fisher, M. Schmidt, M. Möller, *Langmuir* **1997**, *13*, 5368.
[2] T. Duschek, R. Mulhaupt, *Polym. Prepr.* **1992**, *33*, 170.
[3] S. C. Hong, S. Jia, M. Teodorescu, T. Kowalewski, K. Matyjaszewski, A. C. Gottfried, M. Brookhart, *J. Polym. Sci., Part A* **2002**, *40*, 2736.
[4] K. K. Kang, T. Shiono, T. Ikeda, *Macromolecules* **1997**, *30*, 1231. T. C. Chung, H. L. Lu, W. Janvikul, *Polymer* **1997**, *38*, 1495.
[5] T. Tsutsui, A. Mizuno, N. Kashiwa, *Polymer* **1989**, *30*, 428.
[6] S. Kojoh, T. Tsutsui, M. Kioka, N. Kashiwa, *Polym. J.* **1999**, *31*, 332.

Synthesis of Polyolefin Based Materials with Improved Thermo-Oxidative Stability

Marc Dolatkhani,[1] *Henri Cramail,*[1] *Alain Deffieux,*[1] *Maria Rosário Ribeiro,**[2] *Jorge M. Santos,*[2] *João M. Bordado*[2]

[1]Université Bordeaux 1, CNRS, ENSCPB, UMR 5629, Laboratoire de Chimie des Polymères Organiques, 16 Avenue Pey-Berland, F-33607 Pessac Cedex, France
[2]Departamento de Engenharia Química, Instituto Superior Técnico, Av. Rovisco Pais, 1049-001 Lisboa, Portugal
E-mail: rribeiro@alfa.ist.utl.pt

Summary: Polyolefin based materials with chemically bonded phenol compounds have been synthesized. Two unsaturated polyolefins: an ethylene/propylene/5,7-dimethyl-1,6-octadiene terpolymer and an ethylene/5,7-dimethyl-1,6-octadiene copolymer, were used as starting materials. The functionalisation method involves a two-step procedure that consists in the hydrochlorination of the substituted pending double bonds, followed by grafting of the phenol compound. This procedure was found to be more effective for the terpolymer than for the corresponding copolymer. The reasons for this are discussed. A thermogravimetric study, in air atmosphere shows that, poly(ethylene-co-5,7-dimethyl-1,6-octadiene) grafted with 2-*tert*-butyl-4-methylphenol present improved thermo-oxidative stability when compared to a polyethylene reference.

Keywords: antioxidants; EPDM; functionalisation of polymers; non-conjugated dienes; polyolefins

Introduction

Unsaturated polyolefin-based materials, obtained from the polymerisation of olefins with non-conjugated dienes, can be modified by chemical reaction providing a successful route for the synthesis of side-functional polyolefins (PO). By incorporation of polar functional groups it is possible to improve ultimate properties of PO such as adhesion and compatibility with other polymers. Several examples of modification processes of the pending double bonds of unsaturated polyolefins are described in literature and may include: hydroboration followed by oxidation to hydroxyl groups,[1-4] epoxidation,[3-8] bromination,[5] hydroxylation,[9] sulfonation,[10] oxidation to aldehyde and carboxylic acid functions[6,11] and hydrosilylation.[12, 13]

Alternatively the unsaturations introduced in PE main chain may be used for chemical binding of stabilizers in order to inhibit some degradation mechanisms of PE, detrimental to polymer performance. Aromatic functional compounds, such as hindered phenols, are efficient hydrogen donors and can act as radical scavengers preventing thermo-oxidative degradation. Phenolic additives are the most widely used admixed stabilizers for olefin-based polymers. Chemically bound phenols as stabilizers may offer important advantages over admixed additives since they cannot be lost by migration or volatilisation and therefore will guarantee a longer protection of the materials.

In previous works we synthesized olefin-based co- and ter-polymers, using a linear asymmetrically substituted diene, the 5,7-dimethylocta-1,6-diene (5,7-DMO), as comonomer.[14-18] This diene, obtained from terpene fractions, possesses in addition to its vinyl unsaturation a tri-substituted double bond that limits side reactions, like crosslinking or cyclisation, which may occur during copolymerisation. Moreover high amounts of 5,7-DMO can be incorporated in the polymer chains, by careful selection of the metallocene geometry catalyst.

We report now on a two step procedure for the grafting of phenol compounds on these 5,7-DMO based co-and terpolymers, which involves: (1) hydrochlorination of the pendant double bond followed by (2) the grafting of the aromatic compound. The thermal-oxidative stability of an ethylene/5,7-DMO copolymer grafted with 2-tert-butyl-4-methylphenol is also examined.

Experimental

Hydrochlorination reaction: An ethylene/propylene/5,7-DMO terpolymer (8 mol-% of diene units; $\overline{M}_n$ =60 000 g.mol^{-1}, $\overline{M}_w/\overline{M}_n$ =2,) prepared with Et(Ind)$_2$ZrCl$_2$/MAO and an ethylene/5,7-DMO copolymer, (3.5 mol-% diene units, $\overline{M}_n$ =38 000 g.mol^{-1}, $\overline{M}_w/\overline{M}_n$ =2.7) prepared with [(η^5-C$_5$Me$_4$)SiMe$_2$(NtBu)]TiCl$_2$/MAO were dissolved respectively in n-heptane at room temperature and in toluene at 60°C. A ten times molar excess of Et$_3$SiH and SnCl$_4$, relatively to unsaturations was added. The reaction time was 20 min for the terpolymer and 1 hour for the copolymer. During the reaction a white precipitate was formed. At the end this precipitate was filtered and the polymer was precipitated with methanol and vacuum dried.

Grafting of the phenol compounds: Both hydrochlorinated 5,7-DMO-based-terpolymer and

-copolymer were dissolved in n-heptane, at 60 °C. Previously distilled phenol and BF$_3$OEt$_2$ were added to the hydrochlorinated terpolymer and the mixture was stirred for 16 h. 2-tert-butyl-4-methylphenol and BF$_3$OEt$_2$ were added to the hydrochlorinated copolymer and reacted for 20 h. At the end, the mixture was washed with water and concentrated by distilling the solvent. Both polymers were precipitated with ethanol and vacuum dried.

Characterisation of the polymers: ^{1}H and ^{13}C- NMR spectra were recorded on Bruker AC200 and AC250 spectrometers, in different solvents (1,1,2,2-tetrachloroethane-d$_2$, benzene-d$_6$ and dichloromethane-d$_2$). The molar masses of the terpolymers were evaluated by SEC (Varian 5500, UV and refraction index double detection) in THF, at room temperature on the basis of polystyrene standards. Infrared spectra were collected on polymer-KBr pellets with a Perkin Elmer 1600 FT-IR spectrophotometer. Thermogravimetric tests were performed with a Setaram LabSys TG-DTA thermobalance, under an air flux at a rate of 10 K.min^{-1} between 50 and 700 °C.

Results

Hydrochlorination of ethylene/propylene/5,7-DMO terpolymers

The introduction of chloride functions onto polymer chains is an interesting preliminary step for further functionalisation chemistry. Based on data reported by Kennedy, hydrochlorination of unsaturated polyolefins was first attempted in the presence of

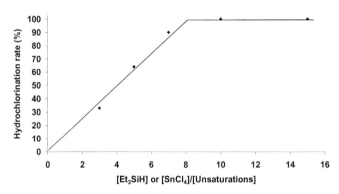

Figure 1. Hydrochlorination rate of a poly(ethylene-co-propylene-co-5,7-DMO) in function of the molar ratio [SnCl$_4$]/[unsaturation]; Solvent hexane; T=25 °C; [SnCl$_4$]/[Et$_3$SiH]=1; [unsaturation]= 3.10^{-2} mol.L^{-1}.

Et$_3$SiH/SnCl$_4$ mixture.[19] The experimental conditions were optimised in order to improve the efficiency of the reaction.

It was found that hydrochlorination rate is controlled by the amounts of Et$_3$SiH and SnCl$_4$ introduced in the reaction mixture. Complete hydrochlorination of the pending double bonds needs, at room temperature, about 10 equivalents (with respect to the unsaturations) of SnCl$_4$ and Et$_3$SiH. It should be noted that this reaction, complete or partial, proceeds with neither formation of gel nor evolution of the initial molar mass of the terpolymer. Under selected conditions the hydrochlorination reaction results in the total disappearance of the ethylenic proton, as shown in the ^{1}H-NMR spectrum (Figure 2).

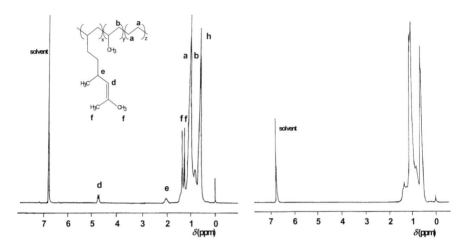

Figure 2. 1H-NMR spectra in deuterated benzene of a poly(ethylene-co-propylene-co-5,7-DMO) (a) and the same polymer after reaction with Et$_3$SiH/SnCl$_4$ (b). Conditions: reaction time 20 min; solvent n-heptane; T = 25 °C; [SnCl$_4$]/[Et$_3$SiH]=1; [SnCl$_4$]/[unsaturation]= 10.

The analysis of the ^{13}C-NMR spin echo spectrum before and after hydrochlorination (Figure 3) supports the formation of tertiary chlorine groups. Two new peaks are observed; one at 72 ppm, characteristic of the quaternary carbon bonded to chlorine, and other at 52 ppm, corresponding to the secondary carbon in α position relatively to the quaternary carbon. This data shows that hydrochlorination reaction of the pending double bonds of the terpolymer is quantitative and regio-selective (Scheme 1).

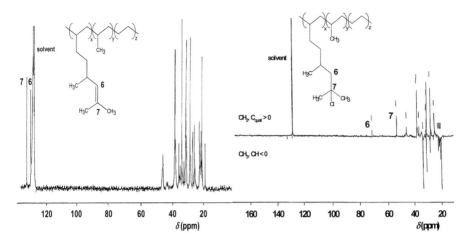

Figure 3. ^{13}C-NMR spectrum of a poly(ethylene-co-propylene-co-5,7-DMO) (a) and the corresponding spin echo spectrum of the same polymer after addition of Et$_3$SiH/SnCl$_4$ (b). Same reaction conditions as in Figure 1.

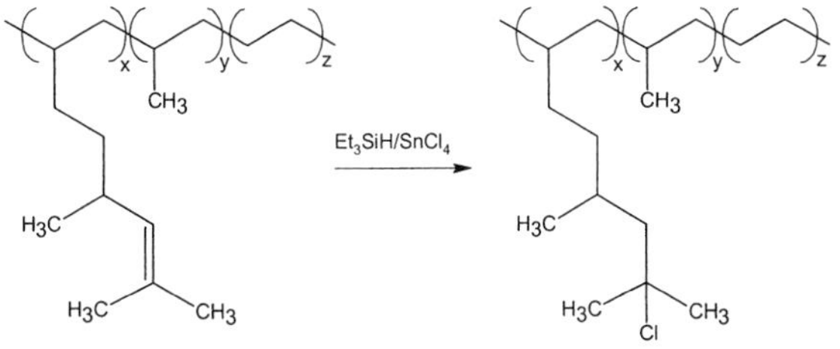

Scheme 1. Hydrochlorination of an ethylene/propylene/5,7 DMO terpolymer.

Hydrochlorination of ethylene/5,7-DMO copolymers

Based on the results presented above the hydrochlorination of ethylene/5,7-DMO copolymers was also performed with a 10 times excess of SnCl$_4$/Et$_3$SiH. To increase polymer solubility reaction was carried out at 60 °C and toluene was used as solvent. The reaction time was increased to 1 h. FT-IR results show that the characteristic band of the substituted 5,7-/DMO double bond, at 842 cm^{-1}, is absent in the reaction product. In the modified polymer new bands characteristic of the C-Cl bond are observed at 653 and 577 cm^{-1}. The ^{1}H-NMR

352

spectrum shows however the presence of residual unsaturations at 4.8 ppm. The ^{13}C-NMR spectrum of the modified polymer confirms the presence of chlorine: new peaks were observed at 52.7 and 72.4 ppm. This spectrum also reveals some other peaks that could be related to side reactions.

In summary, whereas a clean and complete hydrochlorination was achieved for ethylene/propylene/5,7-DMO terpolymers, for the corresponding ethylene copolymers the hydrochlorination is only partial (70 % yield). This fact can be related to the low solubility of the ethylene/5,7-DMO copolymers when compared to EPDM's, and to the use of high reaction temperatures (60 °C), that may favour side reactions.

Grafting of phenol on hydrochlorinated ethylene/propylene/5,7-DMO terpolymers

Friedel-Crafts type reactions allow substitution, in the presence of Lewis acids, of tertiary chlorine functions by aromatic derivatives. Kennedy et al. using boron trifluoride diethyl etherate (BF$_3$OEt$_2$) as Lewis acid, synthesized a bisphenol-terminated polyisobutene, starting from a α-ω-tert-chloropolyisobutene.[20] Based on this data we have performed the phenol grafting reaction in similar conditions: at 55 °C, in presence of an excess of phenol, using the same Lewis acid and n-heptane as solvent (Scheme 2).

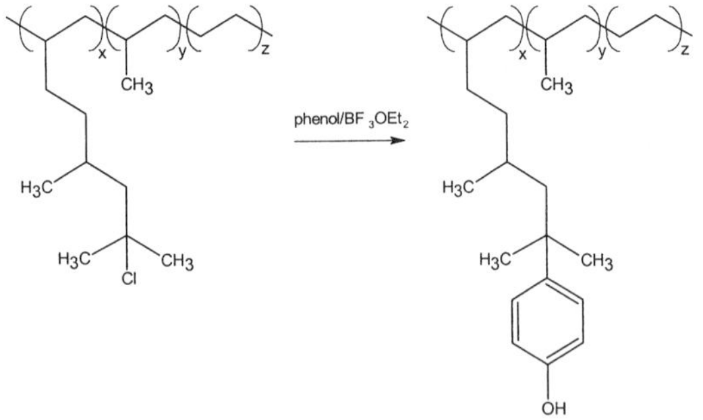

Scheme 2. Grafting of phenol on a hydrochlorinated ethylene/propylene/5,7 DMO terpolymer.

The ^{1}H-NMR spectrum of the modified polymer is shown in Figure 4. The aromatic ring protons of phenol are observed at 6.6 ppm and 7.2 ppm; a broad signal representative of the hydroxyl group proton is also detected at 4.8 ppm. In the chosen conditions, the phenol-grafting rate, calculated from the 1H NMR spectrum, is higher than 75 % with respect to

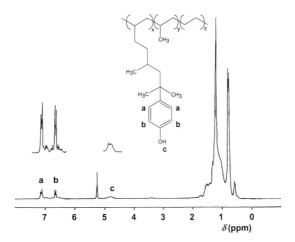

Figure 4. ¹H-NMR spectrum in CD₂Cl₂ of poly(ethylene-co-propylene-co-5,7-DMO) with grafted phenol groups.

hydrochlorinated groups.

Grafting of phenol groups was confirmed by SEC, using UV and refraction index double detection. In fact, the hydrochlorinated 5,7-DMO based EPDM does not have a detectable response in UV but, the SEC chromatogram after phenol grafting shows a strong UV response.

Additional phenol grafting tests were carried out in different conditions (Table 1). A 10 equivalents excess of phenol in relation to tertiary chlorine functions is necessary to avoid chain-coupling reactions.

Table 1. Grafting of phenol on 5,7-DMO-based EPDM. Solvent: n-heptane; reaction time: 16 h; [functions] $= 2.5 \times 10^{-2}$ M; [BF₃OEt₂] $= 0.2$ M.

Functions on the terpolymer	[phenol]/[functions]	T (°C)	Grafting rate (%)	Notes
Chlorine	3	55	/	Chain coupling
Chlorine	7	55	/	Chain coupling
Chlorine	10	55	75	Unimodal SEC peak
Chlorine	15	65	78	Unimodal SEC peak

354

Grafting of 2-tert-butyl-4-methylphenol on hydrochlorinated ethylene/5,7-DMO copolymers

The addition of 2-tert-butyl-4-methylphenol to hydrochlorinated ethylene/5,7-DMO copolymers was investigated (Scheme 3).

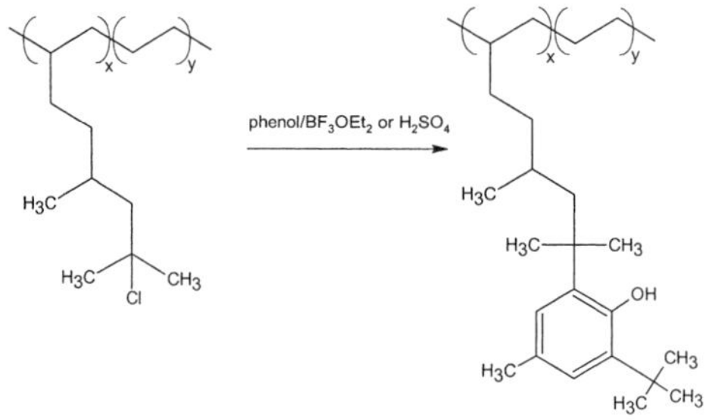

Scheme 3. Grafting of 2-tert-butyl-4-methylphenol on a hydrochlorinated ethylene/5,7 DMO copolymer.

In agreement with the strategy described for the phenol grafting on hydrochlorinated EPDM, the hydrochlorinated ethylene/5,7 DMO copolymer was reacted for 20 h at 60 °C with 2-tert-butyl-4-methylphenol in the presence of BF_3OEt_2. Despite its lower ability to dissolve ethylene/5,7-DMO copolymers, n-heptane was used as a solvent since toluene could compete strongly with phenol, for grafting onto the polymer structure. The ^{1}H-NMR spectrum of the reaction product shows new peaks in the aromatic region, at 6.4, 6.7 and 6.9 ppm, and the presence of non-reacted double bonds. In the ^{13}C-NMR spectrum, the peaks at 52.7 and 72.4 ppm of the hydrochlorinated group are still present although significantly reduced in intensity, confirming the partial substitution reaction. In conclusion, 2-tert-butyl-4-methylphenol grafting takes place, although it does not completely remove chlorine groups from the chains.

In order to evaluate the effect of grafted 2-tert-butyl-4-methylphenol onto the copolymer thermo oxidative stability, a thermogravimetric study, in air atmosphere, was performed.

The thermogravimetric curve of the modified copolymer is compared to that of a polyethylene reference (Figure 5). The phenol-grafted copolymer starts main degradation process (in terms of weight loss) at a temperature near 400 °C, which is almost 50 °C higher than the one observed for homopolyethylene. However a slow degradation process is visible in the range 300 - 400 °C, which can be attributed in part to the elimination of HCl[21] from remaining hydrochlorinated units in the modified ethylene/5,7 DMO copolymer. The presence of chlorine residues in the phenol-grafted copolymers may be detrimental for future applications. An alternative route, that guarantees the total absence of these residues, consists of the direct introduction of phenol groups on the pendant unsaturations of 5,7-DMO based copolymers. Comparison data of direct route vs. indirect route will be published elsewhere.

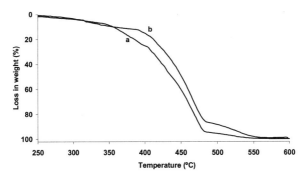

Figure 5. Thermogravimetric curves of: homopolyethylene (a) and ethylene/5,7 DMO copolymer modified with 2-*tert*-butyl-4-methylphenol (b).

The higher stability of the grafted copolymers may be explained considering that thermo-oxidative degradation is mostly due to the formation of peroxy radicals by free radical reactions. In presence of efficient hydrogen donors, such as hindered phenol compounds, these radicals react and form hydroperoxides, preventing therefore the abstraction of hydrogen from the polymer backbone. The fact that in our case the phenol group is covalently bonded to the polymer allows us to expect an increased durability and efficiency of the polymer protection.

Conclusion

Phenolic compounds were chemically bonded to unsaturated polyolefin-based polymers in order to reduce thermal oxidative degradation of these polymeric materials that leads to a loss of physical and optical properties of the polymer.

The synthesis of poly(ethylene-co-propylene-co-5,7-DMO) grafted with phenol and of poly(ethylene-co-5,7-DMO) grafted with 2-*tert*-butyl-4-methylphenol was performed using a two-step procedure involving: hydrochlorination of pending double bond followed by phenol grafting. This route was found to be very effective for the ethylene/propylene/5,7-DMO terpolymer and lead to a phenol-grafting rate higher than 75 %. However the used procedure revealed to be much less efficient when applied to ethylene/5,7-DMO copolymers. This behaviour might be related to: (i) the smaller diene contents and higher crystallinity of 5,7-DMO copolymers when compared to the corresponding terpolymers (resulting in much lower solubility of copolymers); (ii) the higher reaction temperature used for hydrochlorination of the copolymers, which may lead to an increase of side reactions; (iii) and finally the bulkiness of the hindered phenol used, which in one way should be more effective in preventing thermo-oxidative degradation, but may reduce in some extent the efficiency of the grafting reaction.

Compared to polyethylene, the ethylene/5,7-DMO copolymer grafted with 2-*tert*-butyl-4-methylphenol exhibits higher thermal oxidative stability. Long-term durability is also expected since this covalently bonded phenol group will not migrate as happens in stabilization of polyolefins by compounding.

Acknowledgements. The contribution from FCT (POCTI/CTM/2001) and ICCTI/CNRS (423/França) is gratefully acknowledged. MD thanks ADEME and DRT company and JMS the Sub-Programa Ciência e Tecnologia do 2° Quadro Comunitário de Apoio, for their doctoral fellowships.

[1]   T. C. Chung, M. Raate, E. Berluche, D. N. Schulz, *Macromolecules* **1988**, *21*, 1903.

[2]   T. C. Chung, H. L. Lu, C. L. Li, *Macromolecules* **1994**, *27*, 7533.

[3]   S. Marathe, S. Sivaram, *Macromolecules* **1994**, *27*, 1083.

[4]   W. Kaminsky, D. Arrowsmith, H. R. Winkelbach, *Polymer Bulletin* **1996**, *36*, 577.

[5]   M. Hackmann, T. Repoo, G. Jany, B. Rieger, *Macromol. Chem. Phys.* **1998**, *199*, 1511.

[6]   F. Q. Song, D. Pappalardo, A. F. Johnson, B. Rieger, M. Bochmann, *J. Polym. Sci. Part A: Polym. Chem.* **2002**, *40*, 1484.

[7]   M. Arnold, S. Bornemann, T. Schimmel, T. Heinze, *Macromol. Symp.* **2002**, *181*, 5.

[8]   I. Kim, *React. Funct. Polym.*, **2001**, *49*, 197.

[9]   Japan 5603 6508 (1982), Mitsubishi Petrochemical Company, invs.: M. Fujii, M. Masaki; *Chem. Abstr.* **1982**, *95*, 62968.

[10] R. D. Lundberg, *J. Appl. Polym. Sci, Part A* **1982**, *12*, 4623.

[11] P. H. J. Carlsen, T. Katsuki, V. S. Martin, K. B. Sharpless, *J. Org. Chem.* **1981**, *46*, 3936.

[12] X. Guo, R. Farwaha, G. L. Rempel, *Macromolecules* **1990**, *23*, 5047.

[13] T. Uozumi, G. L. Tian, C. H. Ahn, J. Z. Jin, S. Tsubaki, T. Sano, K. Soga, *J. Polym. Sci. Part A: Polym. Chem.* **2000**, *38*, 1844.

[14] J. M. Santos, M. R. Ribeiro, M. F. Portela, H. Cramail, A. Deffieux, *Macromol. Chem. Phys.* 2001, *202*, 3043.

[15] J. M. Santos, M. R. Ribeiro, M. F. Portela, H. Cramail, A. Deffieux, A. Antinolo, A. Otero, S. Prashar, *Macromol. Chem. Phys.*, **2002**, *203*, 139.

[16] M. Dolatkhani, H. Cramail, A. Deffieux, *Macromol. Chem. Phys.* **1995**, *196*, 3091.

[17] M. Dolatkhani, H. Cramail, A. Deffieux, *Macromol. Chem. Phys.* **1996**, *197*, 289.

[18] M. Dolatkhani, H. Cramail, A. Deffieux, *Macromol. Chem. Phys.* **1996**, *197*, 2481.

[19] J. P. Kennedy, B. Ivan; in "Designed Polymers by Carbocationic Macromolecular Engineering. Theory and Practice", Hansen, Munich **1991**.

[20] J. P. Kennedy, S. C. Guhaniyogi, V. Percec, *Polymer Bulletin* **1979**, *8*, 563.

[21] V. V. Korshak, in "The chemical structure and thermal characteristics of polymers", Israel Program for Scientific Translations, Jerusalem, **1971**.

Macromol. Symp. **2004**, *213*, 357-365

Polymerization with Fischer-Tropsch Olefins – A New Challenge in Polymer Science

Ioan Tincul,[1,2] *Siphuma Lufuno,*[1] *Randhir Rawatlal*[2]

[1] Sasol Technology, PO Box 1, Sasolburg, South Africa
[2] School of Chemical Engineering, University of Natal, South Africa

Summary: Co- and ter-polymerization of ethylene with Fischer-Tropsch derived olefins having odd carbon number and branched olefins are investigated. A method of preparation of a robust, high productivity Ziegler-Natta catalyst suitable for ethylene co-polymerization with Fischer Tropsch olefin is reviewed. In supporting titanium tetrachloride, an attempt was made to control the titanium oxidation state. The resulting experimental data were fitted to a sub-sites model that associates titanium oxidation state with catalyst activity.

Keywords: Fischer-Tropsch Olefins; polymerization; Ziegler-Natta catalysis

Introduction

For more than half a century both science and industry have witnessed the unprecedented growth of the polyolefin industry. The discovery of polymerization of ethylene and propylene on transition metal catalysts was a turning point in the successful history of polyolefins. This success was driven continuously by advances in catalyst and process technology, and by affordable olefin resources and versatile market application.

The early work of Natta and his co-workers opened a new era in polyolefin technology. The development of a multitude of heterogeneous catalysts that followed this work made polyolefins the leading polymer family of our times. This was succeeded by a host of homogeneous catalysts. Metallocenes and the "late transition metal catalysts" resulted in further effervescence in polymer science and technology. Process development witnessed numerous solution, slurry and gas phase technologies, combinations of these, and, most recently, the emergence of bi-modal processes.

Oil is the main source for ethylene and propylene while α-olefin comonomers, 1-butene, 1-hexene and 1-octene were produced historically mainly by ethylene oligomerization. A more complex pool is available from the Fischer-Tropsch process. This option is gaining popularity as a major comonomer market source since access to oil is becoming less

 DOI: 10.1002/masy.200450932

affordable. The route to olefins is shorter and a complex mixture of hydrocarbon rich in olefins is obtained. The Fischer-Tropsch process offers the polyolefin industry, in addition to the currently known even carbon number olefins, some less explored possibilities such as odd carbon number olefins and branched olefins, making possible new routes to new polymers such as:

i.   Ethylene co-polymers with odd carbon number olefins

ii.   Propylene co-polymers with odd carbon number olefins

iii.   Ethylene ter-polymers with even and odd carbon number olefins

iv.   Ethylene co-polymers with branched olefins

v.   Ethylene ter-polymers with linear and branched olefins.

Ethylene Polymers with Fischer-Tropsch Olefins

A large number of new polymers with Fischer-Tropsch olefins have been obtained in our laboratories. However, this work is only a beginning in this rather challenging field. The focus of this paper is on ethylene co-polymers with Fischer Tropsch olefins obtained by slurry polymerization of ethylene with Fischer-Tropsch derived olefins using standard polymerization conditions. [1-6] Application properties of some of these polymers having molecular weight and polydispersity index similar to commercial polymers are here reviewed. Ethylene/1-pentene co-polymers are well known. Data are also available in the literature including some of our investigations. [1-2] They have certainly a good balance of properties between ethylene co-polymers with 1-butene and 1-hexene. In the application margins they are at the lower limit where clusters may be identified by NMR techniques.

Ethylene/1-heptene co-polymers have balanced properties between those of ethylene co-polymers with 1-hexene and 1-octene. Co-polymerization rate, branch distribution and crystallizable sequence length average in this class of higher α-olefin make the co-polymer attractive for applications that cover a large field.

Synthesis and application of ethylene co-polymers with 1-heptene have been previously described. [3] Performance range for some of these polymers is here summarised:

- Izod Notched Impact Strength I > 10[C_7]
- Tensile Strength @ Yield (MD) J > -4.4[C_7] + 17

- Modulus $E > -275[C_7] + 850$
- Hardness $H > -10[C_7] + 56$

Ethylene/1-nonene co-polymers are somewhat at the upper margins of application due to their lower co-polymerization rate. However, they have the highest impact strength of ethylene copolymers with C3-C9 α-olefins. Synthesis and application of ethylene co-polymers with 1-nonene have been previously described [3] and performance range is here summarised:

- Izod Notched Impact Strength $I > 13.3[C_9]$
- Tensile Strength @ Yield (MD) $J > -16.67[C_9] + 25$
- Modulus $E > -667[C_9] + 1100$
- Hardness $H > -30[C_9] + 65$

Ter-polymers are a higher step in comonomer application. This was illustrated for co-polymers of ethylene with 1-pentene and other olefins.[1] Synthesis and application of ethylene ter-polymers with 1-pentene and other α-olefins have been previously described.[4-5] Performance is here summarised for some of the polymers described:

- Impact Strength > 60 g
- Tear Strength (MD) > 12 g·μm^{-1}
- Tear Strength (TD) > 3 g·μm^{-1}
- Tensile Strength @ Break (MD) > 25 MPa
- Tensile Strength @ Break (TD) > 25 MPa
- Tensile Strength @ Yield (MD) > 15 MPa
- Tensile Strength @ Yield (TD) > 14 MPa

Ter-polymers of ethylene with branched and linear α-olefins are even more diverse, and this diversity increases further when two branched olefins are employed as co-monomers. Synthesis and application of ethylene ter-polymers with branched and linear α-olefins have been previously described.[6] Performance is here summarised for ethylene ter-polymers with 4- Methyl-Pentene-1 and 3-Methyl-Pentene-1:

- Tensile Strength @ Yield (MD) $J > 240 \rho - 212.4$
- Modulus $E > 700/0.6\rho - 10\,500$
- Impact Strength > 50

Catalysts Used in Fischer-Tropsch Olefin Polymerization

Fischer-Tropsch olefins can be polymerized practically with all known catalysts that polymerise ethylene. However, as for all other catalysts employed in olefin polymerization, it is desirable for the catalyst to have high productivity, easy architecture and polydispersity control, resistance to poisoning and easy incorporation of higher α-olefins. A method of preparing such a catalyst was previously described[7]. A particular support was prepared from partially hydrated magnesium chloride. In supporting titanium tetrachloride, an attempt was made to control the titanium oxidation state. The preparative steps are reviewed:

Step 1 - Ether

- 0.3 to 3 moles per mol $MgCl_2$
- total number of carbon atoms greater than 8

Step 2 – Alkyl Aluminium

- 1 to 6 moles per mol $MgCl_2$
- total number of carbon atoms between 1 and 10

Step 3 – Alcohol

- 0.4 top 4 moles per mol $MgCl_2$
- total number of carbon atoms between 2 and 10

Step 4 – TiCl₄

- 0.5 to 20 moles per mol $MgCl_2$
- wash with saturated hydrocarbons

Catalyst Properties

- Aluminium content: 0.1 to 6 wt%
- Magnesium content: 3 to 15 wt%
- Titanium content: 3 to 15 wt%

Given the complexity of such systems, it is important to develop a fundamental understanding of the catalyst behavior before further work on Fishcher-Tropsch olefin can be attempted. The following section describes the attempts to develop a platform from which a model of the system could be developed.

Modelling the Ziegler-Natta Catalyst System

An investigation was conducted into the oxidation state variation of the active titanium centres in this Ziegler-Natta catalyst for the polymerization of ethylene in the slurry phase. It was found that catalyst prepared according to the method described having titanium at different oxidation states exhibits different activities.[8] This is in line with literature data that discuss the relationship between titanium oxidation state and catalyst activity.[9-13] It is therefore believed that by modelling titanium oxidation state, the catalyst activity may be predicted.

The model of catalyst sub-sites developed by Rawatlal and Starzak [14, 15] was employed in fitting experimental data for the slurry phase homopolymerization of ethylene over a catalyst prepared according to the method described earlier. It was found that the model fit the experimental data quite well in terms of magnitude as well as general trend for the polymer yield and average chain length. Two polymerizing sites are required to fit the data accurately, which is consistent with the experimental observations of previous researchers. The rate equation describing primary site dynamics was written according to Hutchinson [13] by lumping the effect of site transformations which occurs by the various reaction mechanisms [14] into a single parameter β_{st}: [13]

$$\beta_{st}^{q_m q_n} = k_{st,Sp}^{q_m q_n} + \sum_i k_{st,M_i}^{q_m q_n} [M_i] + k_{st,H}^{q_m q_n} [H] + k_{st,Al}^{q_m q_n} [Al] + k_{st,E}^{q_m q_n} [E] \qquad (1)$$

where the $k_{st,i}^{q_m q_n}$ is the rate of transformation from sites of type q_m to q_n as a result of reaction with component-i. Accordingly, the net rate equation on site q_m becomes:

$$\left(\frac{d}{dt} P_*^{q_m} \right)_{rxn} = \sum_n \left(\beta_{st}^{q_n q_m} P_*^{q_n} - \beta_{st}^{q_m q_n} P_*^{q_m} \right) \qquad (2)$$

where $P_*^{q_m}$ is the concentration of catalytic sites of type q_m, usually given on a mol-Ti/g-catalyst basis. Assuming progressive reduction of the sites, this general equation may be re-written in context with the catalyst:

$$\frac{d}{dt} P_*^{4+} = -\beta_{st}^{4+,3+} P_*^{4+}$$

$$\frac{d}{dt} P_*^{3+} = \beta_{st}^{4+,3+} P_*^{4+} - \beta_{st}^{3+,2+} P_*^{3+}$$

$$\frac{d}{dt} P_*^{2+} = \beta_{st}^{3+,2+} P_*^{3+} - \beta_{st}^{2+,1+} P_*^{2+} \qquad (3)$$

Eq. 3 applies in particular to well-mixed batch reactor systems in which mass transfer effects can be ignored. This result is valuable in the parameter estimation procedure, which typically involves data obtained under batch conditions.

Adopting a model of catalyst sub-sites [14, 15] the distribution around a single oxidation state q_m is given by equation (4) and the overall sub-site distribution by equation (5):

$$g_m(t,s) = \frac{f_m(t)}{\delta_m \sqrt{2\pi}} \exp\left(-\frac{1}{2}\left[\frac{s-q_m}{\delta_m}\right]^2\right) \tag{4}$$

$$g_T(t,s) = \sum_m \frac{f_m(t)}{\delta_m \sqrt{2\pi}} \exp\left(-\frac{1}{2}\left[\frac{s-q_m}{\delta_m}\right]^2\right) \tag{5}$$

where

$$f_m(t) = \frac{P_*^{q_m}(t)}{\sum_n P_*^{q_n}(t)} \tag{6}$$

s is the sub-sites index, and δ_m is the sub-sites distribution factor. It is assumed that δ_m does not vary from site to site. The propagation rate constant also varies continuously with sub-site index by a normal distribution in the pre-exponential factor. An Arrenhius relationship is assumed between the kinetic rate constant k_p and the temperature. It is suggested that propagation rate constant distribution spread parameter δ_m be identical for all primary site types m and assumed that activation energy, E_m, is independent of sub-site type but dependent on primary site type. The monomer consumption rate may then be determined:

$$\left(\frac{d}{dt}M\right) = n_{Ti}M(t)\int_{-\infty}^{\infty} k_p(t,s)g_T(t,s)ds \tag{7}$$

where the distribution in the propagation rate constant is given by:[16,17]

$$k_p(s,T) = \sum_m \left[\frac{\overline{A}_p^m}{\delta\sqrt{2\pi}} \exp\left(-\frac{1}{2}\left[\frac{s-q_m}{\delta_{m,p}}\right]^2\right) \exp\left(-\frac{E_m}{RT}\right)\right]$$

$$= \sum_m \left[\frac{\overline{A}_p^m}{\delta\sqrt{2\pi}} \exp\left(-\frac{1}{2}\left[\frac{s-q_m}{\delta_{m,p}}\right]^2 - \frac{E_m}{RT}\right)\right] \tag{8}$$

In total, there are seven parameters to be estimated, namely three site transformation parameters, β_{st}^{4-3}, β_{st}^{3-2}, β_{st}^{2-1}, the sub-sites distribution parameter δ, pre-exponential factors

A_2, A_3, and the number titanium atoms exposed on the catalyst surface, $n_{Ti,s}$.

These parameters were determined using the data obtained from experiments conducted in a slurry phase reactor system charged with heptane and the catalyst. Homopolymerization runs, employing ethylene as monomer, were conducted at a temperature of 80^0C. Hydrogen was employed as a chain termination agent, and TEA as co-catalyst.

The primary measurable of interest was the polymer yield. This was estimated by removing polymer from the reactor and measuring the mass change. The initial oxidation state composition of the catalyst used was found to be 95.734% Ti^{4+}, 4.159% Ti^{3+}, and 0.107% Ti^{2+}.

Results of the regression performed using the model described earlier are presented below. The regression yielded parameters that lie within acceptable limits outlined in the literature.

Table 1. Parameter List.

Parameter	Description	Value	Units
$n_{Ti}M$	Product of moles of titanium and monomer concentration	3.4998×10^{-5}	mol-Ti . mol-C2.m^{-3}
δ	Subsites distribution variance	0.1237	mol-Ti . mol-C2.m^{-3}
A_1	Pre-exponential factor, 3+ site	6793	mol-Ti^{-1}.s^{-1}
A_2	Pre-exponential factor, 2+ site	254210	mol-Ti^{-1}.s^{-1}
$\beta_{st}^{4 \to 3}$	Site transformation parameter, Ti 4 to 3+	7.123×10^{-3}	s^{-1}
$\beta_{st}^{3 \to 2}$	Site transformation parameter, Ti 3 to 2+	0.4176×10^{-3}	s^{-1}
$\beta_{st}^{2 \to 1}$	Site transformation parameter, Ti 2 to 1+	0.6278×10^{-3}	s^{-1}
β_T^1	Termination parameter, site 3+	1.900×10^{-7}	s^{-1}
β_T^2	Termination parameter, site 2+	0.4839×10^{-7}	s^{-1}

Data fits are shown in Figure 1. It was found the best fit was obtained when two polymerizing primary sites were used with a low sub-sites dispersion. It was expected that two primary sites would be required since it is known that ethylene polymerizes both on the 2+ and 3+ oxidation state of titanium. The low sub-sites dispersion index (0.1237) is understood in the context of the model of catalyst sub-sites,[14,15] to mean that the variation in primary oxidation state, which affects the types of sites that are polymerizing, contributes to shape of the yield profile.

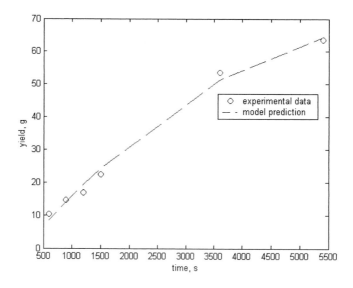

Figure 1. Polymer Yield Profile.

Conclusion

The Fischer-Tropsch process offers the polyolefin industry some less explored possibilities such as odd carbon number olefins and branched olefins, making possible new routes to new polymers. While investigation of these polymers are only at beginning and laborious work is needed to identify all opportunities, work has been initiated in this field by incorporation for a kinetic framework capable of incorporating the effects observed thus far.

The review presented demonstrates that ethylene co-polymers with odd carbon number olefins have in general a balance of properties of the co-polymers obtained with the even carbon number olefins having one carbon unit higher or lower. They extend margin of application of ethylene co-polymers with even carbon number olefin. Ethylene co-polymers and ter-polymers with branched olefins extend the margin of application even more and property change is sharper.

Fischer-Tropsch olefins can be polymerized practically with all known catalysts that polymerise ethylene. A rather robust, high productivity catalyst that is capable of high co-monomer incorporation was developed in this study. A particular support was prepared from partially hydrated magnesium chloride. In supporting titanium tetrachloride, the titanium oxidation state was controlled by the ratio of the alcohol to tri-ethyl-aluminium during catalyst preparation. It is believed that different active centres control olefin incorporation, with titanium oxidation state being associated with catalyst activity. Before incorporating this effect into the co- and ter-polymer systems, an attempt was made to model the homopolymerization of ethylene using a sub-sites model. This kinetic framework developed provides a platform for future studies in this interesting field.

[1]   I. Tincul, D., Joubert *Polyolefins 2 ACS Symposium on Polyolefins*, Napa, CA, **1999**.
[2]   I. Tincul, D. Joubert *IUPAC Symposium Stellenosch*, South Africa **2000**.
[3]   I. Tincul et al. *WO 0032657*.
[4]   D. Joubert et al. *WO 9638485*.
[5]   I. Tincul et al. *WO 9745454*.
[6]   I. Tincul et al. *WO 986441*.
[7]   D. Joubert et al. *WO 9745460*.
[8]   I. Tincul et al. *Ziegler Natta Symposium*, Sorrento, Italy, 2002.
[9]   Soga K, Chen S and Ohnishi R, *Polym Bull*, **1982**, *8*, 473.
[10] G. C. Han-Adebekun, M. Hamba, W. H. Ray *J Appl Polym Sci A: Polymer Chemistry,* **1997**, *35*, 2063.
[11] G. C. Han-Adebekun, J. A. Debling, W. H. Ray *J Appl Polym Sci*, **1997**, *64*, 373.
[12] G. C. Han-Adebekun, W. H. Ray *J Appl Polym Sci*, **1997**, *65*, 1037.
[13] R. A. Hutchinson, C. M. Chen, W. H. Ray *J Appl Polym Sci*, **1992**, *44*, 1389.
[14] R. Rawatlal, M. Starzak *Chemical Technology*, March **2001**.
[15] R. Rawatlal *PhD Thesis*, in preparation, to be submitted **2003**.

Propylene Polymerization in Liquid Monomer with a Ziegler-Natta Catalyst in Bench Scale Reactors. Kinetics and Model to Predict Reaction Yields

J.L. Hernández Vaquero, G. Morrón Lingl, J.M. Perea Gaitán, *
L. Vargas Fernández

Dirección de Tecnología de Repsol-YPF, Carretera de Extremadura N-V km 18.
28930 Móstoles – Madrid, Spain
E-mail: jpereag@repsolypf.com

Summary: The study of new catalytic systems is critical in order to develop improved and cost effective polymerization processes. One of the methods to evaluate the performance of a catalytic system is by means of bench scale reactors. Despite its difference in size with large scale industrial plants, bench scale reactors have proved to be a valuable tool to understand the behavior of the catalytic system during the polymerization. In this work, a method to estimate the kinetic parameters of propylene polymerization over a conventional Ziegler Natta catalyst is evaluated. Thus, it was possible to set up a semi-empirical model to correlate the reaction yield with the polymerization time, the hydrogen content in the reactor and the reaction temperature. This model proves to be useful to evaluate the performance of a catalytic system within the range of normal operating conditions. A brief study on the particle size distribution of the products is also carried out.

Keywords: kinetics (polym.); poly(propylene) (PP); Ziegler-Natta polymerization

Introduction

Since their discovery in the early 1950s, Ziegler-Natta catalysts have centered an intense research activity because of their role in olefin polymerization.[1] Even if these catalysts have been deeply studied over the past decades and that some basic polymerization principles are widely accepted, no unequivocal polymerization mechanism has yet been devised to fully describe their behavior.[2] This is due to the complexity of the catalyst systems employed. Nevertheless, an approach of the kinetic aspects of the polymerization reaction is suitable in order to predict the evolution of the reactive system. It is also important to consider the

© 2004 WILEY-VCH Verlag GmbH & KGaA, Weinheim   DOI: 10.1002/masy.200450933

polymerization kinetics when trying to understand the chemicals processes, either at laboratory scale or large-industrial scale.

There are several ways to evaluate the behavior of a Ziegler-Natta catalytic system during the polymerization process. One of them is by means of small bench scale reactors (1 to 5 liter reactors, so called autoclaves). Because of their size and cost, they are more accessible for research laboratories. Even if they work in a discontinuous way and need special operation and care to provide just a few grams of polymer per batch, bench scale reactors are a very effective tool when trying study the impact of process conditions on the performance of a catalytic system.

Downscaling the process has often been considered to be a difficulty. This is due to the polymerization itself and to its related process conditions. Because of the high exothermicity of the reaction, the equipments need a good temperature regulation system. Because of the extreme sensibility of the catalyst to traces of impurities (such as H_2O, O_2, etc.) any experimental study must be carried out with hermetic inert systems, in order to maintain the high activity of the catalyst. Nevertheless, if used accurately, they prove to be extremely effective and coherent tools for catalyst evaluation.

In this work a bench scale reactor is used to study the behavior of a $MgCl_2$ supported Ziegler-Natta catalyst. Based on the polymerization kinetics, the experimental results lead to the development of a semi-empirical predictive model that correlates the polymerization yield with some basic process conditions: Polymerization time, Temperature, and Hydrogen content in the reactor. A brief study on the particle size distribution of the products is also carried out.

Fundamentals

Polymerization Kinetics

Many authors have studied the kinetic behavior of the Ziegler Natta catalysts during olefin polymerization.[3] Previous works report kinetic models based upon the reaction mechanisms.[4] It must be reminded that the polymerization kinetics of propylene by means of Ziegler Natta catalysts are complex. The activation mechanisms of the catalytic centers by the cocatalyst, and its complexation with internal and external electron donors are still not fully clarified.[5] The presence of multiple catalytic sites must be taken into account, each site having its own propagation rate constant and chain transfer constant.[6] Further more, the catalytic active centers decrease in number during decay because of chemical deactivation.[7] Nevertheless, it is possible to have an approach to the overall kinetic rate by making some assumptions, taking only into account the determining steps of the polymerization.[8] Even if the reaction kinetics are complex, they are limited by the polymer propagation, and the catalyst decay. This last one takes into account the decrease of the number of active sites during the reaction. Thus, through this simplification, it is possible to consider that these two steps describe most of the system kinetics.

Table 1. Kinetic Models as a function of the decay order.

Order (n)	$-\dfrac{dV}{dt}$	$V(t)$	$Y(t)$
1.0	$K_d V$	$V = V_o e^{-K_d \cdot t}$	$Y = \dfrac{V_o}{K_d}(1 - e^{-K_d t})$
1.5	$K_d V^{1.5}$	$V = \dfrac{V_o}{(1 + \dfrac{V_o^{1/2} \cdot K_d}{2} \cdot t)^2}$	$Y = \dfrac{V_o \cdot t}{(1 + \dfrac{V_o^{1/2} \cdot K_d}{2} \cdot t)}$
2.0	$K_d V^2$	$V = \dfrac{V_o}{(1 + V_o \cdot K_d \cdot t)}$	$Y = \dfrac{1}{K_d}\ln(1 + V_o K_d t)$
n	$K_d V^n$	$V = \left[V_o^{1-n} + K_d(1-n) \cdot t\right]^{\frac{1}{1-n}}$	$Y = \dfrac{1}{K_d(n-2)}\left[\left(V_o^{1-n} + K(n-1)t\right)^{\frac{2-n}{1-n}} - V_o^{2-n}\right]$

The single propagation rates of the multiple sites are gathered into a single propagation rate. The overall kinetic rate (V) is then given by the following expression:

$$V = k_p[C_m].[C^*] \qquad (1)$$

Where k_p is the overall propagation constant, C_m is the concentration of monomer and C^* that of the active sites. Many studies have confirmed that the polymerization rate has a 1st order dependence in respect to the monomer concentration.[8-9-10-11-12]

In the same way, it may be considered that the catalyst decay through different chemical mechanisms is summed up by a single deactivation rate. By this way the decay of the catalyst is described by a decreasing number of active sites with time, according to the following expression.[8-9-10-11-12]:

$$-\frac{dC^*}{dt} = K_D \cdot (C*)^n \quad (2)$$

The combination of equations (1) and (2) gives a general expression that takes into account the decay of the catalytic reaction with time:

$$V' = \frac{dV}{dt} = K_d \cdot V^n \quad (3)$$

Where K_d is defined as follows:

$$K_d = \frac{K_D}{K_p^{n-1}} \quad (4)$$

In these expressions, K_D is the deactivation constant and n is the catalytic decay order, V' is decay rate (g PP /g cat.min^2) and V is the polymerization rate (g PP/g cat.min).

Integration of (3) leads to the expression for the polymerization rate (V) against time (t), where V_0 is the initial rate for undecayed catalyst. A second integration leads to the expression of the reaction yield (Y) as a function of polymerization time. The mathematical form of these models changes in relation with the value of the decay order (n), and the results are summarized in Table 1.

The results of the polymerization essays as a function of time are the clue to determine which model fits best to real experience, and thus find out which is the decay order of the catalyst.

Experimental

The polymerization essays were carried out in an autoclave system (Figure 1). The equipment is composed of three sections. The first one corresponds to the raw materials storage and to the purification systems for the streams going to the reactor. The second section includes the measurement vessels and accessories that are used to introduce the catalyst into the reactor. The third section is the reactor itself. With a volume of 4 liters, the reactor is surrounded by a jacket that helps to maintain a constant reaction temperature by means of a cooling control system.

Before beginning the essays, a large amount of the same batch of propylene (polymer grade) and catalyst is stored in order to ensure a constant quality of the raw materials during the hole experimental set up. The raw materials are purified as they are forced to go through molecular sieves and alumina that remove traces of catalyst poisons (such as O_2, CO, H_2O, etc...).

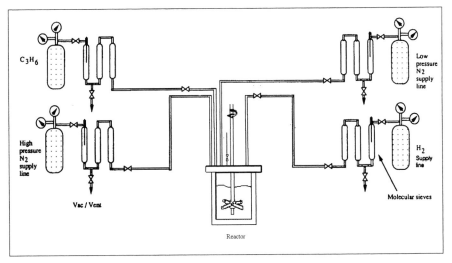

Figure 1. Autoclave system for propylene polymerization.

The propylene polymerization essays were carried out in a liquid propylene (pool) medium. The catalytic system was a conventional Ziegler-Natta catalyst, based on $TiCl_4$ and an internal donor, supported on $MgCl_2$. The use of triethyl-aluminium (TEA) as the catalyst activator, and of an external electron donor to control the polymer stereoregularity, is needed.

Because of the extreme sensibility of the catalyst system to a large range of impurities (poisons), all the operations are done under an inert atmosphere of nitrogen. The reaction operations are performed in successive stages.

At first, fixed amounts of TEA, external donor and catalyst are introduced in an appropriate vessel, in the presence of small amounts of an organic solvent. A precise amount of minutes is then given to the compounds to let them interact and to form the catalytic complex.

There after, the catalytic complex is introduced into the reactor altogether with the amounts of hydrogen and propylene required for the reaction Then the temperature and the pressure are risen until the reaction conditions are reached. The elapsed time of the temperature rise is controlled, and always lasts the same amount of time. The reaction is then maintained during a carefully controlled lapse of time, and it is stopped by turning the temperature down and flashing out the unreacted monomer. The polymer is then recovered from the reactor, dried in an oven and weighed to calculate the amount of polymer produced per gram of catalyst (reaction yield).

During the first part of this work, a set of essays was carried out in order to set the most suitable working conditions. Based on these preliminary results, the most appropriate TEA/Catalyst, TEA/Donor and Catalyst/Donor ratios were used during the rest of the experimental setup. In the second series of essays, the influence of three basic operation variables on the polymerization yield was evaluated, carrying out the essays within a range of the reaction time (from 30 to 120 min), a range of hydrogen content in the reactor (from 0 to 9.5 bars) and a range of temperature (from 50 °C to 80 °C).

Results

Setting the operation conditions

In the first series of essays, a set of reactions were carried out to determine the most suitable conditions for the polymerization reaction. The overall results are shown in Table 2.

Table 2. Operating Conditions (Standard Values).

Bench Reactor Volume	4 liters
Catalyst	20 mg.
Al / Ti molar ratio	1000
Al / External Donor molar ratio	12
Stirring Speed	550r.p.m.
Reaction Time	120 min.
Reaction Temperature	70 °C
Hydrogen Partial Pressure	3,5 bar

Stirring Speed

Because of the nature of the reactive system, based upon polymerization in liquid propylene, problems related to external diffusion of the monomer towards the catalyst active sites are not expected. Nevertheless, the stirring speed in the reactor has been evaluated in order to select a value that ensures a correct dispersion of the catalyst in the reactor, avoiding high shear values that could affect the polymer particle distribution. Three essays were carried out with different stirring speeds (Figure 2). For low speed values, the low polymer yield might be due to a bad dispersion of the catalyst in the reactor. For high speed values, the yield is almost unchanged. In order to avoid fragmentation of the resulting polymer particles due to shear stress at high speed values, the value 550 rpm is selected as a standard for all the essays to be done further on.

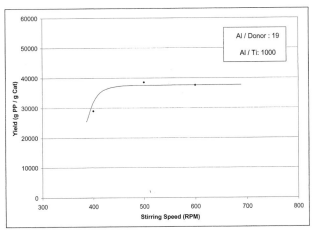

Figure 2. Influence of the stirring speed.

Catalyst content

The amount of catalyst fed into the reactor is limited by the volume of the reactor. It is possible to produce just a few grams of polymer in order to avoid build up and homogeneity problems during the reaction. The essays carried out to determine the amount of polymer to be fed into the reactor (Figure 3) shows that 12 mg is the most suitable quantity of catalyst to be fed into the reactor.

Al/Ti Ratio

TEA has an important effect over the polymerization kinetics. TEA is not only believed to act as the catalyst activator, but also as a scavenger, reducing the impact of the presence impurities that would poison the catalyst. A series of essays were carried out using Al/Ti ratios from 450 to 1200 (Figure 4). From the results, it can be seen that the polymerization yield reaches a maximum when the Al/Ti molar ratio ranges values close to 900. This value is selected as standard for all the essays to be done further on.

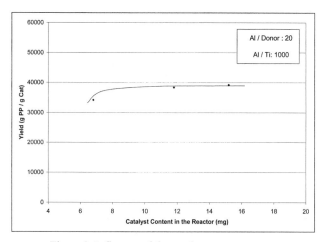

Figure 3. Influence of the catalyst content.

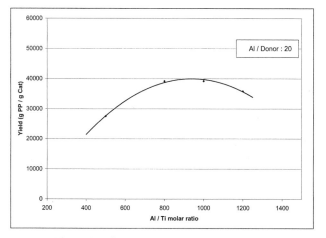

Figure 4. Influence of Al/Ti molar ratio.

Al/External Donor Ratio

During the olefin polymerization with Ziegler Natta catalysts, external donors are generally employed. These are believed to react with non-stereospecific sites, and deactivating them.

By this way, the presence of an external donor within the catalytic system increases the stereospecificity of the reaction and affects the polymerization kinetics. Thus, as the donor content increases the polymer isotacticity increases too, but the catalyst activity decreases. As it can be seen in figure 5, the catalyst yield increases with increasing Al/Donor ratios. From these results, the value 20 is selected as a standard condition for the Al/Donor molar ratio.

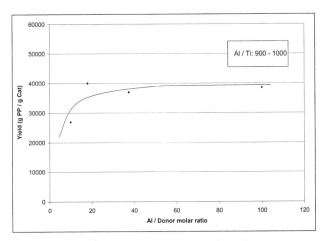

Figure 5. Influence of Al / Donor molar ratio.

Influence of the polymerization conditions

In order to evaluate the influence of the operation conditions on the reaction yield, during the second series of essays, the operating variables are studied one by one within a certain range as the rest of conditions are set to the values of the standard conditions shown in Table 2.

Influence of the reaction time

Several polymerizations were carried out, measuring the reaction yield for a set of reaction times. The experimental results show how the reaction yield increases as the polymerization rate decays with reaction time. The essays were repeated under the same conditions for several hydrogen contents in the reactor. By means of statistical correlations, it was possible to determine that the experimental results fit the best to the non-linear function of the reaction yield with a decay order n=1. This conclusion is coherent for all the results, for the whole range of hydrogen contents in the reactor. The correlation factor (R2>0.9950) gives an estimate of how good the adjustments are. Figure 6 shows the plot of the reaction yield against time for the experimental data and for the mathematical model. The percentage deviations of the experimental points with the mathematical model are in this case below 5%.

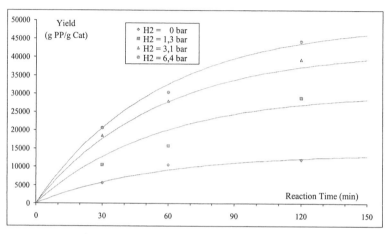

Figure 6. Experimental Results (Yield vs. Time).

These results are also in accordance with the results of previous works.[8-9-10-11-12] Obviously, the decay order for the Ziegler Natta catalysts has been a subject of debate for quite a long time. Even if some authors report that the reaction yield model fits properly to a decay order n=2 or even n=1.5 for reaction times above 2 or 3 hours, from the present results it is possible to see that the decay order n=1 fits properly when the reaction times are below 120 minutes. Here, it is necessary to point out that the decay of the catalyst is an overall parameter that expresses the interaction of several elementary reactions. By that way the value of the decay order does not reveal the basic mechanisms of the catalyst decay.

Influence of the hydrogen content in the reactor

It is well known that the catalyst activity increases with higher hydrogen content in the polymerization reactor.[13] From the experimental runs (Figures 7 and 8), it was possible to determine that the value of the deactivation constant is not significantly affected by the level of hydrogen content in the reactor. This result confirms an assumption undertaken in a previous work.[14] that the hydrogen content in the reactor had little influence on the deactivation constant. On the contrary, the initial polymerization rate grows with the hydrogen content in the reactor, until an upper limit is reached for values above 6 bar (Figure 7). This might be attributed to a reactivation of dormant sites by hydrogen. This is reflected in the previous model by an increase of the initial polymerization rate with higher hydrogen contents, which may be introduced by the following empirical expression:

$$Vo = A\left(1 - Be^{-C \cdot P_{H_2}}\right) \qquad (5)$$

Where P_{H_2} is the hydrogen content in the reactor (bar) and A, B, and C are specific constants. The non-linear correlation of the model with the results gives the values of the model parameters, while the correlation factor allow us to estimate the accuracy of the model ($R2 > 0.997$).

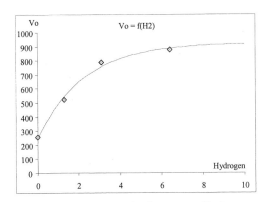

Figure 7. Initial polymerization rate vs. Hydrogen content.

Influence of the reaction temperature on the polymerization yield

From the experimental essays, it can be seen how the catalyst activity increases with increasing values of the reaction temperature (Figure 8).

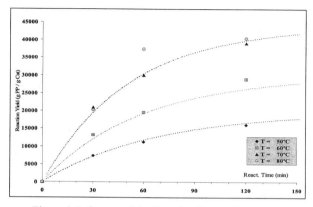

Figure 8. Influence of the Reaction Temperature.

The reaction temperature is assumed to affect both kinetic terms, the initial polymerization rate and the deactivation constant (Vo and K_d), according to the Arrhenius laws:

$$Vo = A_o e^{\frac{-E_a}{RT}} \qquad (6)$$

$$K_d = A'_o e^{\frac{-E'_a}{RT}} \qquad (7)$$

Where Ea and Ea' are the activation energies for each the kinetic parameters.

Figure 9 shows the plot of ln(V) versus the inverse values of the absolute reaction temperature. On the same way, Figure 10 shows the plot of ln(K_d) versus the inverse values of the absolute reaction temperature. In both cases, the values fall nearly on a straight line for reaction temperatures below 75 °C. But in both cases, the values for the reaction temperature of 80 °C are slightly off the Arrhenius plot. This behavior might be due to increasing diffusional effects at the highest temperatures.

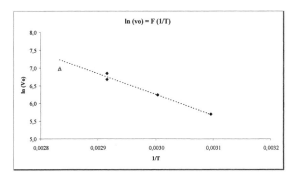

Figure 9. Arrhenius plot of Vo vs. 1/T.

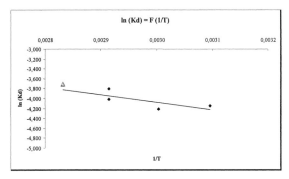

Figure 10. Arrhenius plot of K_d vs 1/T.

As a global result, the polymerization yield is then given by the following expression

$$Y = A\left(1 - Be^{-C[H_2]}\right)\frac{\left(1 - e^{-K_d \cdot t}\right)}{K_d} \qquad (8)$$

Where A, B, C are specific constants, P_{H2} is the hydrogen content in the reactor (bar), K_d the deactivation constant of the catalytic system and **t** the reaction time (min).

In Figure 11, the plot of the experimental Yields is given in relation to the ones estimated by the model for all of the series of essays. As it can be seen, the model fits to the experimental results with high accuracy, with percentage deviations below 10%.

380

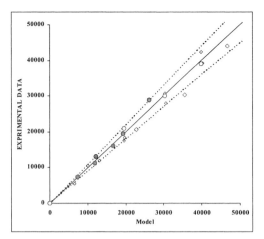

Figure 11. Correlation model vs. experimental results.

Replication phenomenon

Since the first industrial developments, several theories have been established in order to explain the morphology of the Ziegler Natta catalysts and its reaction products.[15-16] As a matter of fact, the final shape and size of the polymer powder depends on how it expands around the catalyst grain: as the reaction proceeds, the polymer particle grows and replicates the physical characteristics of the catalyst particle at a larger scale (shape, texture, particle size distribution, etc.). This is known as the replication phenomenon. Even if the particle growth mechanisms are complex, experimental evidence demonstrates that, during the first polymerization steps, the catalyst particles disrupt into a larger number of small fragments. Even though they are segregated, the catalyst fragments are hold together by the growing polymer that acts as a binder, while being uniformly dispersed inside the polymer particle. As a consequence, as the polymer particle grows around the catalyst particle, the fragments expand outwards, maintaining the same physical shape and characteristics of the initial catalyst particle (Figure 12).

The polymer particle size depends on the initial catalyst particle size, the reaction yield (i.e. the quantity of polymer formed), and the catalyst geometric parameters (volumetric factors).

The very first seconds of the polymerization are critical on what concerns the replication phenomenon. In this phase the catalyst particles are broken apart, and the polymer is formed around them. If the heat that the reaction generates is not properly evacuated and the catalyst activity and mass flow at the particle level are not carefully controlled during these first steps, the polymer fragments are more likely to break, and the replication phenomenon does not occur. Thus, if the prepolymerization conditions are severe, the catalyst particles might segregate in independent particles and not be bounded. In order to observe the replication phenomenon, a prepolymerization should be carried out under mild conditions.

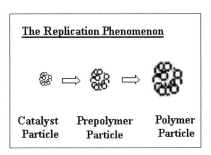

Figure 12. The replication phenomenon.

From the series of essays carried out to study the influence of the polymerization time, the initial catalyst particle size distribution, and the final polymer size distribution were evaluated. As it can be seen from Figure 13, the polymer particle size grows with the increase of the polymerization time. This is due to the increase of the reaction yield with the reaction time, reaching an upper limit value for the largest polymerization times. This behavior is in accordance with the model developed when the decay order is n=1.

It is thus possible to develop a method to asses the final particle size as a function of the initial catalyst size distribution, the reaction parameters and the catalyst physical characteristics. On the same way, it is possible to determine the prepolymer size distribution when the final polymer particle size distribution and the other reaction parameters are known.

In Figure 13, the experimental results correspond to the catalyst size distribution and to the final polymer particle size distribution. By means of a mathematical model, the prepolymer size distribution can be calculated from the final polymer size distribution. As it can be seen in the figure, the prepolymer particle size distribution is the same for all of the essays, meaning that the replication phenomenon takes place, all the products being a replica of the same prepolymer. Nevertheless, the prepolymer particle distribution does not match the one of the catalyst. This is likely to be due to fragmentation during the initial polymerization steps. As mentioned before during the experimental set up, the polymerization begins directly with a controlled increase of the temperature from 25 °C to 70 °C. In order to obtain a more accurate catalyst replication, the initial prepolymerization step should be studied, employing milder conditions.

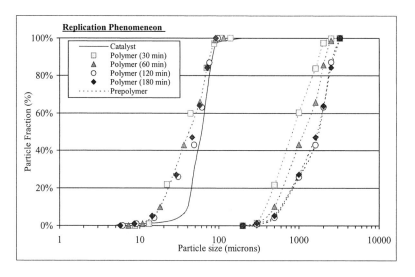

Figure 13. The replication phenomenon.

Conclusions

This work allowed us to prove the suitability of liquid pool propylene polymerization in bench scale reactors to investigate the reaction kinetics. After selecting a set of optimum reaction conditions, and evaluating the impact of the Al/Ti and the Al/Donor molar ratios on the reaction yield, it was possible to model the kinetic behavior of a highly active TiCl$_4$/MgCl$_2$ catalyst with an expression where the initial reaction rate and the deactivation constant are the key variables. From the evaluation, it was possible to determine that the reaction yield increases as the polymerization rate decays with reaction time, a behavior is typical from many Ziegler Natta catalysts. Within the range of time considered, the experimental results fit properly to a kinetic model where the overall deactivation order is n=1. It was also possible to see that the initial polymerization rate grows with the hydrogen content in the reactor, until an upper limit is reached for values above 6 bar. The presence of hydrogen strongly increases the initial activity of the catalyst, by increasing the initial polymerization rate. Within the range of polymerization conditions, the reaction yield may be then given by the following expression:

$$Y = A\left(1 - Be^{-C \cdot P_{H_2}}\right)\frac{\left(1 - e^{-K_d \cdot t}\right)}{K_d}$$

The model predicts the reaction yield as a function of three basic operating conditions: the reaction time, the hydrogen content in the reactor, and the reaction temperature.

In this work it was also possible to study the catalyst replication phenomenon and see that the polymer particles not only replicate the shape and texture of the initial catalyst, but also its particle size distribution.

[1]   E.P. Moore Jr. in "Polypropylene Handbook", E.Moore. Ed. Hanser. **1998**.
[2]   E. Albizzati, U. Giannini, G. Collina, L. Noristi, L. Resconi in *"Polypropylene Handbook"* E.Moore. Ed. Hanser. **1998**
[3]   Peña García B.. Univ. Autónoma de Madrid. Inst. Ciencia y Tecnología de Polímeros. **1992**. *Tesis doctoral*.
[4]   P. Tait, N.D. Watkins, *Comprehensive Polymer Science*, Pergamon Press Ed., **1989**, Vol *4*. pp 533-573.
[5]   I. Kim, S.I. Woo, *Poly. Bull.*, **1990**, *23*, 35.
[6]   L.L. Böhm, *Polymer*, **1978**, *19*, 545.
[7]   Y. Doi, T. Kei, E. Suzuki, M. Tamura, *Macromol. Chem.*, **1982**, *183*, 2285.
[8]   N.F Brockmeier, J.B. Rogan, *Ind. Eng. Chem. Prod. Res. Dev.*, **1985**, Vol *24*, Nº2, pp278-283.
[9]   P. Galli, **1982** *Proceedings IUPAC Macro '82, 28th Macromol. Symposium*, Amherst, MA, July 12-16.
[10] P. Galli, L. Luciani, G.Cecchin, *Angew. Macro. Chem.* **1981**, *94*, 63-89.
[11] I. Kim, H.K. Choi, J.H. Kim, S.I. Woo, *J. Poly. Sci. Part A Poly. Chem.*, **1994**, *32*, 971.
[12] J.J.C. Samson, G. Weickert, A.E. Heerze, K.R. Wesrerterp. *AiChe J.* **1998**, *44*, 1424.
[13] J.J.C. Samson, P.J. Bosman, G. Weickert, K.R. Wesrerterp *J. Polym. Sci.: Part A*. **1999**, *37*, 219.
[14] J.L. Hernandez Vaquero, J.M. Perea Gaitan, L. Vargas Fernández, **2000**. *ECOREP 2000 Conference Proceedings. European Conference on Reaction Engineering of Polyolefins* pp 149 – 152.
[15] Floyd, S., K.-Y. Choi, T.W. Taylor and W.H. Ray, *J. Appl. Poly. Sci.*, **1986**, *32*, 2935.
[16] L. Noristi, E. Marchetti, G. Baruzzi, P. Sgarzi, *J. Polym. Sci. Part A: Polym Chem*, **1994**, *32*, 3047.

Comparative Analysis of the Clathrate Forms of Syndiotactic Polystyrene, Poly (*p*-methylstyrene) and Poly (*m*-methylstyrene)

Vittorio Petraccone, * *Oreste Tarallo*

Dipartimento di Chimica, Università degli Studi di Napoli "Federico II", Complesso di Monte Sant'Angelo, Via Cinthia, 80126 Napoli, Italy

Summary: The crystal structures of the clathrate forms of syndiotactic polystyrene (s-PS), poly (*p*-methylstyrene) (s-PPMS) and poly (*m*-methylstyrene) (s-PMMS) containing guest molecules having widely different steric hindrance are compared in detail. Common features and differences concerning the packing of the chains, the shape and the dimensions of the cavities and the stability of the forms deprived of the guest molecules are pointed out. A new clathrate form of s-PPMS containing CS_2 is also described.

Keywords: clathrates; crystal structure; host-guest system; syndiotactic polystyrene; WAXS

Introduction

Homogeneous catalytic systems, based on titanium or zirconium compounds and methylalumoxane, have allowed producing highly syndiotactic polystyrene and substituted polystyrenes. Some of these polymers (i.e. s-PS, s-PPMS and s-PMMS) are able to co-crystallize with low molecular weight substances to form polymeric clathrates. For this class of molecular compounds some promising applications have been suggested in the field of chemical separations, in particular for water or air purification from volatile organic compounds.

A common feature of all the clathrate forms of these polymers is that the polymer chains assume a minimum energy conformation corresponding to a **s**(2/1)2 helix with a repetition period of 7.7 – 7.8 Å, whose steric hindrance can be very schematically delimited in a prism with a rectangular base. This conformation is characterized by an orientation of the benzene rings generating on two opposite sides of the prism hollows in which properly shaped molecules can nest, a fact that is itself particularly favourable to the formation of clathrate

DOI: 10.1002/masy.200450934

structures. On the other hand, as far as the final shape of the cavities and the packing of the chains in the crystalline clathrates are concerned, appreciable differences can be found among these three polymers. A further relevant difference between the clathrates of these three polymers concerns the removal of the guest molecules at temperatures lower than the glass transition temperature.

The present note analyzes the differences of the various clathrate forms of these polymers and presents structural data for a new clathrate of s-PPMS containing CS_2.

Clathrate Forms of s-PS

In the framework of this class of polymers, the clathrate forms of s-PS have been the first ones studied.[1] In particular, Chatani et al. solved the structures of the clathrate forms containing toluene[2] and iodine.[3] They also emphasized that the clathrate structures of s-PS containing other solvents (i.e. benzene, p-xylene, chlorobenzene and so on) are all similar to each other and are all characterized by centrosymmetric cavities delimited by enantiomorphous chains. Subsequently, the clathrate structure of s-PS containing 1,2-dichloroethane (DCE) has been described.[4] This clathrate form, that is nearly isomorphous to the others already described, represents an example of the extreme selectivity of this type of cavities towards the absorption of guest molecules, since the DCE molecules absorbed into the crystalline phase of the clathrate sample are only in the *trans* conformation.

Structural data for these clathrate forms of s-PS are reported in Table 1. The crystal structure parameters of a metastable polymorphic form of s-PS (δ form), that is nanoporous and can be obtained by removal of guest molecules from clathrate samples by suitable solvent and thermal treatments,[5] are also reported. This form is of particular importance, since it is able to rapidly absorb selectively some organic substances from various environments, also when present at low concentration.[6,7]

Figures 1.a and 1.b show two projections of the crystal structure of the clathrate form of s-PS containing DCE and of the δ form of this polymer, respectively.

Clathrate Forms of s-PPMS

s-PPMS is able to give clathrate structures containing a wide range of guest molecules. The clathrate structures containing *orto*-dichlorobenzene (*o*-DCB),[8] tetrahydrofuran (THF),[9] and benzene[10] have been completely described. New structural data concerning the clatrate form of this polymer containing CS_2 are presented here for the first time.

A distinctive feature of s-PPMS is the fact that it forms two different types of cavities in which guest molecules can be hosted; therefore, the terms α class and β class have been introduced in order to differentiate the corresponding types of clathrates. Figure 2 shows projections of the structures containing *o*-DCB and THF, representative of the two types of cavities. The first type of cavity characterizes the clathrate form containing *o*-DCB, corresponding to the α class. This cavity is very similar to that observed in the clathrate forms of s-PS, and is realized by two enantiomorphous helices correlated by an inversion center. The clathrate form containing THF shows instead the second type of cavity, corresponding to the β class, in which isomorphous chains correlated by a 2/1 screw axis delimit the cavities. In this class, guest molecules may interact since the cavities are not isolated, as in the case of α class cavities, and the distance between contiguous guest molecules is about 4 Å.

Table 1. Crystal structure parameters, number of guest molecules per chain, class and calculated density for a totally empty form for the clathrates of s-PS, s-PPMS and s-PMMS described in literature.

	Ref.	Space Group	a (Å)	b (Å)	c (Å)	γ (deg)	$\rho_{clathrate}$ [b]	n [c]	Class	ρ_{empty} [d]
s-PS/TOLUENE	2	$P2_1/a$	17.5	13.3	7.7	121	1.11	1	α	0.90
s-PS/IODINE	3	$P2_1/a$	17.3	12.9	7.8	120	2.05	2	α	0.92
s-PS/DCE	4	$P2_1/a$	17.1	12.2	7.7	120	1.23	1	α	0.99
s-PS δ FORM [a]	5	$P2_1/a$	17.4	11.8	7.7	117	-	-	-	0.98
s-PPMS/*o*-DCB	8	$P2_1/a$	23.4	11.8	7.7	115	1.13	1	α	0.81
s-PPMS/THF	9	$P2_1/a$	18.8	12.7	7.7	100	1.07	2	β	0.87
s-PPMS/C₆H₆	10	$C222_1$	19.5	13.3	7.7	90	1.05	2	β	0.79
s-PPMS/CS₂	-	$C222_1$	20.0	12.5	7.7	90	1.08	2	β	0.81
s-PMMS/CS₂	13	$Pcaa$	17.8	13.1	7.8	90	1.14	2	β	0.86
s-PMMS/C₆H₆	14	$Pcaa$	17.4	14.8	7.8	90	1.04	2	β	0.78
s-PMMS/*o*-DCB	14	$Pcaa$	17.3	16.4	7.8	90	1.15	2	β	0.71

(a) Crystal form obtained by removing the guest molecules from the clathrates of s-PS (see Figure 1.b).
(b) Density (g/cm³) for a crystal of a clathrate form having the cavities completely occupied by the guest molecules.
(c) Number of guest molecules per chain in a crystal of a clathrate having all the cavities occupied.
(d) Density (g/cm³) for a crystal of a clathrate form completely emptied by the guest molecules.

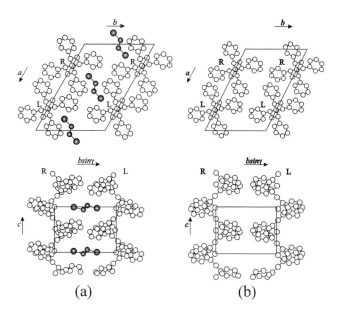

Figure 1. Schematic projections of the clathrate structure containing DCE (a) and of the pure form δ (b) of s-PS. In the lateral projection, only a couple of polymer chains, along the *a* axis, are reported. R = right-handed chain, L = left-handed chain.

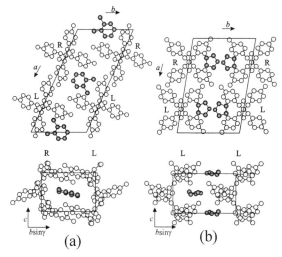

Figure 2. Schematic projections of the clathrate structures of s-PPMS containing *o*-DCB (a) and THF (b). In the lateral projection, only a couple of polymer chains, along the *a* axis, are reported. R = right-handed chain, L = left-handed chain.

The clathrate forms of s-PPMS containing benzene and CS_2 also belong to β class.

The crystalline structure of the clathrate form containing CS_2, presented here for the first time, has been deduced from a comparative analysis of the X-ray diffraction patterns of unoriented samples of the CS_2 clathrate and of the clathrate form containing benzene. The close similarity of the patterns made us assume an orthorhombic structure analogous to that already established for the clathrate form containing benzene. The position of the reflections that characterize the X-ray diffraction pattern is very well interpreted by an orthorhombic cell with axes a=20.0 Å b=12.5 Å c=7.7 Å.

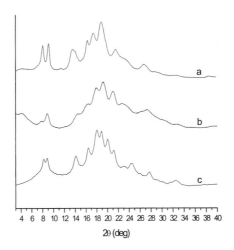

Figure 3. Experimental X-ray powder diffraction patterns of the clathrate form of s-PPMS containing benzene (a) and CS_2 (b). Calculated X-ray powder diffraction pattern for the clathrate form of s-PPMS containing CS_2 assuming a degree of crystallinity equal to 50% (c).

Through molecular mechanics calculations, assuming the same symmetry $C222_1$, we have obtained a structural model, whose simulated X-ray diffraction pattern is reported in Figure 3.c together with the experimental one (3.b). For comparison, in the same figure, the powder diffraction pattern of the clathrate form containing benzene (3.a) is also reported. The agreement between the experimental and the calculated diffraction patterns can be considered sufficiently good in order to classify this clathrate form within the clathrates of s-PPMS without making further refinements.

These last two clathrate forms show some peculiarities. First of all, although they are characterized by cavities very similar to that found, for the first time, for the clathrate with THF (and characterizing the β class), they show an orthorhombic packing of the chains. On the other hand, they represent two unusual examples of chiral crystalline phases in which all the polymer helices assume the same chirality in the lattice. Schematic projections of these structures are represented in Figures 4.a and 5.a in comparison with the corresponding clathrate structures of s-PMMS containing the same guest molecules (see next section). The structural data of the clathrates of s-PPMS are reported in Table 1.

On the basis of X-ray diffraction experiments, as well as thermal treatments, it has been deduced that, for s-PPMS, clathrate forms belonging to the α class can be obtained also with other molecules with a large steric hindrance (such as N-methyl-2-pyrrolidone, o-chlorophenol, o-xylene), while smaller molecules (such as 1,4-dioxane, cyclohexane and cyclohexanone) induce the formation of β class clathrates.[11] It is concluded that, in the case of the clathrate forms of s-PPMS, the steric hindrance of the guest molecules determines the formation of one type of cavity or of the other.[12] Finally, it has to be noticed that, by removal of the guest molecules at any temperature, the clathrate forms of this polymer become completely amorphous or give rise to one or the other stable helical pure form of this polymer[11] that do not show any similarity with the microporous form described for s-PS.

Clathrate Forms of s-PMMS

As far as s-PMMS is concerned, it seems that the solvents that are able to induce the crystallization into clathrate forms are less numerous that for the other two polymers. In particular, only the clathrate structures containing CS_2, benzene and o-DCB, whose schematic projections are represented in Figures 4.b, 5.b and 6, have been described so far.[13,14]

The structural data relative to these forms are listed in Table 1.

It may be noticed that all the clathrate forms of this polymer belongs to the β class despite of the great differences of the dimensions of the three guest molecules[13,14] Moreover, also in this case, by removing the guest molecules from the clathrate forms, no microporous form is generated but one of the pure forms is always obtained.[15]

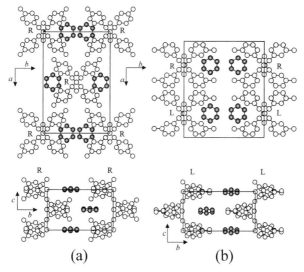

Figure 4. Schematic projections of the clathrate structures of s-PPMS (a) and s-PMMS (b) containing benzene. In the lateral projection, only a couple of polymer chains, along the *a* axis, are reported. R = right-handed chain, L = left-handed chain.

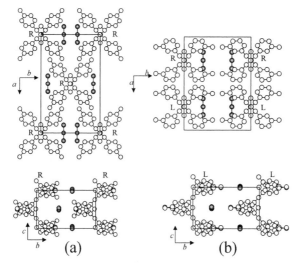

Figure 5. Schematic projections of the clathrate structures of s-PPMS (a) and s-PMMS (b) containing CS_2. In the lateral projection, only a couple of polymer chains, along the *a* axis, are reported. R = right-handed chain, L = left-handed chain.

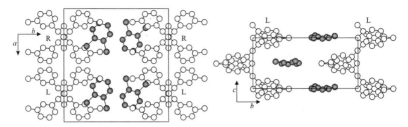

Figure 6. Schematic projections of the clathrate structures of s-PMMS containing *o*-DCB. In the lateral projection, only a couple of polymer chains, along the *a* axis, are reported. R = right-handed chain, L = left-handed chain.

Comparison and Conclusions

Despite the great similarity of the polymers that we have studied, the data presented in the previous sections show that the behaviour of these polymers is quite different with respect to the formation of clathrate forms.

s-PS give rise only to α class clathrates, while s-PMMS forms β class clathrates only. For each of these two polymers, all the known clathrates are characterized by the same symmetry (space group *P21/a* for s-PS and *Pcaa* for s-PMMS) and by an evident similarity in the lattice constants. Moreover, the formation of clathrate structures of these two polymers does not seem to be influenced by the dimension or by the shape of the guest molecule, except for the spacing between the chains forming the cavities, that coincides with the *b* axis of the unit cell for s-PMMS while in the case of s-PS is related mainly to this axis. As a matter of fact, the *a* axis, that depends primarily on the packing of the chains, remains almost unchanged.

The situation is much more complex in the case of s-PPMS. First of all, at variance from the other two polymers, s-PPMS can crystallize in both classes of clathrates. The analysis of the molecules that induce crystallization of clathrate forms in one or the other class show, in an evident way, that the choice is strictly dependent on the dimensions of the guest molecules. Small molecules (having dimensions comparable to the part of the cavity created by the chain itself, such as CS_2, THF and benzene) form β class clathrates, while larger molecules give rise to α class ones.

Moreover, it has to be noted that s-PPMS, within the β class, can crystallize with different symmetries. In particular, benzene and CS_2 clathrates crystallize in the *C222₁* symmetry while the THF clathrate crystallizes in the *P21/a* symmetry. It is interesting to observe that the clathrate forms containing THF and CS_2, presenting cavities almost of the same dimensions (as we can see from the approximately equal values of the *b* axes that, in the case of β class clathrates, are directly proportional to the dimensions of the cavities), differ considerably for the values of the density relative to a completely emptied crystal (see Table 1). This difference of the density is to be attributed, without any doubt, to the different efficiency of the packing of the chains for the two clathrate forms. Consequently, it is surprising that the clathrate form containing CS_2 crystallize with a *C222₁* symmetry rather than with a *P2₁/a* symmetry.

Despite the extreme similarity of these host-guest systems, all the features we pointed out demonstrate that the influence of the factors that control the relative stability of the possible crystalline structures is scarcely predictable. Probably, in our case, an important role has to be attributed to the contribution that the interactions between the chains have in determining the total packing energy of the clathrate structures. We believe that this contribution is particularly relevant in the case of the clathrate forms of s-PS. In this respect, it may be useful to compare the densities of the emptied clathrate forms that we have examined, listed in the last column of Table 1. It is seen that the clathrate forms of s-PS are characterized by densities, relative to the packing of the chains only (without guests), appreciably higher than those of the other two polymers (to be compared with the density of the amorphous phase, always in the range 1.02-1.05 g/cm^3). Moreover, in the clathrates of s-PS the removal of the guest molecules, also in the case in which the cavities are larger, may lead through a negligible rearrangement of the chains to the packing of the δ form, that has a density not too far from the amorphous one. Consequently, the clathrates of s-PS, by removal of the guest molecules at temperatures lower than T_g, do not collapse in an amorphous phase or evolve in the thermodynamically stable γ form, at variance with the behaviours of s-PPMS and of s-PMMS.

Acknowledgements

Financial support from the "Ministero dell'Istruzione, dell'Università e della Ricerca" (PRIN 2002) is gratefully acknowledged.

[1]   (a) A. Immirzi, F. De Candia, P. Ianelli, A. Zambelli, V. Vittoria *Makromol.Chem., Rapid Commun.* **1988**, *9*, 761; (b) V. Vittoria, F. De Candia, P. Iannelli, A. Immirzi *Makromol.Chem., Rapid Commun.* **1988**, *9*, 765; (c) G. Guerra, M. V. Vitagliano, C. De Rosa, V. Petraccone, P. Corradini *Macromolecules* **1990**, *23*, 1539; (d) Y. Chatani, Y. Shimane, Y. Inoue, T. Inagaki, T. Ishioka, T. Ijitsu, T. Yukinari *Polymer* **1992**, 33, 488.

[2]   Y. Chatani, T. Inagaki, Y. Shimane, T. Iijtsu, T. Yukimori, H. Shikuma *Polymer,* **1993**, *34*, 1620.

[3]   Y. Chatani, Y. Shimane, T. Inagaki, H. Shikuma, *Polymer* **1993**, *34*, 4841.

[4]   C. De Rosa, P. Rizzo, O. Ruiz de Ballesteros, G. Guerra, V. Petraccone *Polymer* **1999**, *40*, 2103.

[5]   C. De Rosa, G. Guerra, V. Petraccone, B. Pirozzi, *Macromolecules* **1997**, *30*, 4147.

[6]   G. Guerra, G. Milano, V. Venditto, P. Musto, C. De Rosa, L. Cavallo *Chem. Mater,* **2000**, *12*, 363.

[7]   This emptied form is different from the γ form, which also contains chains in the helical conformation and is obtained by annealing the δ form at temperatures higher than the glass transition temperature of s-PS [*Macromol. Chem. Phys.*, **1995**, *196*, 2795].

[8]   V. Petraccone, D. La Camera, L. Caporaso, C. De Rosa *Macromolecules*, **2000**, *33*, 2610.

[9]   V. Petraccone, D. La Camera, B. Pirozzi, P. Rizzo, C. De Rosa *Macromolecules* **1998**, *31*, 5830.

[10] D. La Camera, V. Petraccone, S. Artimagnella, O. Ruiz de Ballesteros, *Macromolecules* **2001**, *34*, 7762.

[11] A. Dell'Isola, G. Floridi, P. Rizzo, O. Ruiz de Ballesteros, V. Petraccone *Macromol. Symp.* **1997**, *114*, 243.

[12] D. La Camera, V. Petraccone, S. Artimagnella, O. Ruiz de Ballesteros *Macromol. Symp.* 2001, *166*, 157.

[13] V. Petraccone, O. Tarallo, V. Califano *Macromolecules* **2003**, *36*, 685.

[14] O. Tarallo, A. Buono, V. Petraccone, V. Califano *Macromol.Symp.* in press.

[15] C. De Rosa, A. Buono, L. Caporaso, V. Petraccone *Macromolecules* **2001**, *34*, 7349.

RETURN TO: CHEMISTRY LIBRARY

100 Hildebrand Hall · 510-642-3753

LOAN PERIOD	1	2	3
4	*2-HR*	5 *USE*	6

ALL BOOKS MAY BE RECALLED AFTER 7 DAYS.

Renewals may be requested by phone or, using GLADIS,
type **inv** followed by your patron ID number.

DUE AS STAMPED BELOW.

FORM NO. DD 10
3M 5-04

UNIVERSITY OF CALIFORNIA, BERKELEY
Berkeley, California 94720–6000